UITGAVEN VAN HET
NEDERLANDS HISTORISCH-ARCHAEOLOGISCH INSTITUUT TE ISTANBUL

Publications de l'Institut historique-archéologique néerlandais de Stamboul
sous la direction de

Machteld J. MELLINK, J. de ROOS,
J.J. ROODENBERG et K.R. VEENHOF

LXXV

THE OLD ASSYRIAN COPPER TRADE IN ANATOLIA

THE OLD ASSYRIAN COPPER TRADE IN ANATOLIA

JAN GERRIT DERCKSEN

NEDERLANDS HISTORISCH-ARCHAEOLOGISCH INSTITUUT
TE ISTANBUL
1996

CIP-GEGEVENS KONINKLIJKE BIBLIOTHEEK, DEN HAAG

Dercksen, J.G.

The Old Assyrian copper trade in Anatolia / J.G. Dercksen. - Istanbul : Nederlands Historisch-Archaeologisch Instituut ; Leiden : Nederlands Instituut voor het Nabije Oosten [distr.]. - Ill. - (Uitgaven van het Nederlands Historisch-Archaeologisch Instituut te Istanbul, ISSN 0926-9568 ; 75)
Met lit. opg., reg.
ISBN 90-6258-076-9
NUGI 644
Trefw.: Assyrië ; geschiedenis / archeologie ; Anatolië.

Printed in Holland

ACKNOWLEDGEMENTS

The text of this book was submitted as a dissertation, written under the supervision of Prof. K.R. Veenhof, to the Faculty of Arts of Leiden University.

Although a considerable part of the excavated textual material is still hidden in a safe in the Ankara museum, a number of important unpublished texts could be included in the research. I gratefully acknowledge the permission which Profs. E. Bilgiç and T. Özgüç gave me to visit the Museum for Anatolian Civilizations in Ankara in September 1990, where I could search part of the tablets excavated in 1948 (kt a/k) and 1950 (kt c/k) for relevant texts.

In June 1994, I had the pleasure of visiting Münster. Prof. K. Hecker gave valuable advice and provided data from unpublished texts entrusted to him.

Within the framework of a cooperation agreement between Ankara University and Leiden University, Drs. C. Günbattı and S. Bayram each stayed in Leiden for several months during which they worked on the publication of kt n/k texts.

Permission to use and quote texts which they are preparing for publication, was kindly given by Dr. S. Bayram, Prof. E. Bilgiç, Dr. S. Çeçen, Mr V. Donbaz, Dr. C. Günbattı, Prof. K. Hecker, Drs. Y. Kawasaki, Dr. C. Michel, Prof. Ö. Tunca, Ms L. Umur, and Prof. K.R. Veenhof.

Collations of several texts were put at my disposal by Mr V. Donbaz, Dr. C. Günbattı, Drs. Y. Kawasaki, Dr. K. Reiter, and Prof. K.R. Veenhof.

Special tribute is due to Dr. Guido Kryszat for his help and advice.

I am greatly indebted to Ms Helen Richardson, who not only corrected the English text, but made useful suggestions as well. Any mistakes in the text are the result of changes I made subsequently. Ms Gül Demirel provided a fine translation in Turkish of the Dutch summary included in the dissertation.

The text has been printed in 'Times' with supplementary characters taken from the fonts 'Nimrud' and 'Sippar' developed by Dr. W.H. van Soldt.

The research for this dissertation, including the visits to Ankara and Münster, was made possible by a scholarship from The Netherlands Organization for Scientific Research (NWO) during the years 1989-1994.

CONTENTS

INTRODUCTION

The site of Kültepe (ancient name: Kanish), some 20 km north-east of Kayseri, attracted attention when c. 1880 dealers in antiquities in Kayseri started to sell cuneiform tablets. These came allegedly from a ruin called Kültepe and were dubbed 'Cappadocian' after the name of the region. They contained documentary evidence for the commercial activities of Assyrian merchants in Anatolia. In 1893 and 1894 E. Chantre excavated parts of the mound without success. In 1906, H. Winckler conducted a small excavation with the same result. Everything changed dramatically in 1925 when the Czech assyriologist B. Hrozný, after obtaining more reliable information from the local inhabitants about the places where they had found these tablets, dug in the fields at the foot of the mound. He excavated the remains of several houses in which pottery and circa 1,000 tablets and fragments of tablets and envelopes were found. Since 1948, the Turkish Historical Society has conducted excavations at Kültepe almost annually, firstly under the supervision of T. Özgüç, later under that of K. Emre. Many archives were discovered, some of which contained hundreds of texts. At present nearly 15,000 tablets and envelopes, some fragmentary, have been excavated by the Turkish archaeologists, but most are still unpublished. The circa 3,000 texts obtained before the start of the Turkish excavations in 1948 through art dealers have almost all been published.

The lower city of Kanish was located where fields and orchards now surround the mound. This lower city contained the town's commercial district, which was called the *kārum* or 'harbour' according to Assyrian and Babylonian custom. It was inhabited only in the so-called *kārum* period. The lower city is divided into four archaeologically distinct layers. The tablets were found in the two later layers: the majority in level II, and a few hundred in the later level Ib. The presence of Assyrian traders as documented in level II texts, started probably during the reign of the Assyrian king Erišum, well before 1900 B.C., and lasted for about a century. It apparently ended with a local catastrophe during the reign of Puzur-Aššur II. Level Ib ended somewhere during the middle of the 18th century B.C.

Assyria depended heavily upon trade for its survival because, like the rest of Mesopotamia, it was devoid of important natural resources. Raw materials had to be imported, including the metals which were also used as money: tin, copper, silver, and gold. The city of Assur functioned as a centre of transit trade, linking various commercial circuits. Merchants from the South, the Akkadians, brought woollen textiles and copper to Assur to sell these for silver and presumably also for tin. Tin, perhaps originating from Afghanistan, was apparently conveyed to Assur by people living east of it. Little is known of this aspect of trade. The merchants bringing tin into Assur in all likelihood exchanged it for silver and possibly for textiles as well. Assyrian traders exported tin and textiles into North Syria and into Anatolia.

The long distance trade from Assur to Central Anatolia is well documented due to the texts discovered in Kültepe. The merchants involved were private entrepreneurs driven by the desire to make profit. They worked in a business which usually consisted of members of the same family. Tin and textiles were bought in Assur and sent by donkey caravans to Kanish. From there, the merchandise was distributed over Central Anatolia and exchanged for silver and gold, which was sent back to Assur. Frequently, the merchandise was first converted in Anatolia into commodities other than silver or gold, e.g. wool and notably copper. The copper was brought to other towns and sold there for silver or gold. Most of the woollen textiles brought to Anatolia were luxury items and therefore in demand by the local dignitaries. Anatolia possessed many deposits of copper, silver, and gold, but lacked the tin needed to produce bronze. As weaponry and tools were largely made of bronze, this alloy was of strategic importance to the rulers of the various city-states into which Anatolia was divided. The evidence documents how local palaces bought a considerable portion of the tin brought by the Assyrians. For more than a century, these commercial activities resulted in an important flow of silver and gold to Assur.

The study of the texts led to a number of important publications during the last few decades. The grammatical features of the Old Assyrian dialect were treated by K. Hecker (*Grammatik der Kültepe-Texte*, 1968). A detailed investigation into the trade and the economic and political role of the various Assyrian trade colonies was made by P. Garelli (1963). Aspects of trade and terminology were clarified by M.T. Larsen (1967) and K.R. Veenhof (1972). The historical background to the trade settlements and the economy during the Old Assyrian period were analysed by M.T. Larsen (1976). Some scholars presented analyses of a merchant's archive: M. Ichisar (1981) and M.T. Larsen (1982) on Imdilum, and C. Michel (1991a) on two traders called Innāya. Detailed studies have also emerged about geographic aspects (Kh. Nashef (1987), and about two important towns, Purušhattum (Garelli 1989) and Turhumit (Michel 1991b). The legal peculiarities of the Old Assyrian period were treated by B. Kienast (1984) and K.R. Veenhof (1991). The chronology of the level II period has been studied by G. Kryszat (forthcoming). The role which copper played in the Assyrian trade in Anatolia was investigated by E. Bilgiç in 1941 (*Sumeroloji Araştırmalırı* 1, pp.925-934, not accessible to me), and by P. Garelli 22 years later. During the more than 30 years which elapsed since the publication of the latter's *Les Assyriens en Cappadoce*, many new texts have become available for study. Moreover, new evidence on ancient metallurgy and especially on that of copper was obtained recently by other sciences and sheds new light on the data contained in them. Both circumstances warrant a fresh investigation of this aspect of a well-documented trade.

This study has to start with an attempt to establish the origin of the copper which was traded by the Assyrian merchants in Central Anatolia. Although many copper deposits are attested, it is unclear which were actually mined. The texts mention several towns where copper was obtained or which gave their name to a particular quality. In order to locate these towns, the geographical situation of Central Anatolia has to be re-examined. The next step is to define the different shapes and qualities in which copper was traded, both on the basis of a study of the terminology employed in the texts, and of archaeological objects. Recent archaeological and archaeo-metallurgical research may be helpful in solving some of the technical questions. After this basic information is available, the structure of the trade has to be analysed. This leads to questions as to where copper was obtained, who bought it, to which cities was it brought, how and where was it sold, and at which prices. Interrelated are the commercial and political motives of Assyrian and Anatolian traders and rulers, and the role of the *kārum* as the organization of Assyrian traders deserves special attention. Were commodities deliberately exchanged for copper to increase the profit, or are there other explanations why Assyrians accepted this metal? It is also important to establish what role the Anatolian traders and rulers played in the trade, and what the monetary significance of copper was in relation to other metals. I will try to answer these questions by examining a number of Assyrian archives which contain substantial information. Of great help were the numerous unpublished texts which scholars from Turkey and Europe permitted me to use, and some of which I was able to study myself in Ankara. Even with the help of this greatly increased corpus of sources not all questions could be solved. The knowledge of the geography and topography is still limited, and certain procedures still raise questions. As can be seen from the Catalogue and appendices at the end of this book, several words or technical terms are rare and hence difficult to understand. New textual material will furnish more data on the trade in copper, and may lead to corrections of some of my interpretations.

In rendering the original texts, Assyrian words are given in italics, Sumerograms in capitals. The small vertical wedge sometimes used as a word divider (occasionally even between signs of one word) is rendered by a dot, and not by a slash (K. Hecker) or a colon (C. Michel). A slash is used to indicate that the ancient scribe did not write the continuation of a word on the same line. The Sumerogram for silver, 'KÙ.BABBAR', is in this book abbreviated to 'KB'.

CHAPTER 1
THE GEOGRAPHICAL BACKGROUND

1.1 *Scientifically*

1.1.1 *Old workings*

In the course of its geological development, many mineralizations were formed in the Earth's crust in Turkey. The geomorphological processes which determined these formations are complex and will not be discussed here. The abundance of minerals like lead, silver, gold, and especially copper, created a favourable background for a local metallurgy to develop. The first modest use of copper is attested in Çayönü, about 20 km south of Ergani Maden, where some utensils dating from c. 7250-6750 B.C., have been identified as being made of native copper.[1] In their manufacture the techniques of cold working and annealing were used. The heat-treatment of copper, exemplified by annealing, is remarkable because the artefacts date from layers belonging to the Aceramic Neolithic period. The earliest artefact made of copper smelted from ore is a knob-headed pin from Tepe Yahya in Iran, 225 km south of Kerman, dating from c. 4000-3500 B.C.[2] Bronze is first attested in finds from Mesopotamia dating from c. 3000 B.C., a period characterised by a major increase in metalworking, so far as can be judged from the archaeological remains.[3]

The obvious question is where the metal needed for the manufacture of these artefacts originated from. Efforts to trace the provenance of the metal used in the artefact back to the mine from which the ore ultimately derived, called sourcing, have received increased attention during the last decades. A general impression of the metal resources in Turkey can be obtained from the publications by the Turkish Mineral Research and Exploration Institute, MTA (see Map B). Publication MTA 1972 contains a catalogue of the deposits of lead, copper, and zinc in Turkey. The mineralizations noted on the map represent those which can be worked using modern technology, some of which might have been beyond the scope of ancient miners. On the other hand, mines exploited in the past may not have been found or registered, as they are exhausted, or economically unviable today. They are therefore not mentioned. These exploration maps can only give a general outline of the areas where mining of a given mineral is to be expected.

On the basis of the geological data, several surveys have been made with the intention of collecting evidence of ancient workings. The MTA, during its exploratory surveys, assembled many references to old mining activity, and some of the information stored in its

1 Çambel 1974, p.376; Muhly 1989, pp.5-6; Muhly 1993, p.124.
2 Heskel and Lamberg-Karlovsky 1980, p.232.
3 Muhly 1993, p.129.

archives is presented in MTA 1972 and in de Jesus 1980. A German project, the aim of which is to gather (experimental) evidence on the role played by Asia Minor in the development of metallurgy, has led to publications yielding interesting data on possible ancient mines which was obtained in the course of a number of surveys.[4] The sites, which were investigated, were identified by a number preceded by 'TG'. The methods used are those developed by archaeo-metallurgy.

An ancient mining area is generally recognizable by its vegetation and shape which distinguish it from its surroundings. Archaeological remains to be found are slag, remnants of furnaces, tuyères, potsherds, and sometimes a roasting site or part of the ancient mine itself. Slag (see Bachmann 1982) is the natural result of smelting activities, and because is was usually worthless to the smelters, it was discarded. In some cases ancient slag was resmelted at later periods, when better techniques made it possible to extract metals still left in them. When slag is found, ranging from a small quantity to heaps consisting of tons of this material, it should first be identified by analysis. Most ancient smelting slag is ferrous silicate (fayalite), high in iron and low in non-ferrous metals.[5] The composition of the slag depends on several factors (ore, flux, gangue material, charcoal used etc), which may give an indication of the ore itself, or of the method used to smelt the ore. The occurrence on the surface of secondary copper minerals as an indicator of copper smelting slag is ambiguous; copper oxide-carbonites (malachite, azurite) can be the result of weathering.[6] If some charcoal is trapped in a slag, it is possible to establish its approximate date by ^{14}C. Dating with the thermoluminescence method yields a rough indication, and permits only a distinction between mediaeval, classical, and prehistoric slag.[7]

The discovery of the actual mines is often impossible because the pit frequently has been refilled with debris after the ore extraction had stopped, or because it has been destroyed in the course of time. Those mines which have been found provide information on the composition of the type of ore, but unless remains of wood used in fire setting or as mine timber, or pottery are preserved, little can be said about dating them. Wood from a mine near Kozlu was found to date from around 2800 B.C. (^{14}C), thus giving a date for the mine.[8] At an ancient mining site at Yuvalar (TG 138), in the Troad, about 3 km north of Çan, charcoal from a slagheap was dated by ^{14}C to 1760-1260 B.C.[9] On the basis of field observations the copper mining sites of Ergani Station (TG 176C), Derealan (TG 164B), and Karadağ (TG 172) could go back to the Early Bronze Age,[10] but the physical remains of old workings like shafts and adits are not very distinctive before classical antiquity.[11] If a mine had not been exhausted, but remained to be exploited in subsequent times, all traces of former activities might have been destroyed. In the absence of any archaeologically datable

---

4 Literature: Wagner et al. 1983-1984, Pernicka et al. 1984, Seeliger et al. 1985, Wagner et al. 1986, Pernicka 1987, Wagner et al. 1989a, Wagner et al. 1989b, Lutz et al. 1994.

5 Tylecote 1976, p.19.

6 Pernicka et al. 1984, p.597.

7 See Elitzsch et al. 1983.

8 Giles and Kuijpers 1974.

9 Wagner et al. 1986, p.726.

10 Seeliger et al. 1985, pp.639-640.

11 Pernicka et al. 1984, p.570; Pernicka 1987, p.624.

artefacts (written documents are unknown from the mining areas), it follows that for Turkey it is virtually impossible to assign any ancient mine or slag heap to the period around the 19th century B.C.

1.1.2 *Sourcing*

While it is not feasible to identify the very mines that were exploited during the Old Assyrian period by using chronological criteria alone, one may try to establish whether the copper used in manufacturing a given artefact excavated in Kültepe came from a certain mining area. In this case, the chemical and isotope characteristics of the ores and of the artefacts are relevant. Unlike the sourcing of stone artefacts, where the possibility of mixture is excluded, this method is only valid for metals if the object was not made from metal originating from different sources, as might be the case with recycled copper.

There are basically two types of copper ore involved: oxide and sulphide ores. The major deposits in Turkey consist of sulphide ore. The winning of copper starts at the mines, where ore is collected, crushed, sorted, and presumably smelted not too far away from the mining site. In the case of sulphides, the ore has first to be roasted to remove the sulphur, and to turn the ore into a matte. After this treatment, it can be smelted.[12] The raw copper resulting from the smelting, also called the primary smelting product, is transported out of the production area to places where it can be further refined and melted in a crucible to be made into artefacts.

A copper ore contains all other elements besides the main element copper. A number of these follow the copper during the production stages of the metal, and can ultimately become part of the artefact. The theoretical possibility of tracing the origin of the metal back to its ore, starting from the artefact, has long been recognised. The best known project of classifying European prehistoric metal artefacts on the basis of chemical analyses is that of Stuttgart,[13] and thousands of objects have been analysed in the course of this project. The method which should be followed has been treated in some detail in Pernicka 1984. In a quantitative analysis, which takes all elements into consideration, the elements are counted by chemical methods (neutron activating analysis is widely used for this purpose). Since a number of elements may have entered the metal during the processing stages (introduced by flux or otherwise, see on this subject Tylecote et al. 1977), or behave differently during this process, not all can be considered representative of the original copper ore. The Stuttgart method for establishing 'material groups' (*Materialgruppen*) consists of a sequential sorting based on the limit values of the arsenic, antimony, bismuth, nickel, and silver content. Indicative of copper ore are those elements which remain in the copper, independent of the type of ore: gold, silver, nickel, and to a lesser degree cobalt. In the case of raw metal the elements arsenic, antimony, perhaps lead, bismuth, iron, and selenium can be added.[14]

Ores offer the advantage over slag material and artefacts that they certainly have not been mixed. It is thought that by careful sorting (benefication) of the copper minerals, ore

12 For details see de Jesus 1980, pp.26-29.

13 Junghans et al., published in SAM = *Studien zu den Anfängen der Metallurgie*.

14 Pernicka et al. 1984, pp.590-591.

comparable to that smelted by the ancient miners can be selected. The profile or 'finger print' of an ore deposit on the basis of a dozen or more samples is difficult to make with chemical methods only. Because the structure of the ore may show differences within the same deposit, ratios between trace elements may provide a means to identification. Antimony/arsenic and cobalt/nickel ratios were used in Seeliger et al. 1985 (p.658, Abb. 35). The possible importance of the gold/silver ratio to distinguish between sulphidic ores from North-West Anatolia and those from North and East Anatolia has also been noted.[15]

On a different level of investigation, the isotopes of the lead contained in an ore, slag, or artefact of copper or bronze can be taken into consideration. These lead isotopes are determined by the geological age of the ore deposit, and have the advantage over trace elements that they do not change during the processing of the ore and metal. A first study of the isotope ratios in ancient lead was made by Brill and Wampler (1967). The information to be gained from such isotope ratios is, however, only negative. If the ratios of an ore deposit are different from those found in a particular artefact it is clear that this artefact was not made of metal from that deposit. A complication is that also lead isotopes differ within one deposit if this was formed at different times, and the differences in age do not necessarily point to a great geographical distance. There is also a chance that the isotopes may have changed by mixing with fluxes or alloying material containing different isotopic ratios. If the material used for fluxing comes from the immediate vicinity of the mine, it might be of the same geological age and therefore show little difference in isotopes.

A serious problem is the measurement of the amount of lead contained in the copper or bronze artefact, and the question whether lead was part of the copper ore or alloyed to the metal.[16] In bronze the main alloy is tin, considered to have been in the form of cassiterite (tin oxide), which contains very little lead. Lead was presumably added when its content was greater than 1%.[17] The information on the lead isotopes based on several samples taken from a deposit, is expressed in a isotope-diagram. With more ore deposits being investigated, it turns out that the individual isotope-fields occasionally overlap each other.[18] It is perhaps possible to overcome this by combining trace element analyses with the lead isotope ratios as a means of classifying ore groups.

The sourcing of artefacts based on trace element characteristics has been performed on several occasions. In view of the aim of this investigation, which is to establish the provenance of the copper traded by the Old Assyrian merchants, the metallurgical data available should be archaeologically discovered metal artefacts. The excavations at Kültepe, the commercial centre of the Assyrians, yielded many objects made of copper and bronze, and even the remains of work-shops (see § 3.5.1). Artefacts from Kültepe, together with those originating from other sites, have been analysed in Stuttgart and the results were published in Esin 1969 (see Appendix 4). Using the tables in Esin 1969, H. McKerrell

15 Wagner et al. 1986, p.743 and Abbildung 18.

16 See for instance Muhly 1985 and the answer of Gale and Stos-Gale on pp.97-99.

17 Pernicka 1987, p.645 with references.

18 Wagner et al. 1986 Abbildung 19; p.747: 'Gerade die Tatsache aber, daß die seither untersuchten anatolischen, thrakischen und pontischen Erze isotopisch nicht sehr verschieden sind von denen aus der Ägäis, bedeutet andererseits auch, daß für alle Artefakte mit $^{208}Pb/^{206}Pb$-Verhältnissen etwa größer als 2,13 nach wie vor keine möglichen Quellen identifiziert werden können.'

compared the results for Kültepe with those for Horoztepe and Alacahöyük.[19] Considering the silver and lead/tin contents, he concluded that 'basically only one copper metal (was) involved at Kültepe and Horoztepe'. The basis for McKerrell's conclusion seems rather limited when viewed in the light of the criteria established by other scientists for assessing the characteristics of an ore type. The material from Horoztepe has to be dated to the Early Bronze Age (EB III), and is several centuries older than that from Kültepe (19th century B.C.). Artefacts found in Kültepe have been made locally or imported from somewhere else. Because metal was relatively scarce, it was often recycled. If the metal came largely from one source, or from deposits showing the same characteristics, the majority of artefacts would in fact form a unity, as suggested by McKerrell. However, Horoztepe is situated close to mining areas around Tokat, while the evidence from Kanish points to a source west and north-west of the Kızılırmak (see below).

Chemical analysis of small samples taken from copper ingots excavated in Acemhöyük, slightly later than Kültepe, showed the content of nickel especially to be similar to that in coppers found in Tepecik and Tülintepe, near Ergani Maden.[20]

Çayönü is c. 20 km south of Ergani. The native copper used in the manufacture of artefacts excavated there, does, however, not appear to be identical to the native copper which can still be found at Ergani, and a different source seems probable.[21]

A thorough investigation of the Early Bronze Age artefacts of copper and bronze from North-West Anatolia (Troy and surroundings) and of mineral deposits in Anatolia has revealed that the majority of analysed ores cannot have been used in the production of these artefacts. Only the ore from the region of Balıkesir (TG 192) is consistent with the cobalt/nickel ratio detected and with the lead isotope ratio to be found in some of the artefacts.[22] After a detailed investigation, Pernicka concluded that it was impossible to trace the introduction of bronze (tin-bronze) into the Troy civilization back to a local development, because the tin needed was not available in the vicinity.[23]

Some international examples of sourcing come from Iran, Germany, and Canada. Berthoud et al. tried to assign Iranian artefacts from the fourth and third millennia B.C. to known ore deposits, using trace element analysis and establishing the correspondence by the simple overlap of artefact and ore groups. The result was that according to the analyses on trace elements the copper used during Susa A came from Anarak, that from Susa D and Umm-an-Nār (Oman) from Oman, and that from Tepe Yahya from Sheik Ali. No clear correspondence was found for Susa B.[24]

The copper production at Helgoland has been discussed by Schulz.[25] Chemical analysis of discs of black copper found near the island showed remarkable fluctuations of the percentage of nearly all trace elements. These mediaeval discs, which were shown to come from Helgoland, were then compared with five Bronze Age (*kupferzeitlich*) copper

19 McKerrell 1977, p.16.
20 Çukur and Kunç 1990, p.35.
21 Muhly 1989, p.8.
22 Wagner et al. 1986, p.744.
23 Pernicka 1987, p.705.
24 Berthoud et al. 1980.
25 Schulz 1983.

artefacts found in Schleswig-Holstein. Four of these samples could be assigned to ore from Helgoland and the fifth was definitely not made from its ore.

A relatively uncomplicated case is the composition of artefacts used by the Copper Inuit in the western central Canadian Arctic. Because pre-contact Inuit used native copper, which is quite free of impurities, the artefacts could be distingished from artefacts made from copper produced in Europe.[26]

The published attempts to correlate a group of artefacts with ore deposits illustrate the methodological restrictions. The sourcing of the metal used in an artefact is very difficult. Junghans was unable to link metal objects with specific ores and instead defined the regions of metal types. The situation of the Anatolian region is illustrated by the example from Troy. The investigators seem to have a preference for prehistoric artefacts as objects of their analyses, perhaps on the assumption that in the succeeding Middle Bronze Age the quantity of metal in circulation in Anatolia was too big to originate solely from 'local' sources. The underlying thought is that from that time onwards it became necessary to use metal from different ore deposits, occasioned by either the mixing of different ores, or, at a later stage in the production process, by recycling used artefacts. It seems very probable that the artefacts made during the preceding periods also did not always consist of first-use metal, but already contained a mix.

Although copper mines undoubtedly were exploited in Anatolia during the 19th century B.C., it remains difficult to produce positive evidence for this. The sourcing of artefacts from the Old Assyrian period is almost impossible, not only because of the absence of a database containing all necessary information. Metal could have reached the cities from different directions and in more than one form. Only raw copper transported from the smelting area might retain enough of its original characteristics to compare it with the available data on ore deposits, and then it might be more or less probable that it originated from a certain mine. But no raw copper has been found at Kültepe. Because of the impossibility of defining by scientific means the areas where copper traded by Assyrians was mined, it is better to confine oneself at this stage to data furnished by the MTA and subsequent research on the occurrence of copper (and lead-zinc) deposits. De Jesus combined the MTA data with archaeological evidence for mining activity and ancient use of copper and bronze.[27] He formulated a hypothesis to answer the questions of which regions provided copper and of where it was used. Six areas of intensive exploitation of copper deposits were defined (see Map B):

1. Küre, in an archaeologically poorly documented region;
2. Yapraklı, which could have supplied the Central Plateau and especially the sites on the western side of the Kızılırmak around Ankara;
3. Pontic (Tokat), which supplied the Pontic area and after Early Bronze III in part the Central Anatolian plateau;
4. Giresun-Trabzon, which exported copper via the naval trade along the coast;
5. Ergani, which supplied copper to the Mesopotamian plain and the Anatolian interior;

---

26 Wayman et al. 1985; the distinction between native and smelted copper is not always as easy as in this case, see Maddin et al. 1980.

27 De Jesus 1980, p.148 and Map 21.

6. Murgul-Kuvarshan, which supplied eastern Anatolia and the Caucasus.
Which of these areas was or were likely to have provided copper for the Anatolian cities mentioned in the Assyrian texts will be investigated below.

1.2 *Philologically*

1.2.1 *The geographical aspect*

Two ancient cities which played a role in the Old Assyrian trade have been identified: Kanish with Kültepe, and Hattuša with Boğazköy, so that the situation of all other toponyms on Map A remains hypothetical. Several itineraries contain important details for the understanding of the geographical situation in that part of Anatolia which was travelled by the Assyrians. The first is TC 3, 165, where expenses are recorded which a merchant incurred when he travelled with a caravan from Kanish to Purušhattum. This text was last discussed by Kh. Nashef (1987, pp.50-53):
Kanish - Wašhania: (1-4): 'From Kanish to Wašhania I spent 2 minas of tin on guards, on an inn, and on donkey-fodder.'
43-44: 'I paid 20 shekels of silver to porters from Kanish to Wašhania.'
Wašhania - Ninašša: (5-12): 'The palace took 21 shekels of tin as *nishātum*-tax; the gentleman took x minas; the lord of the town took 9 1/2 minas; I spent from Wašhania to Ninašša 1/3? mina of copper on an attorney, and 10 minas of copper on an inn and on donkey-fodder.'
44-47: 'I paid 17 1/3 shekels of silver to porters and to my guides from Wašhania to Ulama.'
Ninašša - Ulama: (13-23): 'The palace took 24 shekels of tin as *nishātum*-tax; the lord of the town took 15 shekels of tin; the gentleman, the *imrum*, and the head of the ... took 10 1/2 shekels of tin; I spent from Ninašša to Ulama 2 minas of copper on an attorney, and 10 minas of copper on a stable, donkey-fodder, and an inn.'
Ulama - Purušhattum: (24-30): 'From Ulama to Purušhattum I spent: the gentleman took 5 minas of copper; the attorney took 10 minas of copper; 10 minas of copper on an inn; 10 minas of copper on donkey-fodder; 30 minas on smuggling.'
in Purušhattum: (31-42): '15 minas of copper for the rent of a house; 2 minas of silver (to pay for) my food and (for) that of the servants; I gave 1 shekel of silver per mina (of silver) and 1 mina of copper per talent (of copper) to the *kārum*-office as *šaddu'atum*-tax. They took 8 *kutānum*-textiles as "five per cent" levy at the *kārum*-office. The deficit of the *biltum*-loads was 12 minas of tin. I gave 2 minas of copper at half a mina of copper per shekel of tin(?) to the *kārum*-office.'[28]

---

28 TC 3, 165: [1] 2 *ma-na* AN.NA *lu ma-ṣa-ra-[tim]* [2] *lu* É *wa-áb-ri lu ú-ku-ul-tí* [3] ANŠE *iš-tù Kà-ni-iš* [4] *a-dí Wa-áš-ha-ni-a ag-mu-ur* [5] 1/3 *ma-na* 1 GÍN AN.NA *ni-is-ha-tim* [6] É.GAL-*lúm i[l*$_5$*-qé* x] *ma-na kà-šum* [7] *il*$_5$*-qé* 9 1/2 *ma-[na* AN].NA *be-[el]* [8] *a-lim*[ki] *il*$_5$*-qé* 1/3 *ma-na* URUDU [9] *a-na ra-bi*$_4$*-ṣí-im* 10 *ma-na* URUDU [10] *[a-na]* É *wa-bi-ri ù [ú]-ku-ul-tí* [11] ANŠE *iš-tù Uš-ha-ni-a* [12] *a-dí Ni-na-ša-a ag-mu-ur* [13] 1/3 *ma-na* 4 GÍN AN.NA [14] É.GAL-*lúm ni-is-ha-tim* [15] *il*$_5$*-qé* 15 GÍN AN.NA *be-el a-lim*[ki] *il*$_5$*-qé* [16] 10 1/2 GÍN AN.NA *kà-šu-um* [17] *im-ru-um ù be-el ha-ṭí-tim*

I assume that all expenditures incurred during the trip were carefully recorded in this document. Leaving aside certain interpretational problems irrelevant to the present investigation, like the meaning of the words in line 17 or of lines 40-42, it can be seen that a trip with a donkey caravan from Kanish to Purušhattum took four days. The average daily distance covered by such a caravan is put at 30 kilometres.[29]

Before giving the proposed location of the cities named in this text, it is necessary to consider a number of recently published or still unpublished itineraries which detail the costs of travelling from Purušhattum to Šalatuar and to Wahšušana. The first is AKT 3, 34,[30] which records the costs of a journey from Wahšušana to Purušhattum:

'From Wahšušana to Šalatuar the road-tax, (costs for) donkey-fodder and inn together amounted for us to 1 mina 55 shekels of copper per (donkey); furthermore they levied 20 shekels (of copper) per donkey at the bridge. The donkey-fodder (cost) 2 1/2 minas of copper in Šalatuar. As far as Purušhattum (the costs) per (donkey) amounted for us to 2 1/2 minas (of copper). They levied 15 shekels (of copper) per donkey at the bridge. (I paid) 1 1/2 minas (of copper) in Purušhattum for donkey-fodder and for food for the servant. I gave Arwanahšu 2 1/2 minas of copper for his expenses.'

H. Sever published two texts recording expenses which were made when an animal called *perdum* was bought in Šalatuar and taken along to Wahšušana and further. That part of kt t/k 1 which is duplicated by the second text, kt t/k 25, runs in translation:[31]

---

[18] *il5-qé-ú* 2 *ma-na* URUDU *a-na* [19] *ra-bi4-ṣí-im* 10 *ma-na* [20] *lu iš-pá-tá-lu lu ú-ku-ul-tí* [21] ANŠE *ù* É *wa-bi4-ri* [22] *iš-tù Ni-na-ša-a a-na* [23] *Ú-lá-ma ag-mu-ur* [24] 5 *ma-na* URUDU *kà-šu-um il5-qé* [25] 10 *ma-na* URUDU *ra-bi4-ṣú-um il5-qé* [26] 10 *ma-na* URUDU É *wa-bi4-ri* [27] 10 *ma-na* URUDU *ú-ku-ul-tí* ANŠE [28] 30 *ma-na* URUDU *a-na pá-zu-ur-tim* [29] *iš-tù Ú-lá-ma a-dí* [30] *Pu-ru-uš-ha-tim ag-mu-ur* [31] 15 *ma-na* URUDU *ig-ri* É-*tim* [32] 2 *ma-na* KB *ú-ku-ul-tí* [33] *ù ú-ku-ul-tí ṣú-ha-ri-i* [34] KB 1 GÍN.TA *ù* 1 *ma-na*.TA URUDU [35] *a-na* É *kà-ri-im ša-du-a-tám* [36] *a-dí-in* 8 TÚG *ku-ta-ni* [37] *a-na me-tù-um ha-<am>-ša-at* [38] É *kà-ri-im il5-qé-ú* [39] 12 *ma-na* AN.NA *mu-ṭá* [40] *bi4-lá-tim* 2 *ma-na* URU[DU] [41] *a-na* 1/2 *ma-na*.TA *a-n[a]* [42] É *kà-ri-im a-dí-in* [43] 1/3 *ma-na* KB *iš-tù Kà-ni-iš a-dí Wa-áš-ha-ni-a* [44] *a-ša bi4-lá-tim a-dí-in* 17 1/3 GÍN KB [45] *iš-tù Wa-áš-ha-ni-a a-dí Ú-lá-ma* [46] *a-ša bi4-lá-tim ù mu-qá-ri-bi4-a* [47] *a-dí-in*.

29 The same distance in Nashef 1987, p.63 n.39. According to the *Encyclopaedia of Islam* IV, p.1011, *sub* KHān, a day's journey was about 20 miles.

30 AKT 3, 34: [1] *iš-tù Wa-ah-šu-ša-na* [2] *a-dí Ša-la-tù-a-ar* [3] 1 5/6 *ma-na* 5 GÍN.TA URUDU [4] *lu da-a-tum lu ú-ku-ul-tí* [5] ANŠE *lu* É *ub-ri-im* [6] *ik-šu-ud-ni-a-tí . a-ha-ma* [7] 1/3 *ma-na*.TA *i-na* [8] *tí-tù-ri-im i-na* ANŠE [9] *i-sú-hu* 2 *ma-na* URUDU [10] *ú-ku-ul-tí* ANŠE.HI.A [11] *i-na Ša-la-tù-a-ar* [12] 2 1/2 *ma-na*.TA *a-dí* [13] *Pu-<ru>-uš-ha-tim* [14] *ik-šu-ud-ni-a-tí* [15] 15 GÍN.TA *i-na* [16] *tí-tù-ri-im* [17] *ni-is-ha-tim ša* ANŠE [18] *i-sú-hu* 1 1/2 *ma-na* [19] *i-na Pu-ru-uš-ha-tim* [20] *ú-ku-ul-tí* ANŠE *ù* [21] *ṣú-ha-ri-im* [22] 2 1/2 *ma-na* URUDU *a-na qá-tí* [23] *Ar-wa-na-ah-šu* [24] *a-dí-in*.

31 Published in Sever 1990a, plates 290 (kt t/k 1) and 291 (kt t/k 25). Kt t/k 1.7-35: [7] ... 3 GÍN KB [8] *a-na* É *wa-áb-ri a-dí-in* [9] 3 GÍN KB . *a-na* É *kà-ri-im* [10] *ša-du-a-tám áš-qúl* 7 *ma-na* [11] URUDU *i-ša-am-ší pé-er-dam* [12] *ni-iš-ú-mu-ú . a-na ki-ra-nim* [13] *áš-qúl* 1 *ma-na a-na iš-pá-tá-li* [14] *mì-ma a-nim i-na Ša-lá-tí-wa-ar* [15] *a-šu-mì pé-er-dí-im ú-ša-qí-il5* [16] *iš-tù . Ša-lá-tí-wa-ar* [17] *nu-ṣa-ma* 2 1/2 *ma-na* URUDU [18] *i-na ša-pá-at na-ri-im* [19] *a-šu-mì pé-er-dí-im áš-qúl* [20] 2 *ma-na* URUDU *a-na* ŠE.AM [21] *i-na Ša-lá-tí-wa-ar áš-qúl* [22] 5 *ma-na* URUDU *a-na* É *wa-áb-ri* [23] 5 *ma-na ša-du-a-tám* [24] *a-na* É *kà-ri-im áš-qúl* [25] 5 *ma-na a-na kà-ší-im* [26] *áš-qúl* 4 *ma-na a-na* ŠE.AM [27] *áš-qúl* 10 *ma-na* URUDU *a-na* [28] *ra-dí-im a-na a-lá-hi-nim* [29] *a-dí-in-ma . a-dí . ša-pá-at* [30] *na-ri-im . ir-dí-a-ni* [31] 1 *ma-na a-na ma-lá-hi-im* [32] *a-dí-in . mì-ma a-nim* [33] *i-na Wa-ah-šu-ša-na* [34] *a-šu-mì . pé-er-dí-im* [35] *ú-ša-qí-il5*.

Šalatuar (7-21): 'I gave 3 shekels of silver to the inn; I paid 3 shekels of silver to the *kārum*-office as *šaddu'atum*-tax; I paid 7 minas of (kt t/k 25 adds: *šikkum*) copper for wine on the day we bought the *perdum*; 1 mina (of copper) for the stable; all this I paid in Šalatuar because of the *perdum*. We left Šalatuar and I paid 2 1/2 minas of copper at the bank of the river (kt t/k 25 has: *ina titūrim*, "on the bridge") because of the *perdum*. I paid 2 minas of copper in Šalatuar for barley.'
Wahšušana (22-35): '5 minas of copper for the inn; I paid 5 minas of copper to the *kārum*-office as *šaddu'atum*-tax; I paid 5 minas to the gentleman; I paid 4 minas for barley; I gave 10 minas of copper for an escort to the *allahhinnum*-official, and he escorted me as far as the bank of the river; I gave 1 mina to the boatman; all this I paid in Wahšušana because of the *perdum*.'

In kt 91/k 424, expenditures are recorded which were incurred on a journey (partly with porters carrying goods) which started in an unnamed locality, and led to Šalatuar and Purušhattum, and from there back to Šalatuar and further to Wahšušana, Tuhpia, and Turhumit.[32] The text runs:
Šalatuar - Purušhattum: (1-14): 'I paid x minas of *šikkum* copper as wages for a porter to get to Šalatuar. I paid in Šalatuar in all, 20 minas of *šikkum* copper on various occasions. I gave 3 minas to the inn (and) I paid 10 minas of copper to porters. I gave 7 minas of copper to Tarkua. I spent 3 minas of copper to get to Purušhattum. In Purušhattum I paid 3 shekels of silver to the *kārum* as *šaddu'atum*-tax. I paid 3 minas of copper as costs to get to *Ušbukatum*, and 3 shekels of silver for an inn in Purušhattum.'
Purušhattum - Šalatuar (15-16): 'I had to pay 1 1/2 minas of copper from Purušhattum to Šalatuar.'
Šalatuar-Wahšušana: (17-20): '10 minas of copper were spent until I left Šalatuar. < > was spent in Wahšušana on an inn.'
Wahšušana - Tuhpia: (20-21): 'I had to pay 1 1/2 minas of copper to get to Tuhpia.'
Tuhpia - Turhumit: (21-22): 'I had to pay 3 minas of copper to get to Turhumit.'
Turhumit - Kanish: (23-25): 'From Turhumit to get to Kanish I had to pay 5 minas of fine copper.'
Kanish - Wahšušana: (25-26): 'I spent 3 minas of copper to get to Wahšušana.'

Further information is contained in kt 91/k 437, which mentions a river crossing or *nēbartum* in the territory of Turhumit.[33]

---

32 Kt 91/k 424.1-26: [1] [ ] *⌜ma⌝-na ší-kam ig-ri ša bi$_4$-il$_5$-/tim* [2] *[a]-dí Ša-lá-tù-a-ar a-dí-in* [3] 20 *ma-na ší-kam i-na Ša-lá-tù-a-/ar* [4] *ú-ša-qí-il$_5$* 3 *ma-na a-*É [5] *ub-ri a-dí-in* 10 *ma-na* URUDU *a-na* [6] *ša bi$_4$-lá-tim a-dí-in* 7 *ma-na* [7] URUDU *a-Tár-ku-a a-dí-in* [8] 3 *ma-na* URUDU *a-dí Pu-ru-uš-ha-tim* [9] *ag-mur* 3 GÍN KB *i-na* [10] *Pu-ru-uš-ha-tim a-na kà-ri-/im* [11] *ša-du-a-tám a-dí-i* 3 *ma-na* / [URUD]U [12] *a-gám-ri-im a-dí Uš-bu-kà-[ti]m*? [13] 3 GÍN KB *a-*É *ub-ri ⌜i-na⌝* [14] *Pu-ru-uš-ha-tim a-dí-in* [15] 1 1/2 *ma-na* URUDU *iš-tù Pu-<ru>-uš-/ha-tim* [16] *a-dí Ša-lá-tù-a-ar ik-šu-dí* [17] 10 *ma-na* URUDU *a-dí i-Ša-lá-/tù-a-ar* [18] *ús-a-ni ga-me-er* [19] < > *[i-n]a Wa-ah-šu-ša-na lu* É *ub-/ri* [20] *ga-me-er* . 1 1/2 *ma-na* URUDU *a-dí* [21] TUR$_4$*-pì-a ik-šu-dí* 3 *ma-na* URUDU [22] *a-dí Tur$_4$-hu-mì-it ik-šu-dí* [23] *iš-tù Tur$_4$-hu-mì-it* [24] *a-dí Kà-ni-iš*^{ki} 5 *ma-na* SIG$_5$ [25] *ik-šu-da-ni* 3 *ma-na* URUDU [26] *[a-d]í Wa-ah-šu-ša-na ag-mur* (by courtesy of K.R. Veenhof).

33 Kt 91/k 437.22-23: [22] 1/2 *ma-na* URUDU *a-né-bar-tim* [23] *i-ša Tur$_4$-hu-mì-it a-dí-in*, 'I gave 1/2 mina of copper for (toll at) the crossing in the territory of Turhumit' (by courtesy of K.R. Veenhof).

These texts provide specific information about the proximity of several cities to a river. There was perhaps more than one river near Šalatuar, over which a bridge was built carrying the road to Purušhattum in one direction, and to Wahšušana in another (AKT 3, 34; kt t/k 1). At some distance from Wahšušana was another river, which was apparently only crossed by boat. The territory up to the river was part of the country of Wahšušana. A river ran through the country of Turhumit as well; the word *nēbartum* could mean a crossing by boat, but also a ford. Little is known about the level of bridge-building techniques mastered in this period. A Hittite letter found at Maşat-Höyük, written some four centuries later, contains details of a bridge. It appears that the foundation on each bank of the river was built of stone, and (part of) the superstructure of wood.[34] Bridges were probably only constructed over rivers (and canals) that were not too wide. Major rivers, like the Euphrates or the Kızılırmak, would have been crossed by boat or at a ford.

The distance between the cities is assessed by the occurrence of an inn in the itineraries quoted above, whereby 30 km was supposedly the average day's journey of a caravan. The road from Kanish to Purušhattum was covered in at least 4 days and led through Wašhania, Ninašša, and Ulama, each city circa 30 kilometres apart from the other. The journey from Turhumit to Purušhattum, frequently made to ship copper to the latter city, also took at least 4 days. The road ran from Turhumit, possibly via Tuhpia, Wahšušana, and Šalatuar to Purušhattum. The use of a boat to cross the river near Wahšušana implies that this river must have been a major one, probably the Kızılırmak.

The city at the major site of Acemhöyük was located in a favourable region where in later periods several large cities were built.[35] Finds of clay bullae with cuneiform inscriptions attest that Assyrian merchants lived in Acemhöyük, but the town's ancient name still remains unknown.[36] The important city of Purušhattum was situated in the Low Country, the plain between Konya and Niğde, called *māt šaplīti* in Hittite times.[37] Acemhöyük is often tentatively identified as ancient Purušhattum, and from the itinerary TC 3, 165 this is possible in view of the distance. But until now no text refers to a river near this city, although only one unknown text would suffice to change this picture. S. Alp suggested identifying Purušhattum with Karahöyük near Konya, where he has been conducting excavations. This seems hardly possible as the distance between Kanish and Karahöyük is nearly double that between Kanish and Acemhöyük, or any other site situated 4 days' journey from Kanish, as Purušhattum must be according to the interpretation of the itinerary TC 3, 165 given above. Level I of Karahöyük contained a town which was almost contemporaneous with the palaces at Acemhöyük,[38] and it is possible that it had economic links with Assyrian merchants. The country of Purušhattum bordered on Ulama (KTS 2, 40.27-30) and was not far from Wahšušana. It can therefore not have been situated too far south of the Kızılırmak, as Wahšušana appears to have been close to this river and lay at a

---

34 S. Alp, *Hethitische Briefe aus Maşat-Höyük*, letter 72. A collection of references to bridges in Hittite texts is given in Otten 1983. Kt n/k 1888.9-12 records the sale of a house and some kind of building near a bridge (masculine; *a-bu-sú ša tí-tù-ri-im*), Donbaz 1988b, p.191.

35 See French 1993.

36 See Tunca 1989 and Veenhof 1993.

37 Garstang and Gurney, pp.64-65. See Garelli 1989 for an outline of the city's commercial importance.

38 Alp 1993, p.193 n.64.

distance of about 2 days from Purušhattum. As the distance from Acemhöyük to the Kızılırmak is circa 70 kilometres, Purušhattum is situated here at Acemhöyük.

OIP 27, 54 provides information about Wahšušana, but because of the character of the details cannot be compared with TC 3, 165. It shows the existence of a route Kanish - Wašhania - Malita - Wahšušana. Wahšušana is likely to be situated north of the line Purušhattum - Kanish, not only because of its intermediate position between Turhumit and Purušhattum, but also on the basis of ATHE 63. The sender of that letter, being in Purušhattum, asks Imdilum to send his textiles, which are brought from Zalpa and Hurama to Kaniš, on towards Wahšušana, '*harrān Tawinia-ma*'. Whatever the exact location of Tawinia is, it is definitely north of Kanish (RGTC 6, p.417; RGTC 6/2, p.167). In this instance, the goods travelled not by the usual road to Wahšušana, if OIP 27, 54 describes this route, but instead (*-ma*) by a northern one. '*Harrān Tawinia*' in this context does not mean 'via Tawinia', but via the road leading to Tawinia. Kt t/k 1 makes it probable that Wahšušana lay near the Kızılırmak.

Šalatuar was situated near a small river. It harboured an Assyrian *wabartum* settlement, which means it held a less prominent place in trade than for instance Wahšušana or Purušhattum where *kārus* were situated, although the reason for this modest place might have been the proximity of these large cities. If Acemhöyük with the Karasu Çay is the site of ancient Purušhattum, this would place Šalatuar north(-west) of Aksaray. This fits in with a passage from kt c/k 445, where it emerges that Purušhattum was the southernmost of the cities mentioned. In this text, Tahašaili entrusted an amount of copper in Wahšušana to a trader who brought it to Purušhattum. On his return, the trader went, after passing Ulama, directly to Kanish, and not to Wahšušana, as there was a blockade in that country (see for this text and the archive it belongs to § 4.3).[39]

Turhumit was a major trade centre and possessed an Assyrian *kārum*.[40] Various attempts have been made to locate this town, mainly on information contained in Hittite documents (see RGTC 6, pp.443-444 and RGTC 6/2, p.175) A location north-east of Kanish for Turhumit can be excluded. The road from Wahšušana to Turhumit led via Tuhpia in kt 91/k 424. Tuhpia was located near Turhumit, as is also clear from Hittite sources.[41] Kt 91/k 437 demonstrates that Turhumit was near a river. It is uncertain whether, coming from Wahšušana and Tuhpia, this river had to be crossed before reaching Turhumit. If the road from Wahšušana to Tuhpia led across the Kızılırmak, and Tuhpia was therefore situated north of this river, one day's journey seems to be insufficient if Turhumit was situated west of the Kızılırmak. The considerable role it played as a copper market makes it likely that Turhumit was the city situated closest to the mining areas west of the Kızılırmak.

39 Kt c/k 445.6-11: [6] ... *a-na Pu-ru-uš-ha-tim* [7] *ú-šé-ri-ib* [8] *a-na-kam ki-ma sú-<ku>-ur-tù-ni* [9] *iš-tù Ú-lá-ma* [10] *a-na Kà-ni-iš*^{ki} [11] *i-tal-kam-ma*.

40 See Michel 1991b for a discussion of Turhumit's commercial role.

41 Otten 1959, pp.356-357.

1.2.2 *The copper-producing regions*

Unlike the often detailed topographical references available of copper reserves and a number of ancient mining sites, the location of most of the toponyms mentioned in the texts in connection with copper trade in Anatolia can only be made with a certain degree of probability.

At present, eleven towns or regions are known which gave their name to certain qualities of copper: Haburata, Ištanuwa, Kunanamit, Sawit, Tapašatta, Taritar, Tišmurna, and Tuhpia, Purušhattum, Wašhania, and Ninašša. The references will be discussed in Chapter 3. The towns or regions where copper was produced were:

Haburata (see p.43; cf. RGTC 4, pp.44-45) does not seem to have any connection with the toponym Habūrā(tum) in Syria or Iraq (see for this city in Old Babylonian texts D.-O. Edzard, *RlA* 4, p.29), apart from the name itself. The texts do not specify where copper thus labelled originated. It is named together with copper from Taritar in two texts. Haburata might be situated west of the Kızılırmak. A location in the Pontic area is less probable, and an identification with classical Cabira (modern Niksar) as proposed by J. Lewy,[42] is impossible. The identity of both names is not 'obvious', as Lewy wants the reader to believe, and Strabo 12,3.30, does not specify which mineral was produced from the nearby mines.

Ištanuwa is attested in only one text (kt n/k 1131; see p.44), where it is spelled Ištunuwa, with 'u' instead of 'a' according to the Assyrian rule of vowel harmony. It is identical with Hittite Ištanuwa (cf. RGTC 6/2, p.56). This has been located near Gordion,[43] but might have been closer to the Kızılırmak.

Copper from Kunanamit (see p.44; cf. RGTC 4, p.73) was exported from or via Sawit in CCT 2, 23. It is possible that Kunanamit belonged to this country.

Sawit (see p.44; cf. RGTC 4, p.96) occurs in two texts. In CCT 2, 23, it was the region from which a consignment of fine copper and Kunanamit copper was transported. This text belongs to the archive of Imdilum, and appears to relate events which happened in the area between Turhumit and Purušhattum. In AKT 3, 56, copper from the 'country of Šawit' was to be transported to Wahšušana, and this makes it likely Sawit was a region bordering on Turhumit.

Tapašatta (see p.44), probably the same as Hittite Tappaššanda (cf. RGTC 6, p.400), is attested only once in an Old Assyrian text (kt n/k 1952). This document records a statement by witnesses before the *kārum* of Hattuš (Boğazköy) that two merchants agreed on the delivery of copper from Tapašatta. The origin of this copper is not given, but could have been a region north or north-west of Boğazköy.

The region of the Devrez Çay near Çankırı has been proposed as the location of Taritar (see p.44-45; cf. RGTC 4, p.115; RGTC 6, pp.406-407).[44]

Tišmurna (see p.45; cf. RGTC 4, pp.121-122) was an alternative to the copper market of Turhumit and was apparently located near that city. M. Forlanini located it at Karaalı,

---

42 Lewy 1958, p.94.
43 Forlanini 1987, p.115.
44 Forlanini 1985, p.50 based on de Jesus 1980.

south-east of Ankara on the basis of remains found there from ancient copper mining which were mentioned by de Jesus.[45] An identification of Tišmurna with Smyrna (= Izmir; B. Landsberger in Garelli, RA 60, p.120 n.1, repeated in Röllig 1983, p.346) is impossible in view of the distance this would involve.

Most of these copper-producing regions or towns were then situated west or north-west of Wahšušana, probably in or close to what de Jesus defined as the Yapraklı Group (Group 2), near Çankırı. If a trader did not obtain the copper there, he could do so in Turhumit. From there, the copper was transported in a south-eastern direction as far as Kanish and Purušhattum. There is only one reference to copper stored in Hattum (VS 26, 56.39), the region north of Kanish. This copper, as well as that from Tapašatta, may have originated from the Yapraklı Group of mines as well.

Note that not a single reference mentions copper brought to Kanish from a region east of it. ATHE 63 does not contain a reference to 'copper stored in Hurama'. Read with Kh. Nashef (RGTC 4, p.60) as first sign in line 17 '*ù*' instead of 'URUDU'. Notwithstanding the uncertainties in locating the mining areas named in the texts, it is evident that copper mined at Ergani near Elazığ did not play a role in the trade conducted west of Kanish. This is in spite of repeated allegations that Ergani was the main source for copper in the Old Assyrian period, mainly based upon the assumption that such a rich mine must have been worked at that time. J. Lewy had located Wahšušana near Niğde.[46] This led him to identify *ha-ra-na(-a)* in KTH 14.18 and 24, interpreted as a toponym, with a village called Tell Charan, '30 kilometres southeast of Erganimaden'.[47] The signs in both lines from KTH 14 (**17**) are, in my opinion, better explained as a part of a personal name.[48] This means that there is no reference to copper mined at Ergani. It is probable that Ergani was indeed worked during this period, but the availability of copper deposits north and west of Kanish, and the costs of transporting the metal from Ergani across the Taurus mountain range to the west, make it very unlikely that copper mined there was brought to the area west of Kanish. Copper from Ergani might have been transported to Northern Mesopotamia and Assur, but the texts from Kültepe do not contain information on this aspect of international trade.

Finally a word on J. Lewy's identification of two areas in the vicinity of Qinnasrīn, south of Aleppo, 'from which copper was exported in the times of the Old Assyrian empire'.[49] The first geographical name is S/Šawit, which Lewy identified with Mount Sawe (*Sa-ú-e*) mentioned in the annals of Tiglath-Pileser III (744-727 B.C.). Sawe would be modern Mount Zāwī, some 60-70 km to the south-west of Qinnasrīn. Although the exact location of S/Šawit remains unknown (see above), it is highly improbable that it was situated in Syria.

---

45 Forlanini 1985, p.50 n.40; de Jesus 1980, pp.237-238 and 240-246 and Map 8.

46 Lewy 1947, p.15.

47 Lewy 1956, p.22 n.95.

48 CAD S, p.97b, Kh. Nashef, RGTC 4, p.51, and Hecker 1993b, p.59, favour the reading *ša Harrana*. In KKS 13b.6, I suggest to read: *Šu-na-bar* <DUMU> *Ša-ha-ra-*[*n*]*a*, cf. the inscription of seal 46 (Teissier 1994 no. 248).

49 Lewy 1952, p.417.

The word *ha-sú-a-am* (see p.39-40) occurs in CCT 1, 42a.22[50] as qualification of 10 minas of copper. It was explained by Lewy as a form of the toponym spelled *Ha-su* in the annals of the same Assyrian king. He identified Hasu with Mount al-Ḥaṣṣ, about 25 km east of Qinnasrīn.[51] But according to the text, the 10 minas of copper were brought to the writer of CCT 1, 42a 'from the City', when he was somewhere in Northern Syria. Although not exclusively, 'the City' usually refers to the Assyrian capital Assur. The copper therefore probably came from Assur, and this precludes a Syrian origin. The word *hassu'um* bears no relation to a toponym. Neither Sawit nor *hassu'um* refer to Syria as the origin of copper. This is in line with the opinion generally held by geologists that Syria lacks copper deposits, although a passage from the annals of the Assyrian king Sargon II (721-705 B.C.) describes the mining and smelting of copper in the Jabal al-Aqra^c in Northern Syria.[52]

1.3 *Conclusion*

It is unlikely that analysis of copper artefacts and ores will permit a convincing identification of the copper deposits from which copper used in Kanish derived. Fortunately, textual evidence is available which contains references to towns and regions connected with the production of and trade in copper. When this evidence is combined with results from archaeo-metallurgical research, the preponderance of geographical details for the region along and west of the Kızılırmak is striking. This situation reflects the trade network operating during part of the 19th century B.C. which served Kanish and the cities west of it, and is not conditioned by the archives studied.

---

50 CCT 1, 42a.21-23: [21] ... 10 *ma-na* URUDU [22] *ha-sú-a-am iš-tù* [23] *a-lim*^{ki} *ú-ša-ak-ší-/dam.*

51 Lewy 1952, p.420.

52 See Muhly 1973, pp.209-210.

CHAPTER 2
COPPER PRODUCTION

2.1 *Introduction*

The primary copper ores are sulphides and arsenides. The most important are the sulphide ores, which consist of copper and iron combined with sulphur. Oxide ores and ores containing native copper are the result of the weathering of sulphide ores, and are usually found near the surface of an ore layer. Further erosion may cause the disappearance of the matrix in which native copper is contained, leaving the copper as nuggets, sometimes the size of a fist. Lumps of native copper and the copper oxides malachite and azurite, which are easily recognizable by their green and blue colours, would have attracted the attention of early miners, probably making these the first copper ores which were smelted. When the oxidized layer of a copper deposit had been exhausted, the subsequent layers consisting of sulphide and arsenic ('grey copper') copper minerals, at greater depth in the rock, would have been mined, requiring a more sophisticated mining and smelting technology.

The extraction from oxide copper ores is relatively simple compared to that from sulphide ores. A concentrate is made by removing unwanted gangue material from the ore, which can then be broken up. Iron containing fluxes, like haematite, are added to facilitate the smelting process, and together with the ore-concentrate are smelted with charcoal. The smelting results in raw copper and slag. The raw copper can be refined by remelting.

Due to their sulphur and iron contents, sulphide ores require a different pyrometallurgical technique to extract the metal. To diminish the levels of sulphur and iron the concentrates are roasted. This is done by exposing the ore-concentrates to air at temperatures between 500° and 700° C. The ore is piled up in heaps over a layer of wood and then ignited. It is left burning for a considerable time, depending on the sulphur content. The sulphur is partially oxidized to sulphur dioxide, which is eliminated as a gas during roasting. The iron sulphides are partially oxidized to iron sulphates and oxides and at the subsequent smelting stage are eliminated as slag. Copper sulphates and oxides are also formed during the roast but they can be reduced back to sulphides during smelting. The product of a roast is called calcine. It is a mixture of oxides, sulphates, and sulphides. The next step is the smelting of the calcine. The furnace charge of calcine is melted at a temperature of about 1200° C usually with silica flux. The silica, alumina, iron oxides, lime, and other minor oxides form molten slag. Copper, sulphur, unoxidized iron and the precious metals form matte. This matte is roasted again if necessary, to reduce the amount of iron, thereby oxidizing iron sulphides and eliminating more sulphur. It is then melted in order to obtain raw or blister copper, named after blisters of sulphur dioxide that are formed on the surface during solidification because of the presence of sulphur and oxygen in the metal. During the melting of the matte the oxidized copper is reduced to raw copper and the oxidized iron is slagged off.

The raw copper (blister, or black copper) obtained after smelting, is a solution and partially also a mixture of copper with other metals and non-metals in different quantities. These impurities can be eliminated from the copper by refining by fire. Most of the impurities possess a greater affinity to oxygen than copper does; for them a selective oxidation by blowing air into the melt is possible. After oxidation, the resulting oxides are slagged off (iron) or evaporate (sulphur). The raw copper, initially dark, gradually takes on the typical reddish colour with a decreasing proportion of impurities. The colour is thus an important factor to assess the purity of the metal. Refining can be performed in crucibles. Refining by fire, however, is not the only way to reduce the degree of impurity; the quality of raw copper already improves by simple remelting.[53] To reduce the amount of oxygen which has dissolved in the copper during melting and refining (oxidation of sulphur above), hydrocarbons (wood) are injected into the molten copper. This step is called 'poling'. The origin of this simple method is unknown; the first reliable documentation for its application comes from Theophilus (12th century AD),[54] but the technique is probably older.

A black layer of scale (copper oxide) forms on the surface when the copper is still hot after melting and pouring it into a certain shape. This gives the product a black, unattractive appearance. To free the copper of the scale, the hot product is submerged in cold water, which makes the scale come off, and the copper gets a reddish colour.

In order to provide a better insight into how copper ore was processed into metallic copper during the time of the Old Assyrian involvement in Anatolia (19th century B.C.), the major results of archeo-metallurgical investigations at several smelting sites from other regions, and often also dating to other periods, will be summarized. These sites, though differing in ecology and types of ore, give an indication of the technological practices that could also have been used in Anatolia, but for which concrete evidence is often lacking.

2.2 *The smelting of oxide ores in the Wạdi Arabah*

2.2.1 *Timna*

The mining area around Timna, in the south-western part of the Wadi Arabah, has been thoroughly investigated during the past decades. Although the amount of copper produced there does not match that of Feinan (see § 2.2.2), the remains of old mining and metallurgy have been left to a large extent undisturbed, due to the isolated and arid nature of this area. According to B. Rothenberg, the beginning of copper production was in the fourth millennium B.C. The only period of large scale industrial copper industry at Timna was during the 14th-12th centuries B.C. (Layers III and II). For about 150 years, expeditions from Egypt in search of copper worked the ore deposits. A short revival of Egyptian activities occurred in the 10th C (Layer I, Iron Age II). More than a millennium later, in

53 Merkel 1990, p.117.
54 *De diversis artibus*, lxvii; in C.R. Dodwell's edition p.125.

Roman-Byzantine and Early Islamic times, the Timna deposits were worked on a small scale.

The deposits consist mainly of oxide ore. Local ores were used for intentional fluxing with haematite, and during a short period (10th-8th centuries B.C.) also with manganese. Charcoal was made mainly of acacia trees growing nearby. During excavation, fragmentary furnace structures, furnace lining, slags, and tuyères were discovered. At the ancient smelting sites some remains of metallurgical significance were found. The interpretation of these archaeo-metallurgical remains together with stratification and other archaeological evidence enabled Rothenberg to postulate production methods and installations specific to certain periods.

In Rothenberg's reconstruction of the chronology of the Arabah, Sinai, and the adjacent arid areas relating to the period c. 4500-2000 B.C. (Sinai-Arabah Copper Age), three major phases of technological and cultural developments can be discerned: c. 4500-3000 (Early Phase), c. 3000-2200 (Middle Phase), and c. 2200-2000 (Late Phase). This interpretation which is based on a reconstruction of the development of copper metallurgy and on the typology of flint tools and pottery, differs from the generally accepted view that most early sites in this region are Aradian Early Bronze II settlements. Rothenberg's dating of the remains from Site F 2 and Site 39 to the second half of the fourth millennium B.C. has been criticized. Muhly considered the metallurgical and archaeological evidence unconvincing. In his view, there is little proof for copper smelting prior to the Ramesside period.[55] Rothenberg defended his chronological framework by publishing further investigations into the typology of flint tools and pottery from Timnian sites.[56] His interpretation is followed here because evidence from Feinan suggests that there copper smelting was practised as early as the fourth millennium as well.

Typical representatives of the Early Phase are Site F 2 and Site 39. Site 39 probably dates into the second half of the fourth millennium B.C. It consisted of smelting furnaces operated on top of a hill, with, at the foot of the hill, a workshop area which also comprised habitation. The furnace investigated was bowl-shaped, a mere hole-in-the-ground with a low stone superstructure. In the proposed reconstruction, its diameter was 20-30 cm, its height including the superstructure 40-50 cm.[57] No tuyères were found, but a large stone at the back of the furnace might have functioned as the basis for bellows. Many early smelting sites in the Arabah were located on top of hills and hillocks, where they were open to the north wind,[58] but it is uncertain whether this natural draught alone was sufficient for a smelting operation. At the very small smelting site F 2, probably the earliest in Timna (Late Neolithic or Early Chalcolithic), fragments of primitive tuyères were found, but this furnace at Site F 2 is located in a wadi bed. The temperatures reached in the furnace at Site 39 did not allow for a separation of the metal from slag and no tapped slag has been found. If smelting occurred in a furnace, the whole of the furnace contents would have been raked out at the end of the smelting operation. However, there are indications that smelting was

55 Muhly 1984, p.287.
56 Rothenberg and Glass 1992.
57 Rothenberg 1990, p.5.
58 Rothenberg and Glass 1992, p.143.

conducted in a crucible, in which case the 'furnace' functioned rather as a charcoal bed.[59] The smelting slag had to be crushed in order to free the metal prills, droplets, and lumps enclosed in a fayalite matrix.[60] The use of iron oxides for fluxing is attested for this period. The total production of copper at a site of this type has been estimated at some tens of kilograms.[61]

In the Middle Phase, smelting occurred in 'cottage industry' workings, close to the houses of the settlement. Compared to the previous phase, an advanced smelting technology was used, which employed better bellows. Together with the use of iron oxide fluxes, a better separation of metallic copper from slag was possible, but, as in the preceding period, this slag contained much copper (2-6%) and had to be broken up into smaller fragments (1-2 cm) to extract lumps of metallic copper. There are indications that some of the slag was tapped out of the furnace.[62]

The Late Phase is a few centuries earlier than the period of the Assyrian involvement in Anatolia. Compared to the Middle Phase, the basic metallurgical technology remained unchanged. The tapped fayalite slag was mostly broken up to extract crude copper prills which were subsequently refined. There was a remarkable increase in the amount of copper produced, at Site 149 (hilltop smelting site, ore crushing workshop at the foot of a hillock) estimated at hundreds of kilograms.[63] The excavator assumes that improved smelting techniques had been caused by an increasing demand for copper in this period, to a large extent resulting from renewed trading contacts with Palestine. It is possible that contemporary bar-ingots found in the Negev and in South Palestine originated from Timna smelting sites of this period.[64]

In the 14th-10th centuries, the time of the main Egyptian mining activities at Timna, the furnace technology was sufficiently advanced to permit a liquid separation of copper and slag. This was the result of larger and better insulated furnaces and the use of blasts of air supplied by pot bellows through tuyères.[65] Site 30 was a walled smelting camp in the centre of the Egyptian copper industry at Timna. Like most of the other New Kingdom sites of the Southern Arabah, it was abandoned by the Ramesside mining expeditions in the middle of the 12th century B.C. The furnaces from this site are tubular and bowlshaped. Interestingly, the furnace bottoms have a slightly concave surface without any intentionally formed cup-, bowl-shaped or other receptacle for the liquid metal.[66] The slag, containing inclusions of metallic copper, was tapped out of the furnace and formed slag plates. These plates were broken up into pieces of various sizes (2-5 cm) to free the enclosed metal, which seems to have been considered an important portion of the smelting product. The freed copper lumps were melted and cast into small ingot-like pieces of copper.[67] At the same time another

59 Hauptmann et al. 1993, p.568.
60 Bamberger et al. 1986, p.1172.
61 Rothenberg and Glass 1992, p.143.
62 Rothenberg and Glass 1992, p.144.
63 Rothenberg and Glass 1992, p.144.
64 Rothenberg and Glass 1992, p.151.
65 Bamberger et al. 1986, p.1167.
66 Rothenberg 1990, p.12.
67 Rothenberg 1990, p.63.

furnace type was operated, dubbed 'intermediate' by the excavator. An example is furnace Z at Site 2. It consisted of two semi-attached, round, rock-cut pits, the smaller of which was the actual smelting hearth, the larger the slag tapping pit.[68] The slag from this furnace was black and had a hole of 10-15 cm diameter in its middle, probably to facilitate the removal of the slag cake weighing 15-25 kg. This 'ring slag' contained relatively little metallic copper.[69]

During the 10th-8th C (Layer I; Iron Age II) only a relatively small quantity of slag was produced, pointing to a short period of Egyptian metallurgical activity. Site 2 revealed two bowl furnaces with a slag tapping pit in front. As usual, the furnace part above ground level was not preserved. The large slag-cakes (15-25 kg in weight) contained almost no metal.[70] At Site 30 the remains of a furnace display a diameter of c. 40 cm for the underground smelting hearth. A shaft above ground level with a diameter of c. 20 cm is reconstructed on the basis of all furnace fragments.[71] Among the remains of furnace lining, 21 fragments of moulds for the casting of plano-convex, oval, or rectangular copper ingots could be identified. These moulds were made of slag-tempered clay and as a result of usage they were heavily sintered on top.[72]

2.2.2 *Feinan*

Surveys and excavations in the Jordanian part of the Wadi Arabah, in the area of Feinan, Khirbet an-Nahas, and Wadi al-Jariya, have provided evidence for extensive mining of the rich copper deposits from the second half of the fourth millennium B.C. to the 13th century AD. In total, some 150,000 - 200,000 tons of slag have been found in the Feinan area, making this the largest ancient smelting site in the Middle East, followed at a distance by that near Bir Nasib in the Sinai. The majority of the slag heaps were produced during 800-400 B.C. (Iron Age IIC) and during the 1st-4th C AD (Roman period). Sporadic metallurgical activities are attested for the first time at the end of the fourth millennium. In the period 2700-2300 (EB II-III), the production of copper was extensive, but traces of mining during the second millennium are rare. The mining of copper which can be dated to 2700-2300 B.C. seems to have continued into the Middle Bronze Age I without major technological change. An indication of this is that from 2700-2300 till the end of the Bronze Age manganese-silicate (knebelite) slag was produced. The Chalcolithic slag, in contrast to that of the Early Bronze Age, is almost free of manganese-oxide. The Iron Age miners were forced to develop new techniques to reach the ore-layers and smelt the ores, and in this period copper mining was practised almost on an industrial scale. The discovery of mining galleries from the EBA and MBA was hampered by the fact that these had been carefully refilled with material after the ore had been mined.[73]

---

68 Rothenberg 1990, p.16.

69 Rothenberg 1990, p.44; see for the copper contents of resp. 'small slag' and 'ring slag' Tables 2 and 3 on p.69.

70 Bamberger et al. 1986, p.1168.

71 Rothenberg 1990, p.46.

72 Rothenberg 1990, p.54 and Fig. 81 and 82.

73 Hauptmann et al. 1985, p.172 and Abb. 10.

In the Chalcolithicum, copper from carefully selected oxide ores of high purity, like malachite, was smelted in crucibles, resulting in raw copper with in general a low level of impurities. The ore deposits do not contain native copper. Pieces of ore from Feinan have been found at a number of Neolithic and Chalcolithic sites in Palestine, and perhaps also in Egypt (Maadi). The discoveries of these lumps of ore are used to corroborate the dating by pottery of Site Wadi Fidan 4 to the Chalcolithic period.[74]

During 2700-2300 B.C. a sharp increase in copper production is attested for the Feinan area, and more than 15 smelting sites with approximately 1,000 tons of slag can be assigned to this period. The smelted copper-manganese ores with a content of c. 25% copper were not as rich as the malachite mined in the preceding period (34.3-51.8%), but the manganese acted as a natural flux which facilitated the smelting process. These ores could be reduced in a one-step reduction process without prior roasting.[75] At site Feinan 9 a battery of some 28 smelting furnaces has been found, situated on top of a hill.[76] These furnaces, consisting of a bowl-formed rear 60 cm in width and 30-40 cm high, with relatively low sides and a semicircular furnace bottom sloping forward, were without a closed front. There are no indications of a dome covering these structures. Remains of tuyères near smelting sites from this period were not found. The slag, originally about 20 cm in diameter, had been broken up into smaller pieces, still displaying a vertical structure containing small prills of copper.[77] In a tentative reconstruction of the metallurgical process used during this period,[78] the furnaces are considered shallow, natural draught smelting furnaces without a shaft, a reconstruction which differs considerably from current reconstructions of such remains as domed shaft furnaces. The temperature of about 1200-1250° C which could be obtained under these circumstances was sufficient for the reduction of the copper-manganese ores. Slag and copper were probably removed together from the furnace after solidification. After the pieces of metal were recovered from the slag they were melted together in a crucible, and then cast into an ingot.

The amount of Middle Bronze Age pottery found is quite small, and it is inferred from this that the metallurgical activities during this period were conducted on a rather small scale.

In the period 800-400 B.C., the ore layers consisting of copper and manganese mineralizations were further worked. To reach the ore deposits shafts reaching a depth of 30 metres were dug. The ore was smelted at two large smelting sites, Feinan and Khirbet an-Nahas. The big slag cakes reportedly had a weight of 40-60 kg. In this period the slag was still crushed to free inclusions of metallic copper.[79] The slag, fragments of tuyères, and remains of smelting furnaces found in the context of the slag heaps from the Iron Age in Feinan and Khirbet an-Nahas display a strong similarity in shape and size to those, a few centuries older, from Timna Site 30 Layer I. They suggest a similar kind of smelting technology and furnace construction at Feinan and Timna.

---

74 Hauptmann 1989, p.122.
75 Hauptmann et al. 1985, p.179.
76 Hauptmann et al. 1985, p.177; Abb. 32 shows sites Feinan 9-11.
77 Hauptmann and Roden 1988, Abb. 4.
78 Hauptmann and Roden 1988.
79 Hauptmann et al. 1985, p.177.

2.3 *Oman, Al-Maysar 1 (c. 2200-1900 B.C.)*

Large slag heaps in the Sultanate of Oman show that in the past a considerable quantity of copper was produced in this region.[80] A number of surveys were conducted, some to explore the possibilities of modern copper mining, others to investigate old workings. Today, nearly 150 copper mines have been identified. The old slag heaps in Oman can mostly be dated to the Early Islamic Period. But in their vicinity, smaller ones dating to the Bronze Age were recovered. All copper deposits worked during the Bronze Age which were not exhausted were presumably mined again during the Islamic Period, and this resulted in the destruction of most evidence of the earlier phases of mining. Remains of copper production during the third millennium B.C., resembling those from Maysar 1, were found at Bilad al-Maaydin, Mazra, Mullaq (c. 10 t slag), Musfa, Bir Kalher (c. 200 t slag), and Bahaysah al-Woqbah (c. 200 t slag). This strongly suggests that the production of copper in Maysar was not an isolated feature, but formed a part of a structural enterprise on a regional scale. On the basis of the slag heaps discovered, the quantity of copper smelted in Oman during the third and second millennia B.C. is estimated at 2,000-4,000 tons.[81]

The German Mining Museum from Bochum conducted a survey during the seventies, and in two campaigns (1980 and 1981), a settlement dating to the second half of the third and first centuries of the second millennium B.C., the so-called Umm an-Nār Period, was excavated near the village of Al-Maysar.[82] Maysar 1 (Samad 5 in the Harvard Archaeological Survey) thrived c. 2200-1900 B.C. The remains of several buildings were excavated in what appears to have been the craft area of Maysar 1 during this period, and traces of metallurgical installations and activities were found. The reconstruction of the production of copper at this site presented in Hauptmann 1985 is based on these data.

Ore consisting of secondary copper mineralizations with a high copper and low iron content, like malachite, chrysocoll, and brochantite, closely connected with silicate gangue and containing small amounts of copper-sulphides and copper-iron-sulphides, was brought into the settlement to be smelted. These ores contained 5-56% copper. Nothing indicates that the ore was roasted, and this seems to exclude the use of sulphide ores. The level of arsenic in copper artefacts and matte from Al-Maysar is high, and it has been suggested that the ores used in antiquity must have contained more arsenic than those which can be found today in that area.

Near the mines, the ore was separated by hand from its gangue material; inside the settlement the ore underwent a further preparation. In smelting furnaces with a content of some 0.01-0.015 m^3 and a height of approximately 40 cm, the ore was smelted in several hours, using iron-oxides (haematite, limonite) as flux. It remains possible that at this period ore was still smelted in a crucible.[83] The slags were removed from the furnace together with the reduced copper and the matte and solidified outside the furnace. At the bottom of the slag, nodules of metal and matte, still containing some slag particles, were formed. These

80 Hauptmann 1985, p.16 Abb. 3.
81 Hauptmann 1987, p.209.
82 Weisgerber 1980 and 1981.
83 Hauptmann et al. 1993, p.568.

pieces of matte and copper were freed from the slag, and supposedly melted without further treatment. By repeated melting, copper and matte were combined into large pieces. The resulting product consisted of metallic copper (raw copper) and a matte with c. 0.57% iron.

The excavations provided direct evidence for the melting of pieces of raw copper in crucibles and the subsequent pouring of the copper into semicircular holes in the floor to form planoconvex ingots. In Maysar 1, House 4, depressions with a diameter of about 15 cm were found in the floor next to two fireplaces.[84] In a trench between both fireplaces, a hoard with 22 fragments of copper ingots weighing more than 6 kg was discovered. The composition of the pieces of raw copper is the same as that of the ingots and shows that the copper in Maysar 1 was not refined or alloyed up till the making of the ingots.[85] A hoard containing 16 complete planoconvex ingots, with a diameter of c. 15 cm or more, was discovered in Al-Aqir in the Wadi Bahla, dating to about 2000 B.C. One of these ingots was the result of ancient swindle; it had a surface of two layers of metallic copper, while the core consisted of slag, which formed half of its volume.[86]

This reconstruction is entirely based on the evidence from Maysar 1 (c. 100 tons of slag), which was only a small settlement where the production of copper played an important role in the local economy. The technology used in smelting the copper ore was presumably representative of that practised at other places in Oman during this period. The amount of copper produced was more than that necessary to satisfy local demand and it is likely that the surplus was used in trade with Mesopotamia.

Copper ingots were also found in Bahrain and Failaka. As this region does not possess copper mines itself, it is evident that this copper was imported. Some, if not all, will have come from Oman. At Saar in Bahrain during excavations, two bun ingots and fragments of several more were found in layers dating to the end of the third millennium. One has a diameter of c. 10.5 cm, the other is slightly larger.[87] A fragment of a planoconvex ingot was found in Failaka. Its diameter was c. 14-15 cm, the height 4.5 cm.[88]

Evidence is growing for an identification of Oman with the country of Magan, which is mentioned in cuneiform texts from southern Iraq dating from the second half of the third and first centuries of the second millennium B.C. In these texts, Magan occurs as a country from which, among other commodities, a considerable amount of copper was imported. After circa 2000 B.C., the trade between southern Mesopotamia and Magan was conducted indirectly via Tilmun (= Bahrain).[89]

2.4 *The smelting of sulphide ores at Mitterberg, Austria (c. 16th century B.C.)*

In the Mitterberg mining area chalcopyrite ores containing few impurities were mined and smelted. An investigation of Bronze Age mine shafts (in German called *Pingen*) near

84 Weisgerber 1981, p.209 and Abb. 38.
85 Hauptmann 1985, p.93.
86 Hauptmann 1987, p.210 and 217.
87 Crawford 1993, p.11, and letter of 18 February 1994.
88 Hurtel and Tallon 1990, Catalogue no. 52.
89 W. Heimpel, *RlA* 7, p.198; for the textual evidence see Heimpel 1987.

Bischofshofen, east of the Mitterberg, revealed that in that period open cast mines were dug, using fire setting and stone tools to mine sulphide ores like chalcopyrite but also *Fahlerze*. At the bottom of one of the shafts (no. 7), ceramic fragments from the end of the Early Bronze Age were discovered.[90] Investigations concentrate on smelting sites dating from the second half of the second millennium B.C. onwards. The sources of information are furnaces, smelting slags, and ingots. The stratigraphy of the smelting sites is not clear enough to assign them a date; however, pottery containing grains of slag showing the same characteristics of (oxidizing) smelting has been found in a well-stratified context, dating to the 16th century B.C. at the Götschenberg close to the Mitterberg. In the same excavation, grains were recovered from older layers, pointing to the possibility of an older, reducing smelting of weathered ores.[91]

The ore was smelted after a light roast. To produce blister copper, the molten matte was oxidized by blowing air into the melt, a method strongly resembling the Japanese *mabuki* process, documented since the 16th century AD.[92] Pieces of metallic copper from the Mitterberg and ingots discovered in Salzburg contained c. 1% sulphur. This means the copper was probably not refined, but traded as blister copper. Even a slight refining operation would have reduced the sulphur content considerably.

A comparison of ingots (*Spangenbarren*) from Obereching (also dated to the 16th century B.C.) with the ores from Mitterberg shows a clear resemblance to these ores in elemental composition.[93] In four depots nearly 500 bars weighing circa 100 kg were found. The average length of the bars is 18 cm; about 90% of these weigh between 170 and 220 grams. The differences in weight are largely due to the purity of the bar: heavy bars contain pure copper, while inclusions of slag result in a smaller weight. The many inclusions, like charcoal and slag, indicate that the blister copper was not melted again or refined at the smelting site, but immediately cast into ingots.

2.5 *Cyprus*

In spite of the discovery of many old workings and tools after the reopening of the copper mines in the Troodos massif in 1920, little is known of metallurgical practices in Cyprus during the end of the third and beginning of the second millennium B.C. This can be largely contributed to the fact that later mining destroyed much evidence from previous periods. Few archaeo-metallurgical remains have been found and in general their dating presents many problems. Where remains of ancient mining could be dated, they turned out to originate mostly from later periods.

At Ambelikou Aletri, in the North-West of Cyprus near the Ambelikou copper deposit, remains of copper metallurgy were discovered dating to c. 19th century B.C. (EBA).[94] It has been suggested that the clay crucible found there was used to remelt copper prills

90 Gstrein and Lippert 1987, p.98.
91 Moesta et al. 1989, n.28.
92 Lewin and Hauptmann 1984.
93 Moosleitner 1988.
94 Merrillees 1982, p.374.

resulting from the smelting of oxide ore with manganese flux, but a different reconstruction of the metallurgical processes applied can not be excluded.

A survey of the Kalavasos area in the valley of the Vassilikos river, near the south coast, revealed a concentration of sites from the 18th-15th centuries B.C. (MBA). This could be related to its location on the route from the copper mines north of the village of Kalavasos to the south. Evidence for metallurgy from this period in this area is very limited. At Laroumena a crucible and pieces of slag were discovered, which might suggest that the copper used was mined in the vicinity.[95]

At present, the oldest installation to smelt copper ore discovered in a mining area was found in Ayia-Varvara - Almyras, where the remains of a smelting furnace, dating to the 7th-5th centuries B.C. were excavated. Slags suggest that this furnace was used to smelt sulphide ores.[96] All other knowledge of metallurgical practices during the second millennium B.C. stems from excavations conducted in ancient settlements which date to the 13th-12th centuries. It is uncertain whether sulphide ores were smelted inside cities like Enkomi, Athienou or Kition; it is more likely that the copper ore was smelted in the mining area, after which the smelting product consisting of matte and slag (the slag itself containing matte and some metallic copper) was transported to the city for further treatment.

2.6 *Turkey*

As in many other mining areas, the search in Turkey for archaeo-metallurgical remains from the second millennium B.C. or earlier is severely hindered by frequent mining and processing of ore on or near the same site in later periods, which very often destroyed the earlier evidence or made it unrecognizable. Remains of copper smelting dating to the middle of the fourth millennium B.C. were discovered at Murgul, in the North-East of Turkey, near the border with Georgia. Five slag heaps contained in all some 200 tons of slag, and charcoal found in the slag was dated by the ^{14}C method. The circular slag cakes, with a diameter of 13-20 cm, 3-4 cm thick, and weighing 1.5-3 kg, are the only metallurgical remains. The fayalite slag with pieces of metallic copper solidified on the bottom of the furnace and was not broken up. On the underside of the slag the impression of the copper regulus was visible. Slag investigations demonstrated that oxide ores had been smelted.[97] The copper production has been estimated at 20 tons at least, which means that the smelting technology at Murgul in the fourth millennium was no longer in the 'trial and error' phase, as at Timna and Feinan, but was performed on an almost industrial scale. This makes Murgul the oldest smelting site discovered thus far.

At Kozlu, some 30 km north of Tokat, and close to the ancient site of Horoztepe, a large rubble fill of an old mine was found, containing several remnants of mine timber. One chunk of wood was dated at about 2800 B.C.,[98] another at c. 4650 B.C.[99] The ore mined

---

95 I.A. Todd, *RDAC* 1988, pp.133-140. A year later, Todd wrote that after this survey 'the evidence for M.B.A. / L.B.A. metallurgy remains as elusive as ever in the mining area', *RDAC* 1989, p.49.

96 W. Fasnacht et al., *RDAC* 1992, pp.59-74.

97 Wagner et al. 1989, pp.655-57; Hauptmann et al. 1993, p.546; Lutz et al. 1994.

98 Giles and Kuijpers 1974, p.823.

was probably oxidic. A large planoconvex copper casting, consisting of copper matte and about 40 cm in diameter, was found 3 km from the Kozlu mine; no date for it is available.

According to E. Kaptan, furnace lining fragments connected with copper smelting were also found in a small slag heap from the Early Bronze Age near Tokat-Almus Bakımlı. A large amount of tuyères was found in another heap at Büklüce village; a piece of charcoal was dated 1887 B.C.[100]

Copper ingots from the 20th-18th centuries B.C were found during excavations at Acemhöyük, west of Aksaray. These ingots are still unpublished, but the results of an analysis of 33 bun ingots were presented by A. Çukur and S. Kunç.[101] The ingots contain particles of impurities and some copper oxide and -sulphide. They were probably cast as blister copper and as such transported to Acemhöyük. Eight copper artefacts contained the same chemical composition as that of the ingots, which indicates that the ingots were processed into artefacts without further refining. These ingots may have been smelted from sulphide ores. Indications of the use of sulphide ores have also been found in some metal artefacts from Early Bronze Age Hassek Höyük.[102]

An important discovery for the history of mining and metallurgy, though not of copper, was made in the Bolkardağ mountains south of Niğde. Two sites are currently being investigated: an old mining complex at Kestel, and opposite it at some kilometres distance, a large workshop and living area at Göltepe.[103] There are indications that the metal mined at Kestel during the third millennium B.C. was tin oxide, although the evidence published by the excavator could not convince a number of specialists.[104] The metal was smelted in crucibles and cast into moulds in Göltepe.

2.7 *Quality and shape of traded copper*

2.7.1 *Archaeological evidence*

During the fourth millennium and the first centuries of the third millennium B.C., copper ore was not only smelted (mainly on a limited scale) at mining sites, but also brought into settlements and smelted there. Ore mined at Feinan and Timna was found in Tell Abu Matar, Bir Safadi, Tell Maqass and Shiqmim; small pieces of slag indicate that ore was smelted at those sites. In Anatolia, remains of copper smelting have been discovered at Nevalı Çorı and Norşun Tepe.[105] The ore, which contained some sulphide, was smelted in crucibles at these sites. This transporting of copper ore seems to have been largely confined to the period up to the beginning of the third millennium. It appears that from c. 3000 B.C. onwards, copper was largely transported out of the mining areas as metallic copper in the form of ingots. Lumps of native copper presumably continued to be transported during the

99 Kaptan 1990, p.77.
100 Kaptan 1990, pp.81-82.
101 Çukur and Kunç 1990.
102 Schmitt-Strecker et al. 1992, pp.112 and 121.
103 Yener and Vandiver 1993a; *AJA* 99 (1995), pp.224-225.
104 Muhly 1993a.
105 Hauptmann et al. 1993.

third and second millennium and perhaps even later. This change from the export of ore to that of metallic copper was determined by a greater demand for copper and an improved smelting technology in the mining areas. The costs of transportation would have played a role as well. It must have been clear that it was cheaper to load a donkey with copper ingots than with copper ore, which contained only a fraction of the amount of metal transported as ingot.

The quality of the ingots found to have been cast and traded during the second millennium B.C. differs according to the degree in which the smelting product was remelted or refined. It ranges from raw copper, still containing important or small levels of impurities like iron, to high quality copper if remelted or refined. These differences in quality are important for the understanding of the terminology employed in the texts. The ingots excavated in Oman have an average copper content of c. 95% and consist of unrefined raw copper.[106] Closest in time to the Old Assyrian period are the ingots from Oman and Acemhöyük. Those from Acemhöyük are important because of their chronological and geographical proximity to the Old Assyrian activities in Anatolia; they too consist of raw copper, which was considered sufficiently clean to be used in the manufacture of artefacts.[107] The Acemhöyük ingots are corroded, which diminishes the copper content of the samples to an average of about 67.3%.

Analyses of some twenty Ulu Burun ox-hide and bun copper ingots from c. 14th century B.C., showed that this copper was smelted from sulphide ores. Primary smelting of sulphide ores generally results in high levels of iron. The relatively low levels (< 0.5%) present in these ingots, suggest they come from a secondary, remelting or refining, operation. Two ingots with 2.14% (bun) and 1.22% (ox-hide) iron either represent unsuccessful refining operations, or primary ingots. Some ingots have particularly dirty metal, which is uncharacteristic of secondary ingots.[108] Ox-hide ingots from Cyprus consist of unrefined blister copper with a sulphide and/or oxide content of c. 1%, an iron content less than 0.5%, and normally c. 98% copper. It is uncertain whether these ingots were later refined, or only remelted and alloyed, in which case a large part of the sulphide and oxygen would be removed.[109] Of the seven ingots from Israel and the Sinai studied by I. Roman,[110] six are the product of primary smelting. Only an ingot found in the sea near Dor (late second millennium) is a secondary cast, produced of copper refined by remelting. This ingot (A 7) still contains c. 1.2% sulphur and c. 0.10% iron, the copper content is c. 98%.[111]

---

106 Hauptmann 1987, pp.211-212.
107 Çukur and Kunç 1990.
108 Maddin 1989.
109 Tylecote 1982, p.94.
110 Roman 1990.
111 Roman 1990, pp.180-181.

2.7.2 *Textual evidence*

The Old Assyrian texts do not contain detailed information on the remelting or refining of copper. Among the texts discovered in the ancient city of Mari, situated on the Euphrates in east Syria and less than a century younger than the Old Assyrian epigraphical material, we find some information on remelting/refining operations. Copper worked in Mari derived largely from Cyprus and this quality was hence called 'Alashian'. The URUDU.KUR, literally 'mountain copper', stems from an amount of copper sent to Mari by Yarim-Lim, king of Yamḫad (the area around Aleppo). To a lesser extent Mari acquired copper from Anatolia and through Babylonia from Iran and Oman. In five texts edited in ARM 25, it is recorded that three copper qualities were 'washed' or refined (*mesû*) The quantities are expressed in minas (m), one mina weighing c. 500 gr:

ARM 25 text	loss:
690: 45 m *mesû* washed again, loss of 2 m, result 43 m *mesû*;	4.5%
692: 18! m URUDU.KUR washed, loss of 3 m, result 15 m *mesû*;	16.7%
693: 12 m URUDU.KUR washed, loss of 2 m, result 10 m *mesû*;	16.7%
691: 42 1/3 m Alashian washed, loss of 8 1/3 m, result 34 m *mesû*;	19.7%
483: [42 1/3 m] Alashian washed, loss of 12 1/2 m; result 29 5/6 m *mesû*;	29.5%

The ancient scribe also recorded the calculated weight losses per mina (introduced in the text by *inūma*). In ARM 25, 691, the weight losses were calculated at 19.3%; in 692 and 693, at 20%. These five references may be compared with experimental evidence obtained during the Timna smelting experiment, 27 and the remelting/refining experiment, 27A-D. During this experiment, a black copper was produced, containing c. 74% copper and 22.2% iron.[112] This was remelted (experiment 27A), resulting in a copper with c. 92% copper and 6.2% iron; this copper was remelted again (27B), resulting in a copper with 93-96% copper and 3.5% iron. A portion from this copper was remelted again (27C), resulting in a copper with c. 96% copper and 2.2% iron. This copper was fire-refined (27D) and contained afterwards almost 100% copper and 0.014% iron. The fire-refining had decreased the level of iron, but the stirring of the molten copper with a wooden stick ('poling') had not decreased the level of oxygen, which caused the copper to have a high porosity and concentration of cuprous oxide. The weight of the raw copper products diminished in this experiment as follows:

remelting (27A) of black copper from (27):	loss of 28.6% weight
remelting (27B) of remelted copper from (27A):	loss of 24.3% weight
remelting (27C) of remelted copper from (27B):	loss of 4.2% weight
fire-refining (27D) of remelted copper from (27C):	no loss in weight

112 Merkel 1990, pp.107-109.

When these weight losses are compared with those from the Mari texts, it is clear that the 'mountain copper' and Cypriote copper qualities subjected to 'washing' were raw copper with different levels of impurities. The amount of already 'washed' copper in ARM 25, 690 may have been the result of a previous, unsuccessful, remelting or refining operation. Such a badly refined copper was not exceptional as is demonstrated by ARM 25, 714 (re-edited by J.-M. Durand in MARI 5, pp.608-609; see also Joannès 1993, p.105a). This text has in line 1: *šum-ma* URUDU *mi-sú za-ku-ú*, 'If the refined copper is (really) pure', with the opposite (line 5): *šum-ma* URUDU *mi-sà-am la ka-aš-du*, 'If the copper is poorly refined', literally 'If the copper (pl.) has not attained washing' (cf. Durand, p.609). The exact meaning of the metallurgical operation expressed by the general verb *mesû* and translated with 'to wash', cannot be established from these texts; it signifies at least that the metal was remelted, but whether a refining and poling operation were an integral part of the 'washing' is unknown.

2.7.3 *Assessing the quality of copper*

Information on the quality of the copper was generally conveyed by its colour and by the shape in which it had been cast, or by the place of origin. As there was no means of establishing the chemical composition of the copper, the quality of the metal had to be discerned by optical inspection. The easiest way is to observe the colour, because the less impurities in the metal, the more the metal shows the typical red copper colour. Raw copper was distinguished on this ground by Agricola in his book *De natura fossilium*, (Basel 1546, Book 8, p.337), as described by V. Tafel: 'Lat. *aes vel fulvum vel rubrum vel caldarium.* Die Bedeutung dieser Bezeichnungen geht aus Agricolas Werk hervor. Hiernach sieht das in Neusohl, Gottesberg, im Hercynischen Walde und in Norwegen erschmolzene Kupfer rot aus (*ruber*). Anderes, wie das in Geier und Schneeberg gewonnene, ist dunkelgelb (*fuscum*). Da es heller ist als das erste, wird es auch weißes (*candidum*) genannt, obwohl es schwarzgelb aussieht. In Saigerhütten wird auch gelbrotes (*in rubo luteum*, hier mit *fulvum* bezeichnet) als das gewöhnlich fallende (*regulare*) hergestellt und ferner braunrotes (*in rubro fuscum*), welches "Kesselkupfer" (*aes caldarium*) genannt wird. (...) Es unterscheidet sich von dem gewöhnlich fallenden dadurch, daß es sich leicht gießen läßt, aber unter dem Hammer zerbricht, während jenes nicht leicht gegossen, aber unter dem Hammer breit geschlagen werden kann. Alle diese verschieden gefärbten Kupfersorten sind demnach sog. Schwarzkupfer, d.h. mehr oder weniger durch fremde Bestandteile verunreinigtes Kupfer.'[113]

Copper which contained little or almost no impurities represented the best quality. This copper could be cold-worked, i.e. hammered into a specific form. If the metal was insufficiently purified and still contained copper oxide, it could only be hot-worked. In Pliny's discussion of various copper qualities, the first is called *regulare*, the second *caldarium*: '34 § 94 XX. Jetzt werden wir uns den verschiedenen Arten und Legierungen des Kupfers zuwenden. Beim Zyprischen gibt es das "*coronarium*" und das "*regulare*",

113 V. Tafel on Agricola, p.437 n.35.

beide sind hämmerbar. Das "*coronarium*" wird zu dünnen Blechen ausgehämmert; in Rindergalle getaucht sieht es bei Schauspielerkränzen wie Gold aus. Gibt man zu je einer Unze Kupfer sechs *scripula* Gold,[114] so erglänzt es feurig wie zarteste Blättchen von Goldflitter. Das "*regulare*" wird ebenso wie das "*caldarium*" auch in anderen Hütten hergestellt. Der Unterschied besteht darin, daß das "*caldarium*" nur gegossen werden kann, unter dem Hammer aber zerbricht, durch den das "*regulare*" verformbar ist - daher von anderen *ductile* genannt - wie das zyprische Kupfer überhaupt beschaffen ist. Aber auch in den übrigen Hütten ergibt sich der Unterschied zum "*caldarium*" durch die Behandlung; denn jedes Kupfer, aus dem im Feuer die Verunreinigungen gründlicher entfernt und ausgeschmolzen sind, ist "*regulare*".'[115] The copper which contains a high percentage of impurities is raw copper, the other sort, which can be hammered, represents the raw copper after remelting or after intentional refining operations.

2.8 *Conclusion*

Most of the copper traded during the Old Assyrian period probably derived from oxide ores. It was often mined in small, localized deposits, and smelted in the vicinity of the mine where the supply of fuel was sufficient. The raw copper resulting from this primary smelting was cast into ingots and transported out of the mining regions. At least two qualities of copper may be expected: one sufficiently refined to be cold-worked, and another containing a higher percentage of impurities making it unsuitable for hammering, but still useful to melt and refine it and cast it into objects. It is likely that lumps of native copper were traded as well. Pieces of ore, containing gangue material, were probably not an object of the metal trade.

---

114 One *uncia* weighed 27.29 gr; a *scripula* was 1/24 of an *uncia* and weighed 1.137 gr.

115 Projektgruppe Plinius 1984, p.12.

CHAPTER 3
THE OLD ASSYRIAN TERMINOLOGY

3.1 *Qualifications of copper*

Old Assyrian uses a rich terminology for copper. Apart from simple *werium*, 'copper', mostly written with the Sumerogram URUDU, there are a series of descriptive adjectives, *dammuqum*, *masium*, *zaku'um*, *watrum*, *lammunum*, *ṣallāmum*, *massuhum*, some reserved for ingots, such as *ṣahhirum*, *šabburum*, and *šikkum*. There is also scrap, *hušā'ū*. In addition one meets various toponyms with a *nisba*-ending or preceded by the relative pronoun *ša*, which designate copper from various places of production and at the same time are indications of quality or form. Certain terms, especially the descriptive adjectives, are rather general in meaning, and it is difficult to link them with specific qualities (degrees of purity). The texts contain a variety of combinations as well, such as *werium dammuqum Haburatāyum* or *werium dammuqum masium*. The following pages offer an analysis of this terminology, in an attempt to narrow down the meaning of the terms and to link them with particular kinds or types of copper (composition, shape) as are known from metallurgical studies and archaeological discoveries. Details from the sources on individual types and designations can be found in the Catalogue.

3.1.1 werium - *'copper'*

The word for copper is *werium*, but it is nearly always written with the Sumerogram URUDU. It is exceptional therefore to find the (plural) forms *we-ri* in KTS 54d.2 (**1**), *we-ri-i-kà* 'of your copper' in Ka 276.8 (Donbaz and Veenhof 1985, p.133), and *e-ru-šu-nu* in the Ilušuma inscription (see next section), where it is written without the initial 'w' although this text is older than the archives from Kanish. Where the Sumerogram URUDU occurs without a qualifying adjective, it may denote a specific quality which the ancient scribe considered unnecessary to write down, or copper in general. It follows that URUDU *per se* can be used to denote copper of one of the known copper qualities, which is reflected in the range of prices available for unqualified URUDU. This complicates modern assessment of the intended quality in a text. In kt t/k 1.11, URUDU is used, but from the parallel passage in kt t/k 25 one learns that this was *šikkum* copper (see above, p.12). In a few instances URUDU is employed as a determinative before an object made of copper. This occurs in URUDU *zirātum* and URUDU *e-li-t[um]* in the school text published by K. Hecker (1993, p.287, kt t/k 76+79.8-9) and presumably also in CCT 5, 34b.29-30 (16 1/3 GÍN URUDU *im-ni-/an* KI *Iš-pu-nu-ma-an*, meaning of *im-ni-an* obscure, '2 *i.*-s'?). URUDU *šaršarretēn* in BIN 6, 227.9 means 'copper in the shape of 2 chains'.

URUDU *per se* does not mean 'copper ore', but the possibility remains that lumps of native copper were in some cases called URUDU.

3.1.2 *Copper of high purity*

3.1.2.1 masium - *'refined'*

The verb *masā'um* (in Old Babylonian *mesûm*), which has the general meaning 'to wash', is used to describe the purification of metal. This 'washing' was done by melting and refining. In Old Assyrian texts the verb itself has generally silver as its object (e.g. in ATHE 38.7 and VS 26, 21.10), but its adjective is used as a qualification of copper, usually written *masi'um* and in the accusative often as *ma-sí-a-am* (BIN 4, 145.5). *Ma-as-am* (in BIN 6, 94.13; KTS 18.9; VS 26, 27.11) is an exceptional form. ICK 1, 85.8 contains the unique writing *ma-sú-im* which is an adjective of the D-stem like many other bulk words, if this writing is not a mistake for *masi'im*.[116] The Sumerian verb LUH and its adjective LUH.HA are employed in the same meaning as *masā'um* and *masi'um*. Contrary to what F. Joannès (1993, p.103) wrote on this subject, the D-stem of *damāqum* is not used in Old Assyrian for 'to refine'.

J. Lewy transcribed the adjective *mazi'um* and translated 'mixed copper' (*Mischkupfer*) in KTB, p.24 ad 5.10ff. By mixed he meant alloyed and *mazi'um* copper was in his opinion bronze (EL II, p.199). A.L. Oppenheim, in his review of Lewy's TC 3/I, suggested the reading *masi'um*, meaning 'washed', or 'refined' (1938, p.134). This reading was accepted by J. Lewy who now translated *masi'um* 'washed' (1958, p.93). It must remain uncertain whether he interpreted this 'washing' in the same way as his wife, Hildegard Lewy, who wrote 'probably native copper that had been freed from impurities by crushing and washing' (1964, p.183 n.1).

The verb 'to wash' occurs with copper in a metallurgical context only in a building inscription of king Ilušuma.[117] In this inscription, relating to the building of the Ištar temple in Assur, the king, living one generation before Erišum I and two or three generations before the time of the bulk of the archives found in Kanish, mentions that he established the freedom (*andurārum*) of the Akkadians and their sons and 'washed' their copper.[118] M.T. Larsen has argued that this 'freedom' may have been established as part of a policy aimed at increasing Assur's economic position by granting certain privileges to merchants from Babylonia.[119] These merchants were instrumental in transporting copper from the South to Assur. Since the established 'freedom' already refers to privileges granted to them, it is unlikely that the 'washing of their copper' would refer again to these privileges, e.g in the

---

116 M. Ichisar (1981, p.324) translated line 8 (*a-na* 1 GÚ URUDU *ma-sú-im*): 'pour un (seul) talent de cuivre raffiné' following CAD M/II, p.235b.

117 See for *im-tí-ZI* in kt a/k 279.20 below, p.43. In BIN 4, 151.14', I read *am-ṣí-ú-ma* ... *al-qé-ú-šu*, see n.129.

118 Grayson, RIMA 1, p.18: 49 *a-du-ra-ar* 50 *A-kà-dì-i* 51 *ù ma-re-šu-nu* 52 *áš-ku-un* 53 *e-ru-šu-nu* 54 *am-sí*.

119 Larsen 1976, p.78.

form of an abolishment of debts of copper, as proposed by K. Balkan.[120] A metallurgical meaning for 'to wash' seems more probable, and Ilušuma apparently boasts himself of having been able to melt and refine copper that was brought to his city by the Akkadians, as a proof of power or of prosperity.[121] The building inscription of king Erišum I, mentions that this king established a 'freedom' on copper as well (Grayson, RIMA 1, p.10.22).

The high prices for *masi'um* (sold for 71-45 shekels of copper per shekel of silver) demonstrate that this copper was of high purity. It belonged to the category of 'fine copper'. This explains why it was equated with *dammuqum*, 'fine' copper several times, as in kt a/k 579.11-14: [10] *šu-ma* (erasure) [11] *ší-im* URUDU *ma-sí-im* [12] *i-ba-ší . té-er-ta-kà* [13] *li-li-kam-ma* URUDU SIG_5 [14] *le-pu-uš*, 'Send me your instructions if refined copper is in demand, and I will obtain fine copper.'

As the purity could not always be the same (see above pp.30-31 on Mari), there are instances of *masium* copper of high quality, *weriam masiam dammuqam* (CTMMA 1, 94.1-2 and OIP 27, 56.33-34).

Copper could be refined by a smith in any city; a major centre for refining was Turhumit (ICK 1, 85.8; kt 89/k 430.15').

3.1.2.2 dammuqum - *'fine'*

J. Lewy interpreted *dammuqum* as 'improved' copper which would be bronze (KTB p.24), similar to his *mazi'um* copper. The translation 'refined' by CAD E, p.322a, shows that it, like Lewy and F. Joannès (1993, p.103), connected the adjective with the factitive sense which the D-stem of *damāqum* possesses. But in Old Assyrian, the adjective for bulk words is often derived from the D-stem without giving a factitive sense to the adjective. This is also the case with *dammuqum*, frequently written with the Sumerogram SIG_5. In CAD D, p.180a, the word is correctly translated 'fine'.

While *masium* is always considered *dammuqum*, the latter comprises more than just *masium* copper. It is used to indicate both raw copper, unrefined but of high quality, and refined copper. Copper from Haburata, Taritar, and Tišmurna could be of this quality. Native copper of high purity might have been called *dammuqum*. Probably the case of 'dirty' *dammuqum* copper, *weriam dammuqam massuham*, in ATHE 38.25-26, is exceptional. The copy of this text does not justify the reading suggested by K. Hecker in *GKT*, p.48 n.1 and adopted by the CAD, which makes in line 25 the sign SIG_5 the continuation of the previous line, reading there *šikkum* / SIG_5 and in lines 25-26 *weriam massuham*. The opposite could be the *weriam dammuqam masiam*, 'refined good copper' in ICK 2, 99.3-4 (or 'good quality copper, i.e. refined'?). As a high level of purity is required for the fabrication of artefacts, copper of lesser quality was often exchanged for *dammuqum*. The expression used is *ana dammuqim ta''urum*, 'to turn into fine (copper)' (AKT 1, 17.31; BIN 4, 64.8; BIN 6, 76.12; CCT 4, 27a.30; kt n/k 1385.27'; KTS 2,

---

120 Balkan 1974, p.33.

121 Likewise Larsen, 1976, p.75. J.D. Muhly rejected in *RlA* 6, p.359b a metallurgical interpretation for this passage.

34.7-9), an activity which often took place in Turhumit. The expression does not seem to be identical with the expression *weriam dammuqum ka"unum*, for which see below, pp.53-57.

The prices for *dammuqum* (sold for 73-50 shekels of copper per shekel of silver) are almost the same as those for *masium*. *Dammuqum* could be twice as expensive as *šikkum* copper (kt c/k 535.18-22).

Like refined copper, fine copper seems to have been usually traded in the form of ingots. The ingot could be broken up, and the copper was then called *šabburum*.

The word *rēštum* in AKT 3, 66.29 is not an adjective to *dammuqum* copper, but means 'fine oil'. Lines 28-30 run: 'Bring me (for a value of) 15 minas of fine copper first-class oil from Kanish or Hahhum.'[122] Although this text is a letter written by Šalim-ahum who asks for copper to be brought to him in Assur in two other texts (AKT 3, 72.42: 40 minas of fine copper, and AKT 3, 74.33-34: 5 minas of refined copper), an alternative interpretation in which both copper and oil must be sent seems unlikely. Oil was measured in *silas* and not weighed in *minas*. The translation of the passage is supported by two other texts, where the weights also express the oil's value in copper. Kt n/k 1502: 'Look out for fine oil from Kanish, send me extra fine (oil) (for a value of) 5 minas (of copper).'[123] In kt 91/k 157, oil was apparently traded for copper: 'Furthermore, be so kind as to send me fully 15 minas of refined copper (for) the fine oil. You also took oil worth 15 minas from PN and you have been doing business with it for 10 years. Give (...) the copper and what has been bought with it.'[124]

3.1.2.3 watrum - *'excellent'*

The word *watrum per se* is used in the meaning 'excellent'. No prices are known for copper of this quality. It also occurs in combination with *dammuqum*, meaning 'extra fine', as in *werium Haburatāyum dammuqum watrum* in Ka 276.3-5, 'extra fine copper from Haburata'. No price is known for this quality.

3.1.2.4 zaku'um - *'clean'*

The adjective *zaku'um* (see for this spelling CAD Z, p.25b discussion section) is not only used for copper, but also for other metals and for *amūtum*-mineral.[125] It occurs independently in RA 58, Sch. 2, where it qualifies copper transported out of Assur. In kt 91/k 390, it qualifies copper of Tapašatta quality. No price is known for this quality.

---

122 AKT 3, 66.28-30: [28] 15 *ma-na* URUDU *da-mu-qám* [29] *re-eš*$_{15}$*-tám ša Kà-ni-iš* [30] *ú ša Ha-hi-im bi*$_{4}$*-lam*.

123 Kt n/k 1502.25-28: [25] Ì.GIŠ.DÙG.GA [26] *ša Kà-ni-iš e-kà li-lik* [27] *ṭá-ba-am wa-at-ra-am* [28] 5 *ma-na šé-bi*$_{4}$*-lam* (by courtesy of S. Çeçen).

124 Kt 91/k 157.21-29: [21] *a-ha-ma* 15 *ma-na* Ì.GIŠ.DÙG.GA [22] URUDU *ma-sí-a-am ma-al-a-nam* [23] *tù-da-ma-qá-ma . tù-ba-lam* [24] 15 *ma-na* Ì.GIŠ.DÙG.GA [25] *ša* PN *a-ta-ma* [26] *tal-qé-ma iš-tù* MU.10.ŠÈ [27] *ta-ma-kàr* . URUDU *ú ší-a-ma-tí-šu* [28] (...) [29] *dí-in* (by courtesy of K.R. Veenhof).

125 For instance tin, Donbaz and Veenhof 1985, p.142. See for *amūtum*-mineral Dercksen 1992, col.796-797.

3.1.3 *Copper of low purity*

3.1.3.1 lammunum - *'poor'*

The opposite of fine copper was *werium lammunum*. J. Lewy translated 'raw copper' (*Rohkupfer*) in EL I, p.134 n.*a*. This is better than the 'literal' translation 'slecht gemachtes Kupfer', as the D-stem is here not used in its factitive sense. CAD L, p.247a, has 'of bad quality', while W. Röllig rendered it 'sehr schlechtes Kupfer' (1983, p.346). The bad quality was caused by the level of impurities it contained, notably iron. In most cases, *lammunum* can thus be regarded as raw copper. Although poor copper was exchanged for fine copper in a number of texts (see above), it was still the object of trade and at times considerable amounts of it were shipped to another city, as in kt 87/k 436, where 14 talents 5 minas of poor copper were brought to Wahšušana and sold for silver. Ingots of poor copper could also be broken up. A sale price of 80:1 is attested for poor copper, which suggests the quality of that particular amount of poor copper was not very bad.

3.1.3.2 massuhum - *'dirty'*

Two words describe copper with impurities, *massuhum* and *ṣallāmum*. The first occurs also with the metals silver and tin, but its exact nature is uncertain.[126] The word *massuhum* is connected with the verb *masāhum* D, 'to make dirty'. Although the unique *ma-sú-im* in ICK 1, 85.8 could be interpreted as alternative writing for *massuhim*,[127] it is better to derive this *ma-sú-im* from *masā'um*, 'to wash', see above p.34. Assuming that the term describes the same feature in all three metals which could be removed by 'washing' or refining, its meaning seems to be that the metals visibly contained impurities and were therefore labelled 'dirty'.[128] B. Kienast translated ATHE 38.26 'Bruch-Kupfer' (following E. Bilgiç, see ATHE p.57). In CAD M/II, p.236b, it is rendered 'of bad quality'. Metal that was impure was not in demand, but nevertheless accepted by some traders, just as *lammunum* copper. For example in BIN 4, 151.11'-15': 'Nobody wanted to accept the copper because it was *massuh*. But I took it on my own responsibility for Ah-šalim's firm.'[129]

3.1.3.3 ṣallāmum - *'black'*

'Black copper' (*ṣallāmum* /*ṣalmum*) was, like 'dirty' and other poor quality copper, a rather impopular object of trade. Black copper coming from Turhumit and stored in

---

126 Silver: see Dercksen 1992, col.800. It was cleaned by 'washing' (*masā'um*). Tin: BIN 4, 15.17; LB 1295.5; TC 2, 42.6, 21.

127 K.-H. Deller in *OrNS* 32 (1963), p.475.

128 See for the relationship with *masākum* *GKT* § 29b.

129 BIN 4, 151.11'-15': 11' ... *ki-ma* URUDU *ma-sú-hu*!*-/ma* 12' *ma-ma-an lá-qá-am* 13' *lá i-mu-ú-šu a-na-ku* 14' *am-ṣí-ú-ma ki-ma* É 15' *Ah-ša-lim al-qé-ú-šu*. The subjunctive depends on *kīma*, '(a tablet) stating'.

Wahšušana was refused in MNK 636.[130] This type of copper may have been covered with scale, which would give it a black appearance. But it is perhaps better to link it with the problem of reducing the level of iron in smelted copper. This word then has the same meaning as black copper in English, being low quality iron-contaminated copper produced from the smelting of oxide ores. Such copper ores were fluxed with iron ores like haematite. Whether haematite was used as a flux, or a weathered chalcopyrite ore containing haematite was smelted, the iron oxide would during the copper smelting result in the growth of solid iron crystals at the copper-slag interface.[131] If these black crystals were visible on the outside of the copper, it was a sign that the iron content was still too high, and that subsequent remelting or refining procedures were needed to remove the iron. Too much iron placed a copper into the category of poor copper. Parallel passages from two texts found in the same archive show that *ṣallāmum* was synonymous with *lammunum*: kt 91/k 514 mentions tablets recording debts by a certain Enna-Suen: '1 tablet concerning 2 talents 40 minas of black copper; a second tablet concerning 15 minas of fine copper, of Enna-Suen's debt'; this Enna-Suen occurs with the same debts in kt 91/k 497: '2 talents 40 minas of poor copper are with Enna-Suen son of Ikua; 15 minas of fine copper are with Enna-Suen son of Ikua.'[132] A person in need of money would even accept lumps of black copper (*kiṣrum*, kt 91/k 422.22-23), presumably pieces of primary ingots.

The verb from which the adjective *ṣallāmum* is derived was translated 'to temper' in CAD Ṣ, p.73, and 'Eisen härten' in AHw, p.1076b, and *ṣallāmum* 'tempered'. Tempering is a low-temperature treatment, which restores some of the ductility to steel (an iron-carbon alloy) which was lost during the rapid cooling of the metal by plunging it into cold water (a hardening process called quenching). This technique is not used with copper. This false interpretation for *ṣallāmum* and *ṣalāmum* (unfortunately repeated in Joannès 1993, p.109b) was based upon two texts. In the first, CCT 5, 13a, a report is given on the attempted sale of *amūtum*-mineral, thought to be iron by the CAD, to an Anatolian official. The passage runs in trancription: [9] *a-dí a-mu-tim* [10] *ša* DUMU *Hi-na-a* GAL *sí-ki-tim* [11] *ú-ṣa-li-mu a-ší-a-tí* [12] *a-mu-tám lá ú-kà-li-im-šu*. Lines 9-11 were translated in the CAD: 'As to the *amūtum*-iron which the son of the *rabi sikkatim* official PN tempered'. However, this so-called 'son of the official' is in fact an Assyrian merchant who owned the *amūtum*. The *rabi sikkatim* is the object of the verb in line 11. Without changing the verb in line 11, I suggest the translation: 'As for the *amūtum*-mineral belonging to the son of Hināya, they have angered the *rabi sikkatim*, and therefore I did not show him the *amūtum*.'[133] The second

---

130 Published in Matouš 1957/58, p.116. MNK 636.10-13: [10] URUDU *ša na-ad-ú* [11] *ṣa-la-am a-na-kam* [12] *ú-la i-ma-ha-ar* [13] *ku-lu* URUDU *ša Tur*$_4$*-hu-mì-it*, 'The copper that is in store, is "black", it is not accepted here (= in Wahšušana). All copper is from Turhumit.' See for this text also Larsen 1980, pp.260-261 and n.34 and Michel 1991b, p.259.

131 T.A. Wertime and J.D. Muhly (eds.), *The Coming of the Age of Iron*, p.466.

132 Kt 91/k 514.5-10: [5] 1 DUB *ša* 2 GÚ 40 *ma-na* [6] URUDU *ṣa-lá-mì-im* [7] *ša-ni-um* [8] *ṭup-pu-um ša* 15 *ma-/na* [9] URUDU SIG$_5$ *ša hu-bu-ul* [10] *En-na-Sú-in*;
kt 91/k 497.10-13: [10] 2 GÚ 40 *ma-na* URUDU *la-mu-nam* [11] *iš-tí En-na-Sú-in* DUMU *I-ku-a* [12] 15 *ma-na* URUDU SIG$_5$ *iš-tí En-na-S*[*ú*]*-/i*[*n*] [13] DUMU *I-ku-a* (both texts by courtesy of K.R. Veenhof).

133 *Ṣallumum* here apparently used for *pānam ṣallumum* (attested in the G- and D-stem, CAD Ṣ, 70b); cf. *ṣulum pāni* 'anger', CAD Ṣ, p.241a, 2'.

text is KTK 6 in Moscow. Unfortunately the reverse with the lines where the verb occurs is damaged: 8' [x x (x)]-*ší-im ù* 9 *me-at* 3 / *ma-na* 9' [URUDU?]x (not WA)-*lim ú-ZA-li-im* . 10' [*um-ma šu-u*]*t-ma* u_4*-ma-am šu-ma* 11' [x x x] *i-kà-šu-du* . *ú-ṣa-áb-ma* 12' [x x *ú-ṣ*]*a-áb-ma a-e-ma* 13' [x x x] . *ú-ZA-lá-am*. The amount weighed in line 8' could be copper, but the context in which it occurs is obscure. In 1984, the CAD (S, p.92b) suggested that the forms in CCT 5, 13a.11 and KTK 6.r 9' and 13' should be understood as forms of *sullumu* 'to make someone act in a friendly manner'. This seems unnecessary in the first text where only the meaning must be changed. The conclusion can be drawn that *ṣallāmum* bears no relation to steel, and that it is impossible for it to denote a metallurgical treatment of copper. On the contrary, black copper needs some treatment like refining.

A text referring explicitly to the iron content of copper is ICK 2, 54 (**2**): '4 talents of fine copper from Taritar (and) 2 1/2 talents of copper that does not contain haematite (*šadwānum*).' In this delivery contract styled as a debt-note, the well-known trader Imdilum presumably entrusted merchandise to one of his servants and charged him to sell it in Turhumit for a specified amount of copper which he had to bring to Kanish. The first amount of copper had to be fine copper from Taritar, the second copper containing no haematite, that is, with a low iron percentage. The most obvious explanation for the use of these different descriptions is, that Idnāya had to obtain or deliver Taritar copper, which seems to have been usually low in iron and was therefore in demand, and in addition a donkey-load of copper whose origin was unimportant as long as it did not contain iron. The Assyrians traded copper mainly as metal and not in lumps of ore, and this precludes J. Muhly's interpretation of this text according to which fine copper would have contained haematite as a natural fluxing agent.[134]

3.1.3.4 ša masā'im - *'which has to be refined'*

The use of the infinitive of *masā'um* is, with the comment on line 1 of AKT 3, 48, taken to mean that this copper needs refining, and was therefore of poor quality. Some *patallum*-ingots consisted of this copper (see below, p.59).

3.1.4 *Referring to the shape of the copper*

3.1.4.1 hassu'um - *'?'*

The qualification *hassu'um* occurs only once. In CCT 1, 42a.22, 10 minas appear to have been brought from Assur (see above, p.17). The word has been interpreted as a *nisba* by J. Lewy and P. Garelli, but its writing does not support this.[135] It is more likely to be the adjective of the D-stem. The verb is perhaps *hussû*, 'to crush or mince'.[136] Less probable is *hašû* from which *huš/sā'ū*, 'scrap', is derived, as the occasional writing with 's'

134 Muhly 1980, p.36.

135 Lewy 1952, p.420; Garelli 1963, p.95; Nashef 1987, pp.28 and 34. In RGTC 4, p.59, Kh. Nashef booked the passage under the toponym Hazu, but expressed doubts about this in his commentary.

136 Said of plants, CAD H, p.257b. But AHw, p.342a, under *hesû*, has '(Drogen mit Tuch) auspressen'.

for 'š' is only attested for the single 's' (GKT § 40g). An interpretation 'copper in pieces' or 'scrap' for *hassu'um* is possible considering the quantity of 10 minas, which might have been intended to pay expenses.

3.1.4.2 kiṣrum - *'lump'*

Black copper occurs in the shape of lumps in kt 91/k 422.22-23. Such lumps were presumably the result of breaking up (primary) ingots.[137]

3.1.4.3 ṣahhirum - *'small-sized'*

The normalization of the word derived from the verb *ṣahārum*, 'to be small', here rendered *ṣahhirum*, is unclear.[138] Whereas the use of *šabburum* was almost exclusively used for high quality copper, denoting that the ingot was broken up into pieces, *ṣahhirum* is only known to describe 'copper' without any further qualification. It evidently refers to lumps obtained by breaking up the ingot, hence the translation 'small-sized'. It must remain uncertain whether there was any distinction between these lumps and *šabburum* copper.

3.1.4.4 šabburum - *'broken'*

Ingots were often broken up into smaller pieces before the copper was conveyed. The verb used is *šabārum* D, of which both the infinitive and the adjective are well attested.[139] The praesens probably occurs in kt n/k 122.23, where I understand *nu-ša-ba-bar-ma* as *nušabbar-ma*, 'we will break (the copper)' because the D-stem of *šapārum* 'to write or send' is unattested.[140] The translation 'in scraps' (CAD E, p.322a) for *šabburum* is correct, although this copper was still unformed. The place where the copper was prepared in this

---

137 The word *kiṣrum* occurs also for a lump of *amūtum*-mineral, see Dercksen 1992, col.797.

138 See also Veenhof 1972, p.431 n.547.

139 Kt n/k 1153.6-10: [6] ... 4 GÚ URUDU [7] *ma-sí-am* 2 1/2 *ma-na* [8] *i-na ša-bu-ri-im* [9] *im-ṭí ku-nu-ki-a* [10] *Hu-dur$_4$-lá na-áš-a-kum*, 'Hudurla is bringing you under my seal 4 talents of refined copper - there was a deficit of 2 1/2 minas from breaking up' (by courtesy of S. Çeçen);
kt 87/k 436.22-25, 29-31: [22] ... 14 *ma-na* URUDU [23] *ig-ri e-re-qí-im iš-ku-nu-nim* [24] 5 *ma-na i-na ša-bu-ri-im* [25] *im-ṭí* (...) [29] *ší-tí* URUDU-*kà* 14 GÚ 5 *ma-na* [30] *a-na Wa-ah-šu-ša-na a-na-ku* [31] *ú-šé-ri-ib-šu*, 'They charged me 14 minas of copper as rent of a wagon; there was a deficit of 5 minas from breaking up. (...) I brought the remaining 14 talents 5 minas of your (poor) copper to Wahšušana' (by courtesy of K. Hecker);
kt 87/k 453.16-21: [16] ... 12 *ma-na* [17] *i-na ša-bu-ri-im* [18] *im-ṭí* (...) [20] *ší-tí* URUDU-*kà* 18 GÚ 7 *ma-na* [21] *a-na Kà-ni-iš as-ri-dam*, 'There was a deficit of 12 minas from breaking up. (...) I transported the remaining 18 talent 7 minas of your (*Haburatāyum*) copper to Kanish' (by courtesy of K. Hecker);
it is unclear whether KTS 38b.1'-2': (*i-na*) [1'] *ša-bu-ri-im* [2'] 36 *ma-na* URUDU *im-ṭí*, describes the deficit established after the copper had been broken (infinitive), or after an amount of broken copper had been received (adjective).

140 Kt n/k 122.21-27: [21] ... *iš-té-ni-iš* [22] 30 GÚ *nu-ma-lá-ma* [23] *nu-ša-*[ba]*bar-ma* [24] *Ta-as-ku-ul* [25] *iš-tí ša bi$_4$-lá-tim* [26] *a-na Wa-ah-šu-ša-na* [27] *ni-ṭá-ra-ad*, 'We will make a full load of (= bring together from various sources) in total 30 talents (of fine copper). We will break (it) in pieces and send Taskul with porters (carrying the copper) to Wahšušana' (by courtesy of S. Bayram).

way for transport seems, as far as the context of these texts allows it, to have been Turhumit. This town is mentioned in two texts as the place where broken fine copper had to be paid.[141] Often ingots of fine and refined copper were broken, but ingots of poor copper could occasionally be broken up as well. A small deficit arose from breaking up, which varies per text:

kt 87/k 453: deficit of 12 minas on a consignment of 1087 minas of *Haburatāyum* copper
= 1.1%
kt n/k 1153: deficit of 2 1/2 minas on a consignment of 240 minas of refined copper = 1%
kt 87/k 436: deficit of 5 minas on a consignment of 845 minas of poor copper = 0.6%

The circumstance that in the last example the deficit is nearly half of that in the other texts dealing with fine copper, may have been caused by the quality of the copper (poor), but the evidence is too small to draw a conclusion from this. The deficits recorded after checking a quantity of transported copper originated from other causes than breaking up, see Veenhof 1972, pp.52-53.

There are several possible reasons why ingots were broken up before transport. It offered a chance to check the ingot's quality and prevent surprises like in one of the plano-convex ingots from Al-Aqir/Bahla in Oman, where the core of the ingot consisted of slag (see above, p.25). Secondly, it enabled the trader to weigh out an amount with greater accuracy, and lastly, it may have facilitated transport.

3.1.4.5 ša šaduīšu - *'native copper'*

The word formation of *ša šaduīšu* is the same as in the Old Assyrian qualifications *ša abnīšu* and *ša mā'īšu* for gold. As *šadu'um* means 'mountain', it is probable that it refers to the form in which the copper was mined or found. This can be either copper ore or native copper. The amounts in the two references are relatively small, 2 and 3 minas. I.J. Gelb, who published OIP 27, 62 where this word occurs in line 42, normalised the form as *šaddu'um*, which would, like *šaddu'atum*, derive from *nadā'um* Š, but its exact meaning was unclear to him. CAD N/I, p.97b, followed Gelb's interpretation by placing this passage under *nadû* Š, 'to have someone make a payment, a deposit'.

It is tempting to relate the qualification URUDU.KUR from Mari texts to this URUDU *ša šaduīšu*. However, the latter is only attested in small quantities of 2 and 3 minas, while 279 minas of 'mountain copper' occur in ARM 25, 20.3, and 67 minas in ARM 25, 712.3. This copper needed refining (see above, p.30), which suggests it was brought to Mari largely as raw copper, whereas the Old Assyrian term is used for a copper that was rare and whose properties are unknown. I interpret *ša šaduīšu* as the Old Assyrian word for 'native copper'.

---

141 AKT 1, 66: [8] ... 3 GÚ 20 *ma-na* [9] URUDU SIG$_5$ *ša-bu-ra-am* (...) [13] *i-na Tur$_4$-hu-mì-it* (...) [16] *i-ša-qal*, and OIP 27, 56: [33] 55 *ma-na* URUDU *ma-sí-am* [34] *da-mu-qám ša-bu-ra-am* (...) [39] ... *i-Tur$_4$-hu-mì-it* [40] *i-ša-qal*.

One of the Akkadian words for 'ore' was *ep(e)rum*, 'earth, dust', in Sumerian SAHAR. It is attested in Old Assyrian only in the case of silver, written with the Sumerogram SAHAR. The word for 'native silver' was KÙ.BABBAR SAHAR.BA, literally 'silver in its dust' (**ina eprīšu*). It occurs in TC 3, 246.1 (amount of 15 shekels), where line 5 of the envelope has [KÙ.BABBAR] *ṣarrupam* instead. Kt k/k 15.1-2, records a debt of 21 shekels of native silver. This word occurs in broken context in CCT 5, 45b.11. The term *epir šadīšu*, 'describing pure metals found in nature' (CAD E, p.189a), is known mainly from texts from the first millennium B.C.

3.1.4.6 šikkum - '?'

A primary smelting product containing many impurities was *šikkum* (conventional spelling). Its quality showed some variety, as a single reference to a good quality *šikkum* (*werium šikkum dammuqum* in CCT 1, 7b.6) demonstrates. In kt c/k 535.18-22, 3 minas of *šikkum* were used to make up for a deficit of 1 1/2 minas of fine copper. The sale prices for *šikkum* (in TC 1, 23 at 140:1 but *dammuqum* at 68:1 and 70:1) also show that its value usually was half or one-third of that of fine quality copper. It was therefore of poor quality, but not exchanged for a better copper. In TC 2, 33.15-16 (**5**), it was even distinguished from *lammunum*.

Although clearly of a lesser quality than refined copper, it was apparently generally accepted as a means of exchange. Expenses en route were paid with *šikkum* in kt t/k 25.5 (7 minas for wine) and kt 91/k 424.1 (for a porter) and line 3 (see transcriptions in § 1.2.1). It is probable that e.g. in the last text most of the payments in copper were actually made with *šikkum*, without mentioning this each time. The parallel passage of kt t/k 25.5 in kt t/k 1.11 has simply URUDU. A monthly ration of 8 minas of *šikkum* was supplied to a merchant's Anatolian wife according to kt 88/k 269, with which she was able to buy food, oil, and firewood.[142]

It is a peculiar feature of *šikkum* copper that it was probably the only copper the weight of which was sometimes expressed in the Anatolian weight system, the *aban mātim*, or 'weightstone of the country', and not in the Assyrian weights. Although copper measured in *aban mātim* is only once identified as *šikkum*, the prices in several other instances indicate that a poor quality must have been meant there (see below, § 3.6.3).

There is not a single occurrence of 'broken up *šikkum*' and this might imply that the ingots were often already in pieces when traded. J. Lewy reached a similar conclusion by deriving the word from the verb *šakākum*, which means 'to harrow', and translated it as 'Kupfer in Stücken' (EL II, p.45 n.*c*), in which he was followed by P. Garelli ('en morceaux', 1963, p.294). The curious translation 'weather-worn' was given in 1958 by the CAD E, p.322a. W. von Soden gave no translation in his AHw (p.1235b). In 1992, CAD

---

[142] Kt 88/k 269.11-17: [11] 8 *ma-na*.TA URUDU *ší-kam* [12] *a-na ú-kul-tí-ša* Ì.GIŠ [13] *ù e-ṣí-ša i-na* ITU.KAM-*im* [14] *Pí-lá-ah-Ištar a-na* [15] *Ta-ta-na a-ší-tí-šu* [16] *i-da-ší-im ù i-na* [17] *ša-tim* TÚG *i-da-ší-im*, '(As long as he is away) Pilah-Ištar will give each month 8 minas of *šikkum* copper to his wife Tatana for her food, oil, and firewood. He will also give her one textile per year' (by courtesy of S. Çeçen).

Š/II, p.441a, stated simply 'copper of poor quality'. *Šikkum* may have possessed a characteristic shape, such as a bar, like the ingots from Obereching.

A difficult passage is contained in kt a/k 279.18-25: [18] [*šu*]-*ma* URUDU 1 [+1] GÚ *ší-kam a-na* [19] *ta-áš-bi-tí-šu* [20] *ú-lá im-tí-ZI* [21] [UR]UDU SIG5 *ta-áš-bi-*[*s*]*ú* [22] *dí-na-šu-um a-pu-tum* [23] *a-ma-kam* URUDU *lá ta-ba-ta-/qá-šu-um-ma* [24] *li-bu-šu* [25] [*lá*] *i-*[*ma*]-*ra-aṣ*, 'Should he not have *m.*-ed the copper, namely 2 talents of *šikkum*, as his final instalment, then give him fine copper as final instalment. Please, do not deprive him of the copper there, lest he gets angry.' The sender of this letter has to leave and he instructs his colleagues about settling a debt for him. They have to pay copper (passage broken) to one of them, who apparently must make the final instalment, and then follows the passage quoted. There are two possible verbs for line 20: either *masā'um* 'to refine' or *maṣā'um* 'to act on one's own responsibility' usually in hendiadys with another verb, e.g. 'to take'. If the first one is meant, then this text contains a reference to the refining of *šikkum* copper. This might be possible because its quality was poor. The subject of the verb would then not be the creditor referred to by the possessive suffix in line 19, but the person entrusted with the task of paying the debt. However, it appears more plausible to assume that both the suffix in line 19 and the subject in line 20 refer to the same individual, namely the creditor. The meaning of the lines is then that fine copper had to be paid in case the creditor had not already taken about 2 talents of *šikkum* on his own initiative. The verb is then *maṣā'um* but the expected form of *laqā'um* 'to take' has been omitted.[143]

3.1.5 *Toponyms with nisba-ending or preceded by* ša

3.1.5.1 Haburatāyum - *'from Haburata'*

Copper qualified by a toponym refers to metal coming from local deposits, or from a local metallurgical industry where raw copper from other regions was refined. Some of the local ore deposits will have been very small and traces of them will often not have survived. A modern example of a localized ore deposit can be found near the village of Kırmızıtarla (near the town of Pötürge, east of Malatya), where lumps of native copper are washed out of the soft ophiolithic bedrocks by streams and rain. Villagers reportedly collect the nodules of native copper each spring.[144]

Alphabetically the first of such a local copper producing town or region in Old Assyrian texts is Haburata. The copper produced there was of fine quality, which is also reflected by the prices attested for it. It was distinguished from refined copper, which suggests it was unrefined. No shape is given, but as it could be broken up, it was probably traded as an ingot. In two texts it is named together with copper from Taritar.

---

143 It is possible that the qualification *šurum*, known for a textile (Veenhof 1972, pp.154-156), was also used for copper. E.g. in BIN 4, 36.16-21: [16] *a-dí-ni . a-wa-tum . ú-lá i-š*[*é*]-*/ra-*[(x)] [17] *i-šé-ra-ma e-t*[*í-q*]*á-m*[*a*] [18] *lu ša* TÚG.HI-*tí-kà . lu* [*ša* UR]UDU-*i-/kà* [19] *šu-ri-im* (...) [21] *a-ba-kam* (collation by K.R. Veenhof).

144 Palmieri et al. 1993, p.586 and Wagner et al. 1989a, p.302.

3.1.5.2 Ištunuwāyum - *'from Ištanuwa'*

There is only one attestation of this quality. This is in a debt-note and the relatively small amount (40 minas) suggests it was of fine quality.

3.1.5.3 Kunanamāyum - *'from Kunanamit'*

The *nisba* of Kunanamit was *Kunanamāyum* (cf. GKT § 57b). copper is distinguished from *šikkum* and refined copper in I 471.6-7 and kt u/k 4, and from fine copper in BIN 6, 133.6-8 // CCT 2, 23.9. This suggests it was of poor quality. The specification *Kunanamāyum* is rare and it is probable that copper of this quality was usually known under the word *lammunum*.

3.1.5.4 ša Ninašša - *'from Ninašša'*

The only reference to copper from Ninašša is contained in kt a/k 461/b.2, which does not detail its quality.

3.1.5.5 ša Purušhattim - *'from Purušhattum'*

Copper 'from Purušhattum' is mentioned in CCT 2, 13, where 10,000 minas were paid at the palace. This suggests that the copper refers to the result of local refining. The fine quality of copper available in Purušhattum is reported in BIN 4, 64.12-13.

3.1.5.6 ša māt Sawit - *'from the country of Sawit'*

The country of Sawit is twice mentioned as the origin of copper: of 'pure' (*zaku'um*) copper in AKT 3, 56.37, and of fine and *Kunanamāyum* copper in CCT 2, 23.6. It apparently was an important mining area producing not only poor copper, but fine (refined?) copper as well.

3.1.5.7 Tapašattāyum - *'from Tapašatta'*

The specification Tapašattāyum is rare. This copper may have been of fine quality referred to as *dammuqum*.

3.1.5.8 Ta/iritarāyum, ša Ta/iritar - *'from Ta/iritar'*

The copper produced at Taritar was usually of fine quality. This appears also from the prices attested for it. In two texts it is named in combination with copper from Haburata. It occurs in relatively small and also bigger amounts, e.g. 2 and 9 talents.

In 1930, J. Lewy offered in EL I, p.29 n.*a*, his reading of CCT 4, 47a.32: *a-na* URUDU *tí-ri ta-ra-im*. He translated with some hesitance '(dem Metall) einen Zusatz geben', by deriving *tiri* from a verb *terû*, 'etwas in etwas hineintreiben' (AHw, p.1388b *ṭerû*). This misreading of the *nisba Tí-ri-ta-ra-im* was destined to have a long life. E. Bilgiç accepted Lewy's transcription, but did not consider *tiri* to be derived from a verb *terû*. Instead, *tiri* would be a copper quality, like *tiri* silver (1954, p.68 n.161). In 1958, CAD E, p.322a, recognized the adjective *tīri*, 'weather-worn'. It was K. Balkan who in 1965 corrected these mistakes by demonstrating that it meant copper from the town Taritari (1965, p.149: 'Mit dieser Lesung wird ein alter Fehler berichtigt.'). M.T. Larsen accepted Balkan's reading, but mentioned the non-existent quality: 'Taritara was the source of much copper (so-called "*tiri*-copper")' (1976, p.91; cf. Veenhof 1989, p.523). This reference to the non-existant *tiri* copper may seem superfluous or at best useful to readers of EL and Bilgiç 1954. However, even in AHw (pp.1388b *ṭerû* G 1 and 1392b *ṭīru* IV) and the *Reallexikon der Assyriologie* (Joannès 1993, p.109a) Lewy's interpretation is repeated.

3.1.5.9 Tišmurnāyum, ša Tišmurna - *'from Tišmurna'*

In Tišmurna fine/refined and poor copper were obtained, usually in relatively small amounts.

3.1.5.10 ša Tuhpia - *'from Tuhpia'*

In the letter kt 85/k 27 (Günbattı 1994), the ruler of Tuhpia mentions two amounts of copper 'from Tuhpia' which he gave to Assyrians. One of these amounts is distinguished from fine copper. Twice in this text, a *kutānum*-textile is compensated for with 50 minas of fine copper, while for another *kutānum*-textile 100 minas of local copper are given. This makes it probable that the copper from Tuhpia was of poor quality and was of half the value of fine copper. Tuhpia copper is also distinguished from 'fine' in KTS 2, 8.3-4.

3.1.5.11 ša Wašhania - *'from Wašhania'*

According to kt m/k 131, copper from Wašhania could fetch a price of 60:1 in Purušhattum, which means that it was of fine quality. An amount of Wašhania copper, apparently obtained there, had to be paid in Wašhania itself according to BIN 4, 113 (studied by Lewy as EL 261).

3.1.6 *Scrap*

The copper bought and sold by the Assyrian merchants was not only raw and more or less refined copper in various ingots or fragments of these, but also in the shape of utensils (sickles, hammers, axes) functioning as concentrators of metal, and as scrap.[145] The words for scrap are *hus/šā'ū* and ZI.IR. The first was derived from *hašû* (AHw, p.335a, *h.* V) by B. Landsberger and K. Balkan (*Belleten* 14, p.242 n.40 'Schrott', rendered 'scraps' by the

145 See Lewy 1950, p.19 and Lewy 1964, p.183.

CAD H, p.262). It also occurs with silver.[146] Landsberger thought that scrap bronze was meant by *hus/šā'ū* and the scrap found e.g. in the late first millennium B.C. shipwrecks from Cape Gelidonya and from Langdon Bay (the English Channel) in fact consisted of bronze.[147] The Old Assyrian texts, however, generally do not specify whether the scrap was copper or bronze. The qualification 'fine' and the exchange ratio of 75:1 suggest that in those instances copper was meant. It must remain uncertain whether this is also the case in the other texts where *hus/šā'ū* occurs.

ZI.IR is attested in BIN 4, 95 and in JCS 41, 3, published by C. Michel and B. Foster in 1989. Both letters were written by Ilia and Lā-qēp to Iddi(n)-Ištar. Although BIN 4, 95 was included in the dossier about Iddi(n)-Ištar's business, line 14 of the newly published text was misinterpreted. As it was considered unlikely that 2/3 *ma-na*.TA represented a rate of exchange, ZI.IR was emended to <*i*>-*ṣé-er* and the interpretational difficulty created by this was hidden behind the translation.[148] ZI.IR is the Old Assyrian writing of Sumerian BA.ZI.IR, which is rendered 'scrap' by J.J. Finkelstein in *JAOS* 90 (1970), p.248. The adjective *šabburum* was never used for 'scrap', but only for pieces of ingots.

It is not surprising that copper implements were recycled considering the permanent value of the metal. Scrap copper was of high quality and after remelting could be cast into new objects. This explains the high prices for scrap (75:1 in KUG 27.28-30). Most recycling of copper will have happened within a small area, like a city or village, where someone would bring old copper implements to a local smith, and have new tools made of them. In that case no (inter)regional trade was necessary for bringing the scrap into the city or village. The evidence contained in the texts from Kanish suggests that, contrary to the finds at Cape Gelidonya and in England and France, there was vitually no trade in scrap metal. Relatively minor quantities are attested for it, ranging from 15 shekels to 30 minas, which resembles the amounts of (old) sickles mentioned as metal (up to 40 minas). Both were largely used as a means of exchange and less as an object of trade, especially when compared to the at times huge quantities of unworked copper.

Another form of scrap may be meant by *tabbu'um* in kt n/k 524.12: 4 *ma-na ta-bu-ú-tim*. It is listed after 40 shekels of silver and 20 minas of *elītum*. It might be derived from the verb *tabbu'um* which occurs in TC 3, 163.8, see below, p.75.

Scrap came mostly from old agricultural implements. It is listed among items brought from Assur to Anatolia, but *hušā'ū* was also brought to Assur in small amounts. In both cases, it probably was to be used as a means of exchange. K. Balkan (1965, p.159 n.1) noticed that 'scrap' in ICK 2, 321.9, was apparently used as a synonym of 'old sickles' in TC 3, 178.9-10. The quantities of *hušā'ū* range from a few shekels to 30 minas. The copper

---

146 CCT 4, 2a, [7] ... 6 *ma-na* [8] KB *hu-ša-e-e* SIG$_5$, '6 minas of good silver scraps' (CAD).

147 *AnSt* 11 (1961), p.27; *Antiquity* 54 (1980), p.107.

148 Lines 14-16 read: [14] URUDU ZI.IR 2/3 *ma-na*./TA [15] (erasure) *ù* URUDU *ša té-zi-bu* [16] *ni-dí-in-ma*, 'We sold the scrap copper at 40 shekels per (shekel of silver) as well as the copper you had left behind.' This was translated: 'Nous avons vendu le cuivre (qui est) sous (la forme de paquets) de 5/6 (*sic*) mine chacun ainsi que le cuivre que tu as laissé', (Michel and Foster 1989, p.40), see also comment to this line on p.41.

object frequently used as a means of exchange was the sickle (*niggallum*).[149] Another object used as a means of exchange was *elītum*, known from two letters recording consignments from Assur (kt n/k 524.11 and TTC 30.18-19). The context in these letters (see the Catalogue) does not allow the interpretation 'top-pack' for *elītum*. Instead, it seems to signify a second-hand item made of copper, which was used to pay expenses en route. In TTC 30, lines 18-19 read: 30 *ma-na hu-ša-e e-li-a-tim*, '30 minas of scrap, in the shape of *e.*-s.' C. Michel's hand copy of this text in RA 80, p.140, shows that nothing is missing in line 19, as M. Ichisar had conjectured ('[x] paquets de dessus', 1981, p.395). URUDU *elītum* occurs also in a lexical text from Kanish (Hecker 1993b, p.289 v 9), and in Ass 13058K (Donbaz 1985, p.16).[150] A badly preserved text contains information on the weight of a copper *elītum*.[151] Although in broken context, the same word probably occurs as well in RA 60, 39, where these items were made of silver and copper.[152]

3.1.7 *Arsenical copper*

The alloying of the relatively soft copper with tin (often 6-15%) results in bronze, a material which by its greater hardness is better able to resist pressure. This effect is also obtained by adding arsenic to the copper. Small quantities of arsenic (up to 2%) offer little improvement over pure copper. The properties approaching those of a medium tin bronze are obtained at circa 4% or more.[153] A considerable number of copper artefacts from the third and second millennia B.C. contain arsenic and are sometimes called 'copper-arsenic bronze', whereby few exceed circa 7% arsenic (most contain 2-6%).[154] The metal smelted from oxide copper ore retains most of the arsenic present below 7%. Artefacts with a high arsenic content show that little refining in a crucible was carried out and that they were cast soon after the metal was melted.[155] According to others, copper with a content of circa 3-10% arsenic apparently was obtained by the deliberate addition of arsenic minerals.[156] Discoveries of high arsenic minerals in the early third millennium B.C. layers from the sites Arslantepe and Norşuntepe suggest that this addition was at that time performed in a crucible by the smith in his workshop.[157]

The use of tin as an alloy increases at the end of the third millennium, probably caused by its increasing availability. The analyses presented in Esin 1969 (see here Appendix 4)

---

149 For instance in KUG 24.1-7: [1] 2 1/2 *ma-na* [2] *ni-ga-li i-na* [3] *Té-ga-ra-ma a-na* [4] É *ub-ri a-dí-in* [5] 1/3 *ma-na ni-ga-li* [6] *a-na ki-ri-im* [7] *i-na Ta-*[*a*]*l-pá*, 'I gave 2 1/2 minas of sickles in Tegarama to the inn. 1/3 mina of sickles in Talpa for a jar.' This text was discussed in Nashef 1987, p.37.

150 Ass 13058K: [1] 1 GÚ 14 *ma-na* U[RU]DU H[I]? [2] *e-li-a-tum* 1/2 *ma-na* [3] 5 GÍN.TA KÙ.BE 2 *ma-na* [4] 6 1/2 GÍN 15 ŠE, '74 minas of copper, *eliātum*, at 35 shekels each, its silver is 126 1/2 shekels and 15 grains'.

151 BIN 4, 214.2'-3': [2'] URUDU *e-li-a-tim* [*ša*] [3'] 5 GÍN.TA.

152 RA 60, 39.6-7: [6] [x *e-l*]*i-a-tum ša* KB [7] [x *e-l*]*i-a-tum ša* URUDU.

153 Northover 1989, p.113.

154 See the percentages of objects made from Cu + As in Table 7, Palmieri et al. 1993, p.596.

155 Tylecote 1980, p.185.

156 Palmieri et al. 1993, p.574. J. Riederer (1994, p.87) considers arsenical copper the accidental result of the smelting of arsenic containing copper ore.

157 Palmieri et al. 1993, pp.574-575.

show that many copper artefacts discovered at Kültepe have an arsenic content ranging from traces to 3.6%. Some of these arsenical coppers also contain tin. On the other hand, nearly all Kültepe bronzes with circa 10% tin contain less than 1% arsenic. This has led to the conclusion that at Kültepe two techniques were used contemporaneously: one using tin, the other arsenic as an alloy to copper. In the last technique, the arsenic could have been added in the form of high grade arsenical copper minerals (cf. Muhly 1993b, p.120a). The word for arsenic is still unknown in the cuneiform texts (Joannès 1993, p.98b), and the awareness of the role it played led to attempts at its identification.

H. McKerrell studied the analyses given in Esin 1969 for Horoztepe and Kültepe and concluded that either arsenic or tin was added to the same basic copper. The arsenic would have been added as high arsenic-content copper, and because the copper ore used probably did not contain arsenic itself, the added mineral would have come from elsewhere. McKerrell suggested that the word *annakum* (often written with the Sumerogram AN.NA), which occurs in the Old Assyrian texts as one of the major articles exported from Assur to Anatolia, does not mean 'tin' as commonly thought, but this 'high arsenic content copper-arsenic alloy'.[158]

McKerrell's interpretation of *annakum* as arsenic alloy was rejected by J.D. Muhly (1983, p.360), who confirmed the meaning 'tin' for it. In Muhly's view, the arsenic contained in the objects of arsenical copper did not result from an addition, but was already present in the copper ore. He argued that McKerrell's study only demonstrated that 'well into the second millennium B.C. there was still an extensive use of unalloyed copper'.

Criticism of McKerrell's hypothesis came also from H. Waetzoldt, who suggested instead that third millennium texts from Mesopotamia and Ebla differentiate between copper with and without arsenic. Together with H.G. Bachmann he offered an interpretation of the Eblaite words represented by the Sumerograms A.GAR$_5$.GAR$_5$, URUDU, and ZABAR.[159] Bronze was called ZABAR, just as in the Mesopotamian texts. Waetzoldt used six arguments why A.GAR$_5$.GAR$_5$ must be different from URUDU: 1) it would be surprising if in Ebla two signs had been used to designate copper; 2) tin was never alloyed to URUDU; 3) three texts contain a passage where A.GAR$_5$.GAR$_5$ was made into an object which was then called URUDU; 4) rings were made of A.GAR$_5$.GAR$_5$ but never of URUDU, and since pure copper is more malleable, this probably was the material from which the rings were made. Tools and weapons were, however, largely made of bronze, as well as of URUDU; 5) A.GAR$_5$.GAR$_5$ is more expensive than URUDU; 6) ZABAR, A.GAR$_5$.GAR$_5$, and URUDU(.GIN) are named consecutively in texts recording the issue of metal. The meaning of A.GAR$_5$.GAR$_5$ is then interpreted as 'refined copper', synonymous with Mesopotamian urudu.luh.ha. Since many artefacts in the Near East were made of arsenical copper during this period, it is assumed that this was also the case in Ebla. URUDU is then considered to be an alloy, and since the sign ZABAR is already used for bronze, URUDU is taken to mean a copper 'improved' by alloying it with arsenic.

The translation 'refined copper' for A.GAR$_5$.GAR$_5$ seems secure because it was alloyed with tin in several texts. However, in ARET 4, 3 (v VI:8-11), 5 shekels of tin

---

158 Eaton and McKerrell 1976 and McKerrell 1977.

159 Waetzoldt and Bachmann 1984.

(AN.NA) and 30 shekels of bronze (ZABAR) were issued to make 6 *bu-di* ornaments, which were often bronze. Similarly, the passage quoted by Waetzoldt in note 10 refers to ZABAR issued to make rings of URUDU. I am inclined to see here in URUDU a designation for metal consisting mainly of copper.[160] This means that it might have been used for arsenical copper as well, but it remains unclear whether in Ebla arsenic was added to copper, as suggested by Waetzoldt and Bachmann, or whether this mineral derived from arsenic containing raw copper, and not an man-made alloy.

One might assume an alloy (refined copper + tin or arsenic) to be more expensive than refined copper. However, amounts of silver paid in exchange for both A.GAR$_5$.GAR$_5$ and URUDU are recorded in ARET 7, 79, and there URUDU is cheaper (37.5:1 in v I.5-7 against 30:1 for 'refined copper' in VIII.3-5; but in ARET 7, 17 v III.2-4, A.GAR$_5$.GAR$_5$ costs 45:1). The metal designated by URUDU seems to have been of less purity. As already noted by Waetzoldt, the use of the word URUDU seems restricted to end products. The exception to this rule is ARET 7, 79 v I, although it might be possible that 5 minas of scrap were bought there.

If URUDU was an alloy, it was probably made in the palace workshop and there was no need to buy it as in ARET 7, 79. One may even go a step further by stating that proof is lacking for the existence of a trade in alloys. Even in the Old Assyrians texts, which contain rich evidence of trade in metals, the alloy bronze apparently was not traded, only objects made of it were occasionally sold.

Finally, Waetzoldt and Bachmann imply that the Eblaites were able to distinguish between copper with and without arsenic. It is true that some objects made from copper with 4% arsenic or more have a thin silvery surface. If this layer is absent, the small percentage of arsenic is invisible to the eye. This was the case with some objects from Peru, and it was only possible to determine by way of analysis whether an object contained arsenic, although copper alloys containing it were frequently used there.[161]

URUDU appears to have been used in Ebla as a logogram for end products with a high copper content. This could be arsenic copper or the alloy bronze, but presumably also unalloyed copper (refined or of lesser quality). The interpretation given by H. Waetzoldt and H.G. Bachmann in which URUDU is in general copper with added arsenic, needs more corroboration from the texts. While it remains likely that arsenical copper was known in Ebla, it is uncertain whether there was a separate word for it in the terminology for metals used in the texts. Moreover, it is still uncertain whether ancient copper workers possessed the skills, which Waetzoldt and Bachmann assume they did, to produce intentionally high arsenic copper alloys on a regular scale.[162]

---

160 In the lexical text TM.75.G.1678 r III.6-7, number 1100 'f' in MEE 4, 319, URUDU is equated with *ga-ba-lum*. Whatever word is hidden behind these signs, it cannot be linked with latin *cuprum*, as done by G. Pettinato and by A. Archi and M.G. Biga in ARET 3, p.394. The word *cuprum* was derived from *aes cyprium* nearly two and a half millennia after Ebla, see Muhly 1973, p.174. In a Hurrian bilingual text *kabali* is rendered in the Hittite version by the Sumerogram URUDU, see E. Neu, *Das Hurritische: eine altorientalische Sprache in neuem Licht* (1988), pp.37-38 and note 113.

161 Lechtman 1980, p.313.

162 Northover 1989, p.117.

The analysis of Kültepe objects demonstrates that arsenic containing copper was used in the manufacture of some artefacts. The arsenic content is not high, which suggests that it derived from the same ore as the copper and was not alloyed to it. The distinction between copper free of arsenic and copper with a low arsenic content presumably was not made, so that there was no reason to have a separate word for high arsenical copper. Even if such a distinction had been made, the copper could have easily been qualified by means of a toponym describing the origin of the metal.

3.1.8 *Conclusion*

In the Old Assyrian terminology, copper was mainly divided into two categories: fine (*dammuqum*) and poor (*lammunum*). The quality of the copper was based on the level of impurity it contained. Iron was the main impurity. The first category included the best quality, refined copper, and copper from Haburata, Taritar, and Tišmurna. It is possible that the copper from Ištanuwa and Tapašatta was of fine quality as well, but the evidence is insufficient to determine this. The unrefined fine copper might be related to the richness of the smelted ore, perhaps including lumps of native copper, or to the level of smelting technology used by those producing these coppers. There is as yet no evidence that copper from Haburata and Taritar was remelted and/or refined, or exchanged for refined copper, before it was sold again. One can infer from this that both coppers consisted of raw copper, sufficiently clean to be cast, or alloyed with tin to make bronze. It is curious that the occurrence of these high quality coppers is relatively rare in the texts, an explanation for this may be that they were implied in many of the cases where *dammuqum* copper is mentioned.

The second category, *lammunum* or poor, consisted of 'black' (*ṣallāmum*) and 'dirty' (*massuhum*) copper, and copper from Kunanamit, Tuhpia, Wašhania, and perhaps also from Ninašša. Poor copper was also called *ša masā'im*, 'which has to be refined'. The main impurity in poor copper was iron.

3.2 *Commercial terminology*

Old Assyrian texts contain two verbs which each describe a specific aspect of trade. The first is *barā'um*, the second is *ka''unum*, for which a metallurgical meaning was given by P. Garelli in the texts where it has copper as its object.

3.2.1 *barā'um* and *barru'um*

A verb *barā'um* occurs in a number of texts dealing with the sale of merchandise.
In the G-stem:
1) BIN 4, 34.6-16: 6 *a-bi a-ta šu-ma . dam-qá-kum* 7 URUDU 10 GÚ *ša e-pu-šu . a-na-kam* 8 *áš-ta-na-me-ma . ma-tum* 9 *em*-ṣa*-at . a-ma-tí ma-tum* 10 *a-na áš-ri-ša li-tù-ur* 11 URUDU *ší-ni-šu ib-tí-ri* 12 *šu-ma li-bi-kà* 13 *té-er-ta-kà . li-li-kam-ma* 14 *lu a-na* KB *lu a-na* 15 KÙ.GI *a-na-kam-ma* 16 *lu-ta-er-šu*, 'My dear father, if it is good in your opinion, concerning every 10 talents of copper which I obtain - here, I keep hearing (that)

the country is starving.[163] When will the country become normal again? The copper has already twice *b*.-ed! If you agree, send me a message and I will exchange it here for silver or for gold.'

2) kt 91/k 350.18-22: [18] *a-na* KB *ša A-šur-na-da* [19] *na-áš-a-ku-nu-tí-ni ku-ta-ni*$_7$ [20] *ša-ma-ni-ma* [21] *ih-da-ma . a-na ba-ra-im* [22] *li-*⌜*ik*⌝*-šu-du-nim*, 'Buy me *kutānum*-textiles for the silver Aššur-nādā is bringing you. Take care, they must reach me to be *b*.-ed' (by courtesy of K.R. Veenhof).

3) TC 1, 29.27-34: [27] ... [K]B [28] *tù-šé-bi-la-nim* [29] *um-ma a-tù-nu-ma la i-ba-/ri* [30] *a-šu-mì té-er-tí-ku-nu* [31] AN.NA *ba-at-qám áš-a-ma* [32] *ù a-ma-kam ib-tí-ri* [33] *ù-ma-am* AN.NA 16 GÍN./TA [34] *ù e-li-iš i-ba-ší*, 'You sent me silver and said: "It must not *b*.!" Because of your instruction I bought expensive tin, and now it has *b*.-ed over there. Today there is tin at a rate of 16 shekels and more per shekel of silver.'

In the D-stem:

4) BIN 4, 2.2-15: [2] ... *a-na* [3] *Ša-lá-tù-wa-ar* [4] *a-na té-er-tí-a e-ta-ra-áb* [5] *ma-ki-ru-ú a-dí-ni lá e-ru-bu-/nim-ma* [6] *um-ma a-na-ku-ma* [7] *lá-áš-pur-ma* AN.NA *iš-tí* [8] URUDU-*i-ni a-ba-ru-im* [9] *lu-šé-tí-qú-nim a-bi*$_4$ *a-ta* [10] *be-lí a-ta i-hi-id-ma* [11] *i-*^{d}UTU*-ši ṭup-pì-i* [12] *ta-ša-me-ú šu-pur-ma* [13] AN.NA*-ki ma-lá ta-le-/e-ú* [14] *lu-šé-ri-bu-nim-ma* [15] *pá-ni-a-ma ab-kam*, 'I just entered Šalatuar for my business; until now no other merchants have arrived, so I thought: "I will write (to Puzur-Aššur), that they bring tin hither to be *b*.-ed together with our copper." My dear father and lord, take care and order them by letter on the very day you read my letter, to bring my tin (into this city) as much as you can (organize); bring it right to me!'

5) BIN 4, 48.5-15, 39-41: [5] *a-na-kam* AN.NA *iš-tí* URUDU*-i-ni* [6] *a-na ba-ru-im lá-šu i-na* ^{d}UTU [7] *ṭup-pì-i ta-ša-me-a-ni* (...) [10] *a-ṣé-er lu-qú-tí-a a-na Tí-mì-il*$_5$*-/ki-a* [11] *le-tí-qú-ma* (...) [13] (...) AN.NA*-ki ú* TÚG*-tí-a* [14] SIG$_5$ *ma-lá ú-šé-tí-qá-ni* [15] *iš-tí a-li-ki li-li-ku-nim* (...) [39] (...) *a-dí* AN.NA [40] *ú-bu-lu-ni-ni sà-ah-ra-ku* [41] *a-na-kam ší-im* AN.NA *i-ba-ší*, 'There is here no tin to be *b*.-ed with our copper. When you hear my letter, let (servants) travel on to my goods in Timilkia. (...) All my tin and fine textiles that I sent across must come to me with travellers. (...) I am wasting my time until they bring the tin. Tin is in demand here!'

6) kt 87/k 422.16-18: [16] ... URUDU*-a-kà* [x x x x x] [17] *ú-ba-ra-ma* KB*-pì* [(x)] [18] *a-la-qé*, 'I will *b*. your copper and collect the silver' (by courtesy of K. Hecker).

7) kt 87/k 453.22-27: [22] *ki-ma i-na da-áš-e-šu* [23] *sí-kà-tù-ni . a-dí u*$_4$*-mì-/im* [24] *a-ni-im* URUDU*-kà* [25] *i-ša-an-tí a-šar ba-ru-um* [26] *ú-ba-ra-šu-ma* [27] *a-na* KB *ú-ta-ar-šu*, 'As there is *sikkatum* in spring, your copper is until the present day ...; I will *b*. it when there is an opportunity to *b*. and exchange it for silver' (by courtesy of K. Hecker).

In 1965, the CAD (B, p.120a) listed text 3, 4, and 5 under *barû* C, with the tentative meaning 'to be available(?), to be on the market(?)' in the G-stem, and 'to put on the market(?), to offer for sale(?)' in the D-stem. W. von Soden (AHw, p.837a) listed texts 4

---

163 Collated by K.R. Veenhof.

and 5 under *parû* II and translated 'to alloy' ('verunreinigen, (Metall) legieren'), but wrote in AHw, p.1547a under *barû* I D: 'aA zT unkl.', with a reference to the CAD. In 1987, K.R. Veenhof proposed to list the forms in the G-stem in BIN 4, 34 and TC 1, 29 under *berû*, 'to be hungry', meaning that the merchandise could be sold (1987, pp.62-63). In 1992, he admitted that the meaning of the D-stem was unclear, and opted for the metallurgical interpretation given by W. von Soden.

Texts 1 (BIN 4, 34), 4 (BIN 4, 2), and 5 (BIN 4, 48) form together with TC 3, 13 and CCT 6, 22a, a group of letters which Buzāzu sent to Puzur-Aššur (see Veenhof 1992). He was on his way to a city with a consignment of copper, but could not enter it because of famine (or rather: problems) in the country. CCT 6, 22a links up with BIN 4, 34 (discussed in Veenhof 1972, pp.312-313), and lines 4-17 read: 'As for what I repeatedly wrote to you in the following wording: "Let them either bring the tin into (the town) via the smuggler's track if that is suitable, or let them make packets and bring (it) piece-meal into Kanish under their clothes." I hear that the country is starving and ... . And the smuggler's track is not appropriate for entering (the town). You must not bring my tin into (Kanish on your own) ..., it should enter it only together with the caravan.'[164] The market apparently was disrupted as a result, and he had to stay in Šalatuar, complaining that copper had in the meantime twice *ibtiri*. This leads him to ask for permission to sell his copper in the city he is staying in (using the D-stem). The juxtaposition of tin with copper in Buzāzu's letters led W. von Soden to propose the meaning 'to alloy' for this verb, which has the supposed meaning 'to blaspheme' in the Erišum inscription (see Grayson, RIMA 1, p.21 Erišum 1.70: *ú-pá-ru-ú*). Proof is lacking for the alloying of copper with tin by Assyrian merchants, which makes Von Soden's and Veenhof's interpretation unlikely. The Assyrians did not trade bronze, but only its two components copper and tin. Buzāzu's wish is to put the tin and copper on the market at the same time, perhaps as a conditional sale. K.R. Veenhof (1992) rejected the possibility of a conditional sale, arguing that Old Assyrian would use the verb *emādum* D for this. If line 11 means 'copper has already twice been up for sale', Buzāzu must refer to other merchants trying to sell their copper, because up till now he had not received permission to sell his copper there. But the translation must be different if the subject in line 11 is the copper from his own consignment. A translation 'to get hungry' for this passage and the one in TC 1, 29 was proposed in 1983 by K.R. Veenhof, meaning that an opportunity to exchange silver and copper for something for profit had been missed.[165] A translation 'to be for sale' or 'to get hungry', however, is impossible in kt 91/k 350.21 in view of a similar expression in BIN 4, 52: ^{15} *šu-a-tí a-na a-lim*^{ki} ^{16} *ṭù-ru-sú-ma a-na* ^{17} *ba-ra-im a-na* ^{18} *Kà-ni-iš li-ik-šu-dam*, 'Send himself to the City and let him reach me in Kanish for *b.*' In AHw, p.109a, this passage is listed under *barû* I 'to see'. A translation 'for inspection' or the like might fit in both texts.

---

164 CCT 6, 22a.4-17: ^{4} ⌜*a*⌝*-dí . ša áš-ta-pá-ra-ku-nu-tí-ni* ^{5} *um-ma a-na-ku-ma .* AN.NA-*ki* ^{6} *lu ha-ra-an . sú-ki-nim . šu-ma* ^{7} *na-ṭù-ú . lu-šé-ri-bu-ni*[*m*] ^{8} *lu ri-ik-sí . lu-pí-šu-ma* ^{9} *a-na Kà-ni-iš . i-na sú-na-tí-šu-nu* ^{10} *lu-uš-té-ri-bu-nim a-ša-me-ma* ^{11} [*m*]*a-tum . em-ṣa-at-ma . i-ṣé-er* ^{12} [x x]-*tí-šu . a-na a-wi-lim* ^{13} [*ú-k*]*à-ú-lu . ú ha-ra-an* ^{14} [*sú-ki-ni*]*m . a-na e-ra-bi*$_4$*-im* ^{15} [*lá na-ṭ*]*ù-ú .* AN.NA-*ki* ^{16} [x x] *lá tù-šé-ra-ba-nim* ^{17} [*iš-tí*] . ILLAT-*tim-ma . le-ru-ba-am.*

165 Veenhof 1987, pp.62-63.

The forms in the D- and G-stem, with the exception of TC 1, 29, are then all derived from the same verb. It does not have a metallurgical meaning, but a commercial, so that the proposal by the CAD is correct.

3.2.2 *ka''unum*

A specific use of the verb *ka''unum* with copper and silver as object led in previous studies to various interpretations. E. Bilgiç (1954, p.55) translated CCT 4, 27a.9-10 'In *T*[*išmu*]*rna* ist kein Kupfer zu verwirkli[chen]', giving the verb here a commercial meaning.

P. Garelli (1963, pp.177-178), in a discussion of the economic and financial activities related to the *kārum*-office, translated the verb in CCT 1, 19b.2 '*convertir* (?)', and connected it with a metallurgical operation, like the refining or hammering of the copper. In view of CCT 4, 27a, however, he did not exclude the possibility that the verb might have a commercial meaning. On the basis of this text, Garelli equated *ka''unum* with the expression (copper) *ana dammuqim ta''urum*, 'to exchange (it) for (copper) of fine quality', which has a clear commercial significance. Upon the assumed equation of *ka''unum* with *ana dammuqim ta''urum*, he interpreted CCT 4, 27a.31-33 (*uka''in*) as an exchange of fine copper, possibly valued in silver. Garelli apparently preferred a commercial meaning for this use of *ka''unum*.

In 1971, CAD K, p.170b rendered the verb (translating CCT 4, 27a.5) as 'to establish (probably referring to refining)'. W. von Soden listed in AHw, p.440a the lines in CCT 4, 27a under the heading 'für jemand sicherstellen'. M. Ichisar (1981, pp.314-317) followed the interpretation favoured by Garelli and gave as the translation of *ka''unum* in CCT 4, 27a 'rendre convertible'. The rate in line 32 was rendered as 164 shekels of copper per 60 shekels (?). The last to discuss CCT 4, 27a was C. Michel (1991b, pp.258-259). She too followed Garelli's interpretation and translated *ana ka''unim* in line 23 by 'de convertir'. Lines 31-33 were interpreted as giving the rate of exchange of 164 shekels of raw copper per 60 shekels of refined copper.

This means that only the CAD thinks it possible that *ka''unum* has a metallurgical meaning, all other authors prefer a commercial interpretation. The texts where *ka''unum* occurs with copper as object are:[166]

1) BIN 4, 64.3-13: [3] *iš-tù a-na Tur$_4$-hu-mì-it* [4] *e-ru-bu* 10 GÚ URUDU *i-na* [5] *ba-áb-tí-kà am-hu-ur* [6] *ší-tí ba-áb-tí-kà a-na* 20 *u$_4$-me* [7] *ú-za-kà-ma a-na* SIG$_5$ [8] *ú-ta-ar* ANŠE.HI.A [9] *ù-šé-ri-a-ma* [10] URUDU SIG$_5$ *pá-ni-a-ma ša ú-kà-nu* [11] *a-šar ma-lá-ki-im mì-li-/ik* [12] *a-ša-me-ma* URUDU *i-na* [13] *Pu-ru-uš-ha-tim ṭá-áb*, 'I have received 10 talents of copper from your outstanding claims since I entered Turhumit. I will collect the remainder of your claims within 20 days and convert it into fine copper. I have led donkeys

[166] In CCT 2, 31a.14, read *ma''uqāku-ma* from *muqqu*. Lines 13-17: [13] *a-dí* URUDU *ša tù-na-hi-dí-ni* [14] *ma-ú-qá-ku-ma i-na* [15] *ša-am-ší e-ru-bu* [16] *e-ru-qá-tim ú-lá a-bu-uk* [17] URUDU *er-sú*, (to Puzur-Aššur from Buzāzu:) 'As for the copper about which you gave me instructions, I was feeble and did not bring the wagons on the day I entered (the city), but the copper is ready!' These lines are parallel to TC 3, 15.3-9, where Buzāzu writes in line 7: *am-ta-ra-aṣ-ma*, 'I fell ill.'

(hither) and take a decision (what to do with) the first fine copper that I will *k*. I hear that the (quality of the) copper is good in Purušhattum.'

2) CCT 1, 19b.1-5: [1] 4 GÚ 28 *ma-na* [2] URUDU *ša kà-ú-nim* [3] *i-na* É *kà-ri-im* [4] *i-na Pu-ru-uš-ha-tim* [5] *a-lá-qé*, 'I will collect in the *kārum*-office in Purušhattum 4 talents 28 minas of copper to be *k*.-ed.', parallel to:

3) CCT 1, 22a.9-11: [9] [x]+2 GÚ 25 *ma-na* [10] *ša kà-ú-ni-im* [11] *A-šùr-na-da i-lá-qé*, 'Aššur-nādā will collect 2+x talents 25 minas to be *k*.-ed.'

4) CCT 4, 27a.18-34: [18] *ki-ma* URUDU *i-na Tí-iš-mur-na* URUDU *a-na* [19] *kà-ú-nim lá-šu-ú a-dí ma-lá ù* [20] *ší-ni-šu a-ṣé-er A-mur-Ištar* [21] *ù Pu-šu-ke-en*$_6$ *e-tí-iq-ma* [22] *um-ma a-na-ku-ma i-na Tí-iš-mur-na* [23] URUDU *a-na kà-ú-nim lá-šu* [24] URUDU *a-na Tur*$_4$*-hu-mì-it lu-šé-ri-/ba-ma* [25] *a-na* SIG$_5$ *lu nu-ta-er* [26] *um-ma šu-nu-ma a-wa-at kà-ri-im* [27] *lá tù-šé-ra-ba-am ù té-er-tí-ni* [28] *a-na A-mur-Ištar a-dí ha-am-ší-šu i-lik* [29] *um-ma a-na-ku-ma* URUDU *lu-šé-ri-ba-am* [30] *a-na* SIG$_5$ *li-tur*$_4$ *um-ma šu-ut-ma lá na-/ṭù* [31] *lá tù-šé-ra-ba-am* 1 GÚ URUDU SIG$_5$ [32] *i-na Tí-iš-mur-na* 2 2/3 *ma-na* 4 [GÍN.TA] [33] *ú-kà-in*, (to Imdilum from Uzua:) 'I went repeatedly to Amur-Ištar and Pūšu-kēn because there is in Tišmurna no copper to be *k*.-ed, saying: "There is no copper to be *k*.-ed in Tišmurna; let me bring the copper to Turhumit and we will exchange it for fine (copper there)." But they answered: "There is a *kārum* ruling: you are not allowed to bring it in(to the city)." Well, up to 5 times a letter by us was sent to Amur-Ištar, in which I said: "Let me bring the copper in(to the city) so that it can be converted into fine (copper)." But he said: "It is not the appropriate moment to bring it in(to the city)." I have *k*.-ed 1 talent of fine copper in Tišmurna at a rate of 164:1.'

5) TC 2, 36.38, 42-48: [38] *a-na kà-ú-nim* URUDU *lá-šu* (...) [42] *iš-tù a-na* URUDU *kà-ú-nim ni-zi-zu* [43] 40 *u*$_4$*-me lá ma-ṣí* 10 GÚ URUDU SIG$_5$ [44] *nu-kà-in* URUDU *a-dí* 1 ITU.KAM *i-za-ku-ma*, (to Imdilum from Uzua:) 'There is no copper to be *k*.-ed (...) After we started to *k*. copper, (it turned out that) 40 days were not enough. We *k*.-ed (only) 10 talents of fine copper. (This) copper will become available within 1 month.'

6) kt a/k 977.5-21: [5] ... 15 *ma-na* [6] [URUD]U *ma-sí-a-am ší-im* 1 *ku-ta-ni-a* [7] *[i]š-tù Tur*$_4$*-hu-mì-it i-na* [8] *e-ra-bi*$_4$*-kà-ma . a-ṣé-er ša ki-/ma* [9] *ku-a-tí . tù-šé-ba-lá-ma* [10] URUDU . *i-Kà-ni-iš i-ša-qú-lu-/ni-ma* [11] ⌜1⌝ *ṭup-pá-am ha-ar-ma-/am* [12] *[š]a hu-bu-li-kà* [13] *[š]a* 6 GÚ URUDU *lá-mu-nam* [14] *[š]a kà-ú-nim ha-bu-lá-tí-/ni* [15] *[a]-da-šu-nu-ti* [16] *šu-ma* URUDU*-i . tám-tí-ri-ma* [17] *[a]-dí* 4 *ša-na-tim* [18] ⌜*ù*⌝ 6 *ša-na-tim lá tù-uš-/té-bi*$_4$*-lam* [19] ⌜*a*⌝*-li ni-na-mu-ru-/ni* [20] 30 *ma-na* URUDU *ma-sí-a-am* [21] *ta-ša-qá-lam*, 'The moment you arrive in Turhumit you will send to your representatives 15 minas of refined copper as proceeds of one of my *kutānum*-texiles, and they will pay me the copper in Kanish, and then I will hand over to them a certified tablet concerning your debt, which says you owe 6 talents of poor copper to be *k*.-ed. Should you be slow and not send me my copper in 4 or even 6 years time, then you must pay me 30 minas of refined copper wherever we will meet.'

7) kt n/k 122.8-11, 14-21: [8] ... 10 GÚ URUDU SIG$_5$ [9] *ší-im* 50 TÚG.HI.A [10] *i-na Tí-ša-mu-ur-na* [11] *am-hu-ur* (...) [14] *lu ší-im* TÚG.HI.A [15] *ú* AN.NA *lá-mu-num* [16] *i-ba-ší a-na-kam* [17] *a-na kà-ú-nim* [18] *lá-šu a-šar e-ku-ší-im* [19] *nu-kà-áš-ma i-na* [20] *ma-at Hi-ru-ùh* [21] *nu-kà-an-ma*, 'I received 10 talents of fine copper as proceeds of 50 textiles in

Tišmurna. (...) There is poor copper available as payment for the (other) textiles and for the tin. Here, there is no (fine copper) to be *k.*-ed. We will travel as far as necessary and *k.* (copper) in the country of Hiruh' (by courtesy of S. Bayram).

8) kt n/k 194.12-15: [12] 6 *me-at* 75 *ma-na* [13] *a-na* SIG$_5$ *kà-i-šu-ma* [14] KI URUDU *ša li-bi$_4$-kà* [15] *šé-bi$_4$-lam*, '*k.* 675 minas for fine (copper) and send it with the copper you owe me' (by courtesy of S. Bayram).

9) AKT 2, 39.4-8, 15-19: [4] 10 GÚ URUDU PN *ub-lam* [5] ŠÀ.BA 7 1/2 GÚ *ma-sí-um* [6] 20 LÁ 2/3 *ma-na* URUDU *ma-sí-a-am* [7] *a-na* É *kà-ri-im a-na* [8] *kà-ú-nim ni-dí-in* (...) [15] 1 GÚ 30 *ma-na* URUDU *ší-ku-um* [16] *ù* 18 *ma-na ša* [17] É *kà-ri-im ù-kà-i-nu-ni-a-*[*t*]*í-ni* [18] ŠÀ.BA 9 *ma-na* [19] *ša-du-a-tám ni-dí-*[*i*]*n*, 'PN brought 10 talents of copper; 7 1/2 talents of it was refined (copper), (of which) we gave 19 1/3 minas to the *kārum*-office to be *k.*-ed (and sold the rest). (...) 1 talent 30 minas was *šikkum* (copper). We gave 9 minas as *šaddu'atum*-tax out of 18 minas which they *k.*-ed for us at the *kārum*-office (11 minas given to PN, remainder 1 talent 28 minas sold).'

10) kt 87/k 417.3-6: [3] 8 GÚ URUDU SIG$_5$ 9 1/2 GÚ *lá-mu-nu-um* [4] PN *ub-lam* 5 *ma-na* SIG$_5$ *im-ṭí* [5] 5 5/6 *ma-na* SIG$_5$ *lu e-ri-ib-tum* [6] *lu ša kà-ú-nim*, 'PN brought me 8 talents of fine copper and 9 1/2 talents of poor (copper); there was a deficit of 5 minas of fine (copper), 5 5/6 minas of fine (copper were paid) as entrance charge and (as costs) for *k.* (remainder 7 talents 49 minas)' (by courtesy of K. Hecker).

11) kt 89/k 394.20-25: [20] URUDU *la-mu-un-ma* [21] *ú-la áš-pur-ra-kum* [22] URUDU *da-mu-qám* [23] *ú-kà-an-ma* [24] *a-ša-pá-ra-ma* [25] ANŠE.HI.A *šé-ri-a-*[*am*], 'I did not write you because the copper is poor. When I can *k.* fine copper, I will send you a message and you must send me donkeys' (by courtesy of Y. Kawasaki).

The verb is used in these texts with the following constructions:

a) copper *ša ka''unim* - 'to be *k.*-ed' in 2, 3, 6, 10. The modality is uncertain and the subject undetermined, and it is translated as a passive. In 6.13 the object is poor copper, in 10.5 fine copper.

b) preposition with infinitive: *ana ka''unim* - with *laššu* in 4.23 and 5.38, both with 'copper' as object, and in 7.17-18; with *izēzum* in 5.42, object 'copper'; with *tadānum* in 9.7-8, object refined copper.

c) fientic with as object fine (*dammuqum*) copper in 1.10, 4.31-33, 5.43-44, 11.22-23, and implied in 10.6; and with poor copper in 6.13-14; no qualification in 2, 3, 7.21.

In text 1, the sender of the letter apparently is collecting poor copper, which he will subsequently convert into fine (copper). By mentioning that he possesses donkeys and that the price of copper in Purušhattum is good, he is suggesting to its owner that he should have it brought to that city. The object of *ka''unum* is fine (*dammuqum*) copper, resulting from *ana dammuqim ta''urum*. The meaning 'to secure (by means of conversion)' seems here appropriate for *ka''unum*, which is not identical with *ana dammuqim ta''urum*, as there the object is poor copper. The verb *ta''urum* in the last expression can be used theoretically both in a commercial and in a metallurgical sense. There are, however, no clear references to an Assyrian merchant bringing poor copper to a copper smith wishing to have it refined and so change it into high quality copper. P. Garelli's interpretation of TC 3, 10 is incorrect, see

§ 3.4.3. Such a visit to a smith is only attested when silver has to be refined. The expression therefore does not have the meaning 'to turn into fine copper' by metallurgical means. Only the commercial interpretation seems possible, whereby poor copper was exchanged for fine.

Also in text 4, *ka"unum* is not the same as *ana dammuqim ta"urum*, as in P. Garelli's interpretation, but rather describes its result, whereby the fine copper was *k*.-ed as *dammuqum* or fine copper without change of quality. The rate given in line 32 means either that 1 shekel of silver was paid per 164 shekels of fine copper, or that 164 shekels of poor copper were paid for each 60 shekels of fine copper. In the first case, fine copper would have been bought with silver in Tišmurna, which is very rare and difficult to prove, but still possible. In the second case, 'to *k*. fine copper' implies that poor copper was exchanged for fine. The rate of exchange then applies to the two qualities, as in C. Michel's interpretation.[167] Although this is an attractive solution, the assumed rate of fine copper in silver appears to me a more plausible explanation.

It is apparent from text 5, that the action of *k*. took some time. As in text 4, copper was in short supply. More than forty days were needed to 'secure' 10 talents of fine copper, which would be available within 1 month. M. Ichisar (1981, p.346) translated lines 42-44a: 'Depuis que nous avons placé le cuivre à rendre convertible, 40 jours ne se sont pas passés, nous avons (encore) rendu convertible 10 talents de bon cuivre', suggesting that the fine copper was made convertible for exchange for something else. An exchange for silver probably would take place later, but can here be excluded as fine copper was at this stage of the trade itself the desired commodity.

Text 6 contains a reference to '6 talents of poor copper to be *k*.-ed'. The debtor owes 15 minas of refined copper for 1 textile, and his creditor offers a settlement for this debt: as soon as the debtor enters Turhumit, he will send these 15 minas to his representatives, who will pay the copper in Kanish to the creditor. A bond concerning the debt of '6 talents of poor copper to be *k*.-ed' will then be released to the debtor's representatives. If he does not pay now, but postpones it for 4 or even 6 years, he will have to pay twice the amount. The handing over of the bond means that the debt was fully paid. The 15 minas of refined copper can hardly be the equivalent of 6 talents of poor, which would make refined copper 24 times more expensive than *lammunum*. This suggests that the 15 minas constitute the final instalment given in refined copper although the debt consisted of poor. It would be attractive to see in the bond a document stipulating that poor copper will be converted into refined copper. However, the obligation seems to be that 6 talents of poor copper, or its equivalent in fine or refined copper, must be paid for an unspecified number of *kutānum*-textiles. This means that poor copper could also be the object of *ka"unum*, here in a contract.

In text 8, *ka"unum* is used as a synonym for *ta"urum*. The object *-šu* refers to the 675 minas in line 12, apparently consisting of poor copper, and a translation 'to exchange' for *ka"unum* seems suitable here. Nevertheless, text 1 shows that *weriam dammuquam*

---

[167] Michel 1991b, p.259.

ka''unum is not identical to *ana dammuqim ta''urum*. The difference between both expressions was apparently not so great, and they were mixed up in text 8.

In texts 9 (**19**) and 10, the verb is used in connection with the *kārum*-office and seems to refer to the payment or a fee in refined/fine copper (9.6-8 and 10.6). Another meaning is contained in 9.16-17, where *ka''unum* describes the payment of 18 minas of copper to merchants by the *kārum*-office. The qualification of this copper is unspecified, but as it is used to pay the *šaddu'atum*-tax on 9 talents (9 minas for 7 1/2 talents of refined + 1 1/2 talent of *šikkum* copper?), it probably was fine copper. The reason for this payment is unclear.

The small amounts paid in 9 and 10 may represent fees which a trader paid to the *kārum* when he wished to participate in a copper transaction organized by it, or in acknowledgement of its role in securing the deal. In 2 and 3 the translation would be 'I shall take copper which has to be *k.*-ed in the *kārum*-office.' This could refer to the result of a collective *kārum* action or of an operation by an *ellutum*: it is certain what amount the palace will pay in future, but it is not yet available and has still to be 'secured'. M.T. Larsen (1976, p.339) used this very word for *k.* in text 3. An expenditure of 18 minas seems insignificant, but might have been part of a greater amount like the nearly 4 1/2 talents mentioned in texts 2 and 3.

In conclusion, *ka''unum* is likely to mean 'to secure', by contract as in text 6, and in practice, in exchange for merchandise or by exchanging poor copper for fine, thus resembling *epāšum* which is often used in the same context.

3.3 *Ingots*

Primary ingots, cast at the smelting site, consisted of either relatively pure or impure metal. Copper of high purity and to a lesser extent also bronze, were cast by the coppersmith into moulds to form secondary ingots. Several types of such moulds have been discovered in workshops (see p.73 and Appendix 2). The metal poured into them may have been other than copper, but as this was the metal most widely used, I consider these moulds as matrixes for casting copper ingots. Judging by the number of bar moulds, copper was frequently cast into a bar. Moulds for such bars were discovered both in private houses and in workshops. Thin bars could have been bent to make torcs or *Ösenhalsringe*. The Assyrian word for 'bar' was *urākum*, thus far attested for *aši'um*-mineral (kt 89/k 206.7) and for silver (BIN 4, 217.10). Another type, found only in the level II workshop, is represented by a rectangular mould to form a plaque. The weight of such plaques seems to have been more or less identical, circa a couple of kilograms. These plaques served a double purpose: made of copper or bronze they could be hammered into a sheet, and ultimately an artefact. It was just as easy to break off a piece from a metal plaque as from a bar, if a smaller amount of the metal was required. A plaque (*lē'um*) of tin weighing c. 4 1/2 kg is mentioned in AKT 3, 49.[168] A small number of lead and tin ingots of irregular shape were found in workshops,[169] but no lumps of copper have not been found in the ruins of the

168 AKT 3, 49.23-24: [23] 9 *ma-na* . 7 GÍN AN.NA [24] *le-a-am*, '9 minas 7 shekels of tin, a *lē'um*.'
169 Özgüç 1986, pp.77-78.

houses excavated at Kültepe. A number of metal lumps, allegedly bronze, were found in a jar during the 1950 excavations.[170] Although moulds for bars and plaques were discovered, the excavations did not yield any copper ingot of such a type. Instead, several disc-shaped pieces of metal have been found, which, however, were interpreted as weights by the excavator. One might be a copper ingot:

- kt b/k 246; D. 9.2 cm, Th. 1 cm, weight c. 596 gr (see p.250).

The majority are of lead:
- kt e/k 137; unpierced; Özgüç 1986, Pl. 130, 4; D. 9.1 cm, Th. 1.9 cm, weight 1012 gr
- kt e/k 142; perforations on the edge; Özgüç 1986, Pl. 130, 3; D. 7.3-7.9 cm, Th. 0.9-1.1 cm; weight 575 gr
- kt j/k 152; pierced in the centre; Özgüç 1986, Pl. 130, 1; D. 6.1 cm, weight 50.75 gr
- kt j/k 153; pierced in the centre; Özgüç 1986, Pl. 130, 2; D. 6 cm, weight 59.95 gr
- kt u/k 95; Özgüç 1986, Pl. 130, 7; D. 3.4 cm, Th. 0.8 cm, weight 80 gr
- kt 76/k 175; 2 notches in the outer edge; Özgüç 1986, Pl. 130, 5; D. 3.2 cm, Th. 0.2 cm, weight 19.4 gr
- kt 82/k 180; Özgüç 1986, Pl. 130, 6; D. 7.6 cm, Th. 1.1 cm, weight 405 gr

It is difficult to determine the original weight of these objects because some of the metal is lost as a result of corrosion. Their weight is tentatively given in Appendix 5. The copper object found in a grave might be an ingot because a disc was a common form. The copper ingots found in the palace at Acemhöyük were 'of flat round shape',[171] and it is possible that an ingot was presented as funerary gift. It could be a primary or a secondary, that is of remelted metal, ingot, if the smith used moulds for discs as well. Plano-convex or bun shaped ingots are thought to have formed at the bottom of the furnace,[172] but the evidence from Oman (see p.25) and Timna (p.21) suggests that such ingots were (usually?) formed in circular depressions in the ground and not in the furnace.

Old Assyrian terminology possessed several words denoting types of ingot. *Kabbārum* is only attested in the damaged text Ka 435,[173] where I read: 2' [x x] *ša*!(GA)*-áp-tám* 3' *Ah-ša-lim . nu-ú-a-a-e* 4' *ú-kà-li-mu-ma* 5' *šu-lu-ša-ú-um* 6' *a-na kà-ba-ri-im* 7' *ú-kà-lu-ni* 8' *a-ni-i* 30 *ma-na* 9' *ša-áp-tám* 10' *a-na ṣú-ha-ri-š*[*u*] 11' *ú-sà-ni-iq-ma* 12' *i-na ma-hi-ri-im* 13' *ša-na-ú-um a-na* 14' URUDU . *ṣa-hi-ri-im ip-šu-*[*ur*]*-/ma*, '(That) Ah-šalim offered wool for sale to the Anatolians, and that they offer to pay (for) three (units of wool) one (unit of copper in the form of a) *kabbārum*. Just now he transferred 30 minas of wool to his servant and he sold it on the market at a rate of two (units of wool per unit of) small-sized (pieces of) copper.' The meaning is clear: wool was sold for copper and the first offer was

170 Özgüç 1953b, p.123: 'II N-O P/20: In der Küche (...), nahe am Herd war eine Schnabelkanne, in der 5 kleine Tafeln, kleine flache Steine und ungeformte Metallstücke (Bronze) waren. Dies ist der erste derartige Fund.'

171 *AJA* 86 (1982), p.560.

172 Tylecote 1976, pp.30-31; Muhly 1983, pp.358-359 § 3.17.

173 This text was published in Donbaz and Veenhof 1985, p.152.

at a ratio of 3:1 for copper in ingots. Now, a better exchange ratio could be realized of 2:1 against small-sized copper. Not only was the price offered for the wool 50% higher, but also the quality of the copper received as proceeds could be better assessed. While a complete ingot might contain impurities not visible on the outside, one broken up in small pieces would easily show the impurity the metal might contain. The word *kabbārum* might be derived from the verb *kabārum*, 'to become fat, heavy, thick' and refer to its shape. It would then have the meaning 'thick'. On the other hand, it might be related to the word *k/gubāru*, mainly attested in texts from Old Babylonian Ur, where it refers to copper ingots.[174] These *k/gubāru* ingots could, according to UET 5, 678.1-2, weigh 3 and 4 talents, or 90 and 120 kg, a piece. This is an astonishing weight, considering that the Late Bronze Age oxhide ingots from the Mediterranean weighed 30-38 kg, and it is thought that they were formed by tapping the whole content of a furnace.[175]

The second type is *kakkartum*, which denotes a round object. A word *kakkaru/kakkartu* is known with the meaning 'round loaf of bread', and in mid-second millennium B.C. texts it is used for a disc weighing 1 talent and usually of copper (CAD K, p.49b; in Hebrew and Ugaritic *kkr*). The Old Assyrian references are AKT 2, 59: ^{8} ... 5 ^{9} *kà-kà-ra-tum* ^{10} 57 *ma-na* ^{11} *šu-qú-ul-tum* ^{12} É *A-mur-Ištar* ^{13} *i-ba-ší-a*, '5 *k*.-s (together) weighing 57 minas remain in storage in Amur-Ištar's house'; I 482 reverse: $^{5'}$ *kà-kà-ra-tum i-ba-ší-a ma-⌈lá⌉* $^{6'}$ *té-er-tí-kà lu* KB *lu* KÙ.KI *a-ša-a-ma* $^{7'}$ *ú-šé-ba-lá-ku-um*, 'There are *k*.-s in storage; according to your instructions I will purchase (= sell the *k*.-s for) silver or gold, and send it to you' (by courtesy of K. Hecker); and kt n/k 492: ^{9} ... *a-na* ^{10} 8 *ma-na* KB . *kà-kà-ra-tim* ^{11} 8 GÍN KB *ni-ik-bu-ús*, 'From 8 minas of silver, (proceeds of?) *k*.-s, we deducted 8 shekels of silver' (by courtesy of C. Günbattı). These texts, in contrast to those mentioning *sadālum* (see below), do not specify from which metal a *kakkartum* was made. Such a specification would have been superfluous when it was always or usually the same metal. Since this was neither silver nor gold according to I 482, it must have been copper or tin. An argument in favour of copper is that Amur-Ištar (AKT 2, 59) was frequently obtaining copper in exchange for merchandise (see § 4.6.1). The discs stored in his house together weighed 57 minas, that is on average 11 minas 24 shekels (c. 5.5 kg) a piece. The fact that such additional information on their total weight is given, suggests that individual discs varied in weight.

The third word is *patallum*, which occurs in I 792, $^{2'}$ ... 24 *pá-tá-li* $^{3'}$ URUDU *ša ma-sá-im* (...) $^{5'}$ ... 24 *pá-tá-li* $^{6'}$ *ù ší-im* GUD, '24 anklets of copper yet to be refined (...) 24 anklets and the price of the oxen' (by courtesy of K. Hecker). G. Kryszat suggested to derive *pá-tá-li* from a Hittite word *patalli*, 'anklet'. It apparently refers to an ingot bent into a shape resembling that of an anklet. The first 24 consisted of copper which still had to be refined, which indicates that a *patallum* was (at times?) a primary ingot.

The word *sadālum* was used for copper. It bears no relationship with the word *šandalum*, a metal vessel, as suggested by the CAD,[176] because the sibilant is always

174 AHw, p.295a; CAD K, p.481b. See for Ea-nāṣir's correspondence Leemans 1960, 39ff.

175 Tylecote 1980, p.189, and 1982, p.195.

176 CAD S, p.11a, but no reference to this word in Š/I, p.374a *šandalu*. AHw, p.1000b read erroneously *sad/talium* in TC 2, 62.

written with the sign SÁ, and the second vowel must be long, since it is not subjected to the rule of vowel harmony. Two texts mention this type of ingot: kt 87/k 439: 4 5 *me-at sá-da-li* 5 *i-na* URUDU *ša* É.GAL-*lim* 6 *A-šur*-SIPA *il*5-*qé*, 'Aššur-rē'ī received 500 ingots from the copper (owed by) the palace' (by courtesy of K. Hecker); and TC 2, 62: 5 12 *sá-da-li* URUDU 6 *ší-im* ANŠE *iš-tí* 7 *Da-ku-ku*, 'Dakuku owes 12 copper ingots as price of the donkey.' The mentioning of the word 'copper' after *patallum* and *sadālum* implies that both words were used for ingots of other metals as well.

Of the four words to denote 'copper ingot', the shape is more or less certain only in the case of *kakkartum* and of *patallum*. A *sadālum* might be a bar ingot, a *kakkartum* a disc-shaped ingot, and a *kabbārum* one of the two. The copper discs excavated in Acemhöyük and possibly the copper disc from Kanish, might be equated with *kakkartum*. One or more of these words may have been used as synonyms for the same type of ingot. See for URUDU *sí-ra-tum*, which K. Hecker tentatively gave the meaning '*Kupferstangen*',[177] p.239 sub *zirum*.

The question can be raised whether there was some form of weight standardisation for ingots of a certain shape. If a plaque or disc of copper, or a bar of a certain size, had more or less a standard weight, it would facilitate estimating an amount of copper. The exact weight could be established at a later stage.[178] A lead ingot from level Ib weighing 520 gr (Emre 1971, p.161 and Pl. 15, 2) represents more than 1 mina (~500 gr ± 40 gr, *RlA* 7, p.509), which might be intentional. The fact that in the text mentioning *patallum* and in both texts mentioning *sadālum* the quantity is expressed by simply giving the number of ingots instead of their weight points to a more or less customary weight and size for this type.[179] The contents of two bags are given in AKT 3, 40: 'I went the Erādi's house and there were 2 bags with his seals. In one 40 minas (of copper) were stored, in the other about 2 talents.'[180] Since the copper presumably consisted of broken up ingots, the contents of both bags were assessed by using a standard capacity for the bag (see below, p.66), or by means of a clay seal mentioning contents and weight.

As appeared from the texts quoted in § 3.1.4.4, the copper ingots were frequently broken up before being transported. The adjectives used for such copper ingots in pieces are *šabburum*, 'broken up', and *ṣahhirum*, 'small-sized'. The advantage of an approximate 'standard' weight in assessing an amount of copper was of course lost once the ingot had been broken up.

177 Hecker 1993, p.290.

178 Likewise Zaccagnini 1986, p.414.

179 The information contained in TC 2, 62 that a donkey was sold for 12 *sadālum* is insufficient for establishing the weight of a *sadālum*, because the prices in copper for a donkey vary, and the quality of both donkey and copper are often uncertain (7 minas in JCS 41, 3.33; 10 minas in BIN 4, 172.15-18; 25 minas in ATHE 37.22-23; 26 minas in ATHE 37.18-19). It is therefore merely a guess when its weight is put at 2 minas.

180 AKT 3, 40.20-25: 20 *a-na* É *E-ra-dí a-li-ik-ma* 21 2 *il*5-*lu ku-nu-ku-šu* 22 *i-ba-ší-ú i-na iš-té-en* 23 40 *ma-na ša-pí-ik* 24 *i-na iš-té-en ki-i* 25 2 GÚ *ša-pí-ik*.

3.4 *Transport*

3.4.1 *Donkeys and porters*

In Assur, a merchant did not only buy merchandise, but also the means of transport in the form of donkeys with their harness and the bags necessary to carry the wares. After such a purchase was made, the merchant would not set off immediately, but wait till enough other merchants had finished their preparations to make up a caravan. When the formation of the caravan was completed, the merchants together set out for Syria and Anatolia.[181] A good donkey could carry a load of at least 75 kg,[182] and the amount of copper carried by a donkey varies from 2-3 talents, or c. 60-90 kg.[183] Its ability to carry such loads, combined with the adequate availability of these animals, made the donkey, called *emārum* (*ṣallāmum*,'black donkey'), the primary means of transport for the long distance trade between Assur and Anatolia. Upon arrival in Kanish, they were either sold together with the merchandise, or first used to carry goods to other destinations and then sold there if necessary. A fair number of donkeys must have been available and used for transportation inside Anatolia.

Local porters were employed within Anatolia as an alternative to donkeys. They were called *ša biltim*, 'he who (carries) a load', indicating what a man could carry, namely one talent (*biltum*), about 30 kg. In contrast to the street-porters of the Middle East and North Africa during the past centuries, who worked in old cities with narrow winding streets where no beasts of burden or carts could be used,[184] the porters attested in the Old Assyrian texts carried goods from one town to another. In kt n/k 482, two Assyrians travelled from Kanish to Tišmurna to sell textiles to the palace: 'After I travelled with Ibizua from Kanish - notwithstanding I had reached an agreement about the textiles, the palace in Tišmurna waited with buying the textiles and for that reason I had to stay until the present day. I will send Taskul on his way with the porters *straightaway* and settle the remainder of your business, and then I will leave (too).'[185] Though the writer of this tablet does not mention it, it must

---

181 Information on the movements of a large Assyrian caravan comprising 300 men and 300 donkeys is contained in ARM 26, 432.

182 Veenhof 1972, p.24. The standard load of a donkey in ancient Rome was fixed at 200 pounds, or 65.5 kg, *Der Kleine Pauly* 2, p.371.

183 2 talents: RA 58, Sch. 4.3: 1 ANŠE 2 GÚ URUDU *na-ší*, 'One donkey carries 2 talents of copper'; 2 talents 30 minas: AKT 3, 56.25: 5 GÚ URUDU 2 ANŠE *lu-ma-li-a-ma*, 'I will load 2 donkeys with 5 talents of copper';
2 talents 37 minas: kt a/k 425: [1] 2 1/2 GÚ 7 *ma-[na]* [2] URUDU SIG_5 ANŠE [3] *ṣa-lá-ma-am a-na* [4] *T.* [5] *áp-qí-id-ma*, 'Two talents 37 minas of fine copper (and) a black donkey, I gave to T. (and he brought it to Šalatuar)';
2 talents 56 minas: TC 1, 108: [1] 2 GÚ 56 *ma-na* [2] URUDU ŠÀ.BA 11 *ma-na* [3] *hu-sa-ú* 1 ANŠE [4] *ṣa-la-mu-um*, '2 talents 56 minas of copper, including 11 minas of scrap, and 1 donkey.'

184 See article *ḥammāl* in *Encyclopaedia of Islam* 3, p.139.

185 Kt n/k 482.3-15: [3] *iš-tù Kà-ni-iš* [4] *i-na a-lá-ki-a iš-tí* [5] *I-bi-zu-a ki-ma a-na* TÚG.HI.A [6] *a-mì-ig-ru* TÚG.HI.A *i-na* [7] *Tí-iš-mu-ur-na* É.GAL-*lúm* [8] *a-na lá-qá-im ú-hi-ir-ma* [9] *a-ší-a-tí a-dí* u_4-*mì-im a-nim* [10] *ak-ta-lá Ta-as-ku-ul* [11] *ù-šé-ri-iš iš-tí* [12] *ša bi-lá-tim* [13] *a-ṭá-ra-ad-ma ú ší-tí* [14] *té-er-tí-kà ú-za-kà-ma* [15] *a-ta-lá-kam* (by courtesy of C. Günbattı).

be understood that the textiles were carried by porters. The party consisted of the two Assyrians in charge, the writer of this letter (Suen-rē'ī) and Ibizua, and an unspecified number of porters who were supervised by Taskul, a (half-)Anatolian. The textiles would be sold eventually, and then the porters would not just walk back empty-handed, but carry the proceeds from this sale, most likely copper, to Kanish or another city. If Tišmurna was situated in the vicinity of Ankara, the distance to Kanish was at least 350 km. These porters were then apparently employed for almost a month by the same firm, that had to provide them with food (and lodging?) and pay them their wages upon conclusion of the journey. It seems unlikely that a firm would have a permanent number of porters on stand-by; rather they were recruted when needed from among those locally available. In contrast to the passage just quoted, in TC 3, 165, the porters coming from Kanish were apparently dismissed in Wašhania since they received their wages in that city, and the new party was paid in Ulama, meaning that in this case the men did not travel more than a couple of days (see below). The terms on which they were contracted included a payment at the end of the agreed trip, and food (and lodging?) during the journey at the expense of the firm.[186]

There is little direct evidence as to what a porter actually cost. It seems that he was paid according to the distance over which he carried goods rather than to the time he was employed. Frequently, the amount of 1 shekel of silver or its equivalent in copper are mentioned as wages for a hireling, e.g. in kt 89/k 232: 'If he refuses to give the silver, hire (some) hireling(s) for silver, about 1 shekel (each as wages).' Bearing in mind the usual equivalence of 1 mina of good quality copper with 1 shekel of silver, the following texts recording payments of hire are also of interest: kt a/k 431: 'I gave 1 mina of good quality copper to Hadāni as hire,' and KTP 26: 'I hired a hireling for 1 mina of good quality copper.'[187] In general, only the sum paid to an unknown number of porters is mentioned, as in BIN 4, 178: 'I paid 7 shekels of silver to those who brought the copper,'[188] and TC 3, 165: 'I paid 1/3 mina of silver from Kanish to Wašhania for porters; 17 1/3 shekels of silver I paid from Wašhania to Ulama for porters and for my guide.'[189] This text lists expenses made while travelling from Kanish to Purušhattum; in the latter city 2 minas of silver were

---

[186] Food to feed porters: from an unpublished letter from the Rosenberg collection cited by J. Lewy (1956, p.68 n.289), lines 29-34: [29] ANŠE *a-na* 30 LÁ 1 *ma-na* [30] URUDU *ta-dí-in* ŠÀ.BA [31] *lu da-tum ša ha-ra-nim* [32] *lu ú-ku-ul-tí* ANŠE *lu ú-ku-ul-tí* [33] *R. ù ša bi$_4$-il$_5$-tí-kà* [34] *a-lá-qé*, 'A donkey has been sold for 29 minas of copper; out of this (copper) I will take (= pay) the road-tax, fodder for a/the donkey, and food for R. and your porter';
TC 2, 53 reverse 8'-10': [8'] ... 1 1/6 GÍN *a-na* [9'] *ší-ri-im a-na ša bi$_4$-lá-tí-im* [10'] *ša ú-ṭá-tám iz-bi$_4$-lu-ni-ni*, '1 1/6 shekel (of silver) for meat for the porters who carried the barley' (in obverse 6-7, a total of 23 sacks of *aršātum* and 18 sacks of barley were bought).

[187] Kt 89/k 232.26-28: [26] ... *šu-ma* KB *ta-dá-/nam* [27] *lá i-mu-wa ag-ra-am* [28] *a-na* KB 1 GÍN *a-gu$_5$-ur-ma* (by courtesy of Y. Kawasaki); kt a/k 431.8-10: [8] 1 *ma-na* [9] URUDU SIG$_5$ *ig-ri a-na* [10] *Ha-da-a-ni a-dí-in*; KTP 26.2-3: [2] ... 1 *ma-na* URUDU SIG$_5$ [3] *ag-ra-am a-gu$_5$-ur-ma*.

[188] BIN 4, 178.1-4: [1] 7 GÍN KB [2] *a-na ša* URU[DU] [3] *ub-lu-ni-ni* [4] *a-dí-in*. See also kt n/k 432.12-16: [12] URUDU *ma-lá i-ba-ší-ú* [13] *ša bi-lá-tim* [14] *ta-ru-ma a-ṣé-ri-a* [15] *a-na Wa-ah-šu-ša-na* [16] *áb-kam*, 'Bring porters along and transport all the available copper to me, to Wahšušana' (by courtesy of C. Günbattı).

[189] TC 3, 165.43-47: [43] 1/3 *ma-na* KB *iš-tù Kà-ni-iš a-dí Wa-áš-ha-ni-a* [44] *a-ša bi$_4$-lá-tim a-dí-in* 17 1/3 GÍN KB [45] *iš-tù Wa-áš-ha-ni-a a-dí Ú-lá-ma* [46] *a-ša bi$_4$-lá-tim ù mu-qá-ri-bi$_4$-a* [47] *a-dí-in*.

paid for food for 'me' and for the servants (*ṣuhārū*). It is more likely that the porters were dismissed after reaching Ulama because servants had come from Purušhattum to carry the merchandise further, rather than to assume that porters and servants are here identical. Thus, not only porters were hired to carry goods, but also servants belonging to the firm were used.[190]

The evidence on the number of textiles carried by a porter is difficult to interpret because the weight of individual textiles is often unknown, and may have differed. In BIN 4, 63, a porter carries 11-12 *makūhum*-textiles: 'I. <is bringing (you)> 2 minas and 23 shekels of refined silver under my seal; my dear brother, buy good *makūhum*-textiles which are (perfectly) finished on the inner side, for 12 or 13 shekels each, and hire a porter and send him with the first person leaving.'[191] The load consisted of 12 (at 12 shekels a piece, 1 shekel short), or 11 textiles (at 13 shekels a piece, all silver spent). But a *makūhum* is a rare textile (see Veenhof 1972, p.169) and may have had a different weight than the more current *kutānum*-textile. The textiles (*ṣubātum*) mentioned in kt n/k 200 weighed on average 1 mina 56 1/2 shekels.[192] The number of porters employed in kt n/k 506 is ambigious: 'Aššur-ṭāb brought 41 *kutānum*- and 10 *šurum*-textiles. (...) We paid 18 1/3 shekels of silver to the porters who came with Aššur-ṭāb.'[193]

Sometimes the political situation or weather conditions made it difficult to find porters willing to go outside the city, as in kt n/k 84: 'They are attacking the country of Wahšušana and the porters refuse to go'[194] and CCT 3, 49b: 'Concerning your copper, I and Adad-ṣulūlī have sent for porters, but because *glazed frost* is on the roads, they just refused to go. I have just sent a message to Ilī-wēdāku for donkeys; if he does not bring donkey(s) to me, if really my own donkeys have recovered strength, I will load my own donkeys.'[195]

---

190 RA 58, Goudchaux 2.19-22: [19] ... *i-na* [20] ANŠE.HI.A *lu ṣú-ha-ru* [21] 1 GÚ.TA *ù* 2 GÚ [22] *li-iz-bi$_4$-lu-nim*, '(Bring copper) with donkeys or let the servants carry 1 each or 2 talents hither.' Also in TC 3, 165.

191 BIN 4, 63.16-26: [16] 2 1/3 *ma-na* 3 GÍN KB [17] *ṣa-ru-pá-am ku-nu-ki-a* [18] *I.* <*na-ší*> *a-hi a-ta* [19] *ma-ku-hi . dam-qú-tim* [20] *ša qé-er-ba-am . šál-mu-/ni* [21] *ša-am-ma . lu* 12 GÍN./TA [22] *lu* 13 GÍN.TA [23] *šu-qú-ul-ma ša bi-il$_5$-/tim* [24] *a-gu$_5$-ur-ma* [25] KI *a-li-ki-im* [26] *pá-nim-ma . ṭur$_4$-da-šu.*

192 226 1/2 'túg' weigh 7 talents 20 minas (by courtesy of S. Bayram).

193 Kt n/k 506.7-8, 22-24: [7] 41 TÚG *ku-ta-ni* 10 TÚG *šu-ru-tim* [8] *A-šùr*-DÙG *ub-lam* (...) [22] 18 1/3 GÍN KB *a-na ša bi-lá-tim* [23] *ša* KI *A-šùr*-DÙG *i-li-ku-ni-ni* [24] *ni-iš-qú-ul* (by courtesy of C. Günbattı).

194 Kt n/k 84.10-13: [10] *a-na ma-at Wa-ah-šu-ša-na* [11] *i-ma-qú-tù* [12] *ú ša bi-lá-tim a-lá-kam* [13] *lá i-mu-ú* (by courtesy of S. Bayram).

195 CCT 3, 49b.3-15: [3] *a-du-um* URUDU-*i-kà a-na* [4] *ša bi$_4$-lá-tim a-na-ku ù* [d]IM-*ṣú-lu-li* [5] *ni-iš-pu-ur-ma* [6] *ki-ma sú-ur-pì-li-nu* [7] *ha-ra-na-tim ú-kà-lu-ni* [8] *a-lá-kam lá im-tù-ú-nim* [9] *a-na* ANŠE.HI.A *a-ṣé-er* [10] *Ì-lí-we-da-ku*[!] *áš-ta-pár* [11] *šu-ma* ANŠE *lá uš-té-ri-am* [12] *i-a-ú-tum-ma* ANŠE.HI.A [13] *sú-ri li-id-ni-nu-ma* [14] *i-a-ú-tum-ma* ANŠE.HI.A [15] *e-ma-a*[*d*]. The word *surpilinnu* is known only from this text.

3.4.2 *Wagons*

A wagon (*ereqqum*) was used for the transport of the most common bulky goods, like tin, textiles, barley, wood, straw, and copper.[196] Traces of the use of wheeled vehicles were found on stone blocks in a main street in the commercial district of Kanish, and according to the excavator the streets generally were wide enough for a wagon.[197] The *ereqqum* was a four-wheeled wagon drawn by donkeys or oxen. The texts scarcely mention this vehicle, which must have been a regular feature in the Anatolian transport system, and when they do, sometimes add words like *parsum* and *makriš*[198] that are difficult to interpret beyond the assumption that they denote parts of the wagon. Primitive four-wheeled wagons, drawn by oxen, were found in Alacahöyük (2500-2300 B.C.), and though these particular scale models had cultic significance, they probably were modelled on existing tranport wagons. In the Old Assyrian period oxen (*alpum*) were also used as a draft animal for a cart or wagon, as is implied in VS 26, 20: 'Buy everywhere wood and reed, as much as the oxen can pull (in a wagon).'[199] Evidence for the use of donkeys to draw an *ereqqum* comes from BIN 6, 94: 'Send me donkeys, bags, their harness and wagons,'[200] and indirectly from kt e/k 46, where lines 15-16 mention the purchase of a donkey and an *ereqqum*.[201] One could argue that in this text only one donkey was bought because another was available, making two donkeys to draw the wagon. The wagon was chiefly used for loads to Purušhattum. It was also possible to transport copper by wagon from Turhumit to the south, as in KTS 3b: 'About the copper which you left behind in Turhumit in Amria's house, write to Amria, and write (also) for my (copper), as much as you left behind, (...) As for the copper, give written order to load it on a wagon. Do not expect donkeys; send

---

196 Add the following Old Assyrian evidence to Littauer and Crouwel 1979, p.64:
textiles: CCT 1, 15a.7-14: [7] ŠUNIGIN 1 *me-at* 48 TÚG [8] 7 *lu-bu-*[*šu*] [9] 4 *na-ru-qá-*[*tum*] [10] 7 *i-lu* 4 1/2 GÍN [11] KB *a-na e-re-qá-/tim* [12] *áš-qúl mì-ma a-nim* [13] *a-na I.* [14] *áp-qí-id*, 'In all 148 textiles, 7 *lubušum*-textiles, 4 sacks, 7 bags, 4 1/2 shekels of silver I paid for wagons - all this I entrusted to I.';
tin and textiles : kt n/k 490.5-13: [5] *a-ma-lá té-er-tí-ni* [6] AN.NA *ù* TÚG.HI.A [7] *ša a-na Ú-ṣú-ur-ša-Ištar* [8] *e-zi-bu-ú* [9] *a-na A-hu-*DÙG *ù* [10] *En-um-A-šur pì-iq-da-ma* [11] *i-na e-re-qá-tim* [12] *li-dí-ú-nim-ma* [13] *lu-ub-lu-nim*, 'Entrust according to our instructions the tin and textiles which I left behind with U. to A. and E., and let them load it on wagons to carry it' (by courtesy of C. Günbattı);
barley: kt b/k 665.8-11: [8] 10 GÍN KB *a-na* [9] *e-re-qá-tim* [10] *ša ú-ṭí-tim* [11] *áš-qúl*;
a wagon-load of firewood: kt a/k 537.1-4: [1] 1 1/2 GÍN KB *a-na* [2] ⌜*e*⌝*-re-qí-im ša e-ṣí*! [3] [*áš*]*-qú-ul* 1 1/6 GÍN KB [4] *a-na e-re-qí-ma áš-qúl*;
a wagon-load of tree-trunks: KTK 94.7-8: [7] *e-re-qá-am* [8] *ša ki-is-na-t*[*im*] (K.R. Veenhof in *BiOr* 27, p.368b).
a wagon-load of straw: BIN 4, 169.12-13: [12] 2 GÍN 2 *e-re-qé-en* [13] *ša tí-ib-ni-im.*

197 T. Özgüç (1959, p.97): 'In the main street in square P/20 at a depth of 4.80 m. one can see the traces of chariot wheels on 6 andesite blocks which were lying in a row (Pl. 17,1). The width of the streets in level II permits the use of wagons.'

198 See for this word Balkan 1979.

199 VS 26, 20.7-10: [7] *e-ṣí . ù qá-nu-e* [8] *tí-ša-a-ma ša* GUD.HI./A [9] *ša-da-dim* [10] *li-iš-du-du-nim.* See also CAD Š/I, p.23b.

200 BIN 6, 94.8-12: [8] ANŠE.HI.A [9] *i-lá-tim* [10] *ú-nu-sú-nu ù* [11] *e-re-qá-tim* [12] *ṭur*$_4$*-da-am.*

201 Kt e/k 46.15-16: [15] 3 1/2 *ma-na* AN.NA *ší-im* ANŠE [16] 1/2 *ma-na* AN.NA *ší-im e-re-*[*q*]*í-im*, '3 1/2 minas of tin price of a donkey [c. 35 shekels of silver]; 1/2 mine of tin price of a wagon [c. 5 shekels of silver].'

with Aya both donkeys that are stabled.'[202] In kt 87/k 436, 14 talents 5 minas of copper were sent by wagon to Wahšušana, presumably from Turhumit. In kt c/k 254, a wagon was rented to bring 20 talents of copper from Wahšušana to Purušhattum (see § 4.3.2). Loads of copper by wagon to Purušhattum are also attested in CCT 4, 47a: 'You said: "I will exchange your copper for that of a better quality and bring it together with my copper to Purušhattum." [...] Give 2 wagons and bring my copper together with your copper,'[203] and TC 2, 18: 'Because there was no demand for copper in Purušhattum, we did not hire a wagon and did not bring it into Purušhattum. Send to us donkeys, then we will bring his copper into Purušhattum (when it is again in demand?).'[204] Kanishites brought, perhaps by wagon, 5 talents of copper in CCT 2, 40a: 'Men from Kanish are bringing you 3 sacks weighing 5 talents of refined copper, under my seal. Give them 6 shekels of silver as their wages.'[205]

Although only a single text mentions the renting of *ereqqū* from the local palace, it is likely that this happened in many other instances where the palace is not mentioned. It is described in TC 3, 3: 'And I spoke to Wawali: "Ask the palace for wagons so that they are ready when the servants arrive," *in the first place*, he refuses to give me (them) and (furthermore) he will not speak to the palace about the wagons, so that we cannot dispatch the wagons. I am lonely here, take care to send your messsage quickly to me, because I want to bring the copper to Purušhattum before the harvest takes the country in its grasp.'[206] How rented wagons or animals were returned to their owner is never specified. The best solution would have been to send them back with a return freight. If rented from an Assyrian or Anatolian merchant, they could be returned to his representative in another city. Data are lacking, but it can be imagined that also the palace received the wagons back via middlemen stationed in other cities, or via a kind of mail service between cities, which was probably using wagons (*sikkum*, see below). Whether palace functionaries in charge of the wagons had a special title is unknown. A title *rabi ereqqātim* or 'head of the wagons' does not exist.[207]

---

202 KTS 3b.7-21: ^{7} *a-šu-mì* URUDU *ša i-na* ^{8} *Tù-ur-hu-mì-it . be-et . Am-ri-a* ^{9} *té-zi-bu-ni* ^{10} *a-na Am-ri-a* ^{11} *šu-up-ra-am* ^{12} *ù i-a-am ma-lá* ^{13} *té-zi-bu šu-up-ra-am* (...) ^{15} ... *a-šu-mì* URUDU ^{16} *šu-pu-ur-ma i-na* ^{17} *e-re-qí-im li-dí-ú-/nim* ^{18} *a-na* ANŠE.HI.A ^{19} *ú-za-⌜kà⌝ lá i-ba-ší* ^{20} 2 ANŠE *ša i-za-za-ni* ^{21} *iš-tí A-a ṭù-ur-da.*

203 CCT 4, 47a.4-10: ^{4} ... *um-ma a-ta-ma* ^{5} URUDU-*a-kà . a-na da-mu-qí-im* ^{6} *ù-ta-ar-ma iš-tí* ^{7} URUDU-*i-a . a-na Pu-ru-uš-ha-tim* ^{8} *ù-ba-al* [ ] *e-re-qá-tim* ^{9} 2* *dí-in-*[*ma* URUDU]-*i* ^{10} *iš-tí .* URUDU-*i-kà . bi*$_4$-*il*$_5$ (collation by K.R. Veenhof).

204 TC 2, 18.7-15: ^{7} ... *ki-ma* URUDU-*i-ú* ^{8} *i-na Pu-ru-uš-ha-tim* ^{9} *ma-ah-ṣú-ni i-re-qám* ^{10} *lá ni-gu*$_5$-*ur-ma* ^{11} *a-na Pu-ru-uš-ha-tim* ^{12} *lá nu-šé-ri-ib* ^{13} *e-ma-ri-i šé-ri-a-ma* ^{14} URUDU-*ú-šu a-na Pu-ru-uš-ha-tim* ^{15} *lu nu-šé-ri-ib.*

205 CCT 2, 40a.4-9: ^{4} 3 *i-lá-tum* 5 GÚ ^{5} URUDU *ma-sí-am šu-qúl-ta-ší-na* ^{6} *ku-nu-ki-a Kà-ni-ší-ú* ^{7} *na-áš-ú-ni-ku-nu-tí* ^{8} 6 GÍN KB *ig-ri-šu-nu* ^{9} *dí-na-šu-nu-tí.*

206 TC 3, 3.9-25: ^{9} *ù a-na Wa-wa-li* ^{10} *um-ma a-na-ku-ma* ^{11} *e-re-qá-tim iš-tí* ^{12} É.GAL-*lim e-ri-iš-ma* ^{13} *a-dí ṣú-ha-ru i-ma-qú-tù-/ni-ni* ^{14} *lu er-sú-wa* ^{15} *a-pá-ma lá i-da-nam* ^{16} *ù a-na* É.GAL-*lim a-dí-i* ^{17} *e-re-qá-tim lá i-qá-bi-ma* ^{18} *e-re-qá-tim lá nu-šé-ša-ar* ^{19} *a-na-kam we-da-ku* ^{20} *i-hi-id-ma té-er-ta-kà* ^{21} *ar-hi-iš li-li-kam-ma* ^{22} *lá-ma ma-tám e-ṣa-dum* ^{23} *iṣ-bu-tu*$_4$ URUDU *a-na* ^{24} *Pu-ru-uš-ha-tim* ^{25} *lu-šé-ri-ib*. The adverb *a-pá-ma* (collated by K. Reiter) occurs also in CCT 1, 12a.3.

207 P. Garelli (1963, pp.217-218) read this title in BIN 6, 258.8, but the text is an inventory list and line 8 reading [ ]-*ri-ga-tum* probably refers to some item.

For the packing of copper leather bags (*illum*) were used. An *illum* contained in general 1 talent, or 30 kg, making it synonymous with the half-pack, *muttatum*, or *šuqlum*. 'One talent of tin' frequently was also 62.5-65 minas in a standard packet (*šuqlum*), which contained ideally 65 minas. So also an *illum* with copper had the ideal size and weight for donkey transport. The three words denote the same bags with copper in texts dealing with Puzur-Aššur's legacy: *illum* in OIP 27, 57.13, *muttatum* in CCT 6, 30a.5, and *šuqlum* in I 633.16 (but *illum* in line 11).[208] The amount of copper stored in an *illum* varied from only 40 minas (AKT 3, 40), and 44 minas (kt 89/k 286), to more than 1 talent (AKT 1, 30), 100 minas (CCT 2, 40a.4, see above), and about 2 talents or 120 minas in AKT 3, 40.25.[209] The two last examples suggest that the maximum contents were not always identical. Bags for packing, *illātum*, are mentioned also in BIN 4, 31.20. If copper was to be used as money, it could be put in smaller packages, such as *huršiānum*.[210]

A porter could carry 1 talent of copper;[211] a donkey 2-3 talents (see above). By using a wagon, one had the opportunity of transporting considerably more at the same time, depending on the size of the wagon. In kt c/k 263, 20 talents, or 600 kg of copper were loaded onto a single wagon; if packed in bags, about 20 bags would have been involved.[212] This single wagon could transport the same as 8 donkeys or 20 porters. In kt 87/k 436, probably 14 talents 5 minas were shipped in one wagon (see above, p.65). All three alternatives involved men on foot, as in the case of donkeys or donkeys pulling a wagon, several persons were needed to guide the party. When the travelling time is considered, the wagon represented the slower option of the three.

Comparing the costs for the use of porters, donkeys, and wagons for transport of goods in general, is difficult because of the absence in the texts of basic data that would permit the calculation of how much would have been paid for each alternative when a given amount was to be transported over a given distance. The available data do not specify how

---

208 See Matouš 1969 for the complexities surrounding this legacy.

209 AKT 1, 30.6-8: [6] *i-na* 3 *i-lá-tim* [7] 3 GÚ 7 1/2 *ma-/na* [8] [URUDU *is*]-*ni-qú*, 'They established that the 3 bags contained 3 talents and 7 1/2 minas of copper' (same affair in VS 26, 152);
AKT 3, 40.21-25: [21] 2 *il*$_5$*-lu ku-nu-ku-šu* [22] *i-ba-ší-ú i-na iš-té-en* [23] 40 *ma-na ša-pi-ik* [24] *i-na iš-té-en ki-i* [25] 2 GÚ *ša-pí-ik*, 'There were 2 bags under his seal; in one, 40 minas were stored, in the other about 2 talents';
kt 89/k 286.5-10: [5] 44 *ma-na* URUDU *ší-kam* [6] *i-lu-um . ku-nu-ku-a* [7] *a-dí-i* (...) [9] *a-tù-ra-ni . i-na* [10] É [*be-tí*]-*kà . li-ib-ší*, 'Let 44 minas of *šikkum* copper, (in) a bag under my seals, remain in your house until I return' (by courtesy of Y. Kawasaki).

210 Kayseri 4695: [1] 12 *i-la-tum* ŠÀ.BA [2] 11 *i-la-tum ša ší-ki-/im* [3] 1 *i-lu-um ša . ma-sí-im* [4] *a-ha-ma* 15 *ma-na* [5] *ma-sí-um* 15 *ma-na ší-ku-um* [6] *hu-ur-ší-a-ni* [7] *ša qá-ti-a*, '12 bags, consisting of 11 bags with *šikkum* copper and 1 bag with refined copper; furthermore 15 minas of refined (copper) and 15 minas of *šikkum* (copper) in packages, for my *expenses*.' Copper is taken out of a *kursinnum*, a 'sack', in the meaning 'capital' or 'investment' in *Athenaeum* 47.4 and 18, see Gelb 1969, pp.122-123.

211 In the passage from RA 58, Goudchaux 2, 1 GÚ.TA, '1 talent each', is considered the amount that could be carried by one man, while the 2 GÚ, 'two talents', refer to the load of a donkey.

212 kt c/k 263.18-19, 21-26: [18] *ur-kà-num* 6 GÚ *ma-sí-am* [19] *ù* 14 GÚ *ší-kam* (...) [21] *i-na e-re-qí-im . a-na* [22] *Pu-ru-uš-ha-tim A-lá-hu-um* [23] *ù Wa-li-a . ub-lu-šum* [24] 21 *ma-na ig-ri e-re-qí-im* [25] 32 1/2 *ma-na . a-qá-tí-šu-nu* [26] *a-dí-in*, 'Subsequently, 6 talents of refined (copper) and 14 talents of *šikkum* (copper), (...) Alāhum and Walia brought to him in Purušhattum by wagon. 21 minas (I paid for) the rent of the wagon; 32 1/2 minas I gave (them) for their expenses.'

long the distance was, or how many days the item was hired.[213] The cheapest way was the use of one's own donkeys, but when these were not available, the means of transport had to be rented. There are some data on the costs of renting a donkey and hiring a porter, but as in the case of the *ereqqum*, it is unknown what distance was covered at those prices.[214] In view of the greater carrying capacity of a wagon, its hiring, whether this included one or two donkeys needed as draft animals, was always cheaper than hiring donkeys for transport, so firms regularly transporting bulky goods, would in general have possessed their own wagons.[215]

3.4.3 *The* sikkum

The word *sikkum* occurs in four letters from the archive of Puzur-Aššur (son of Išar-kitt-Aššur; see § 4.5). Tin was transported by means of *sikkum* in ATHE 66 and CCT 2, 40a. Lines 9-14 of the first text, a letter to Iddi(n)-Abum and Kulumāya from Puzur-Aššur, are clearly in disorder. In view of the content of the next letter, the intention is that tin must be given to porters from Kanish, presupposing in lines 12b-13 a sentence beginning with the preposition *ana* before the genitive of the gentilic, followed by the form *dinā-ma*, and then the next sentence *ina sikkim liššiūnim*. The translation is then: 'As for the tin, hire porters, that they carry it to Ninašša. I hear that they do not let travellers go to Wašhania. If my tablet has caught up <with> the men from Wahšušana and Kanish, who brought (news of) the sworn agreement, (then) give (the tin to the Kanishites), that they carry it in a *s.*

---

213 It would appear from A.A.A. 1, 3.21'-25', that the costs of transporting 10 talents of copper from Turhumit to Wahšušana amounted to nearly 10% of the load. This is impossible when compared with the data on costs collected in § 1.2.1. The incomplete context suggests the talent was paid for a variety of expenses. Lines 21'-26' read: [21'] *a-wa-sú a-na* 10 GÚ URUDU SI[G$_5$] [22'] *ni-ig-mur-ma* ŠÀ.BA 1 G[Ú (x)] [23'] *ša iš-tù Tur$_4$-hu-[mì-it]* [24'] *a-dí Wa-ah-šu-ša-na* . [x x x] [25'] *lu ta-ší-a-tum* . [x x x] [26'] [x x] *ší i* [x x x].

214 Kt u/k 5.19-20: [19] ... 2 LÁ 1/4 GÍN KB [20] *ig-re-e e-ma-ri-im*, '1 3/4 shekel of silver, rent of a donkey', and TC 3, 164.14-17: [14] 4 GÍN KB *iš-tù Za-al-pá* [15] *a-dí Kà-ni-iš ig-ri* [16] ANŠE *áš-qúl* 2 GÍN KB [17] *a-tí-ib-nim áš-qúl*, 'I paid 4 shekels of silver for the rent of a donkey from Zalpa to Kanish; 2 shekels of silver I paid for straw.'

215 We possess the following data on prices for the use of an *ereqqum*:
Rent of 1 wagon (from Wahšušana to Purušhattum): 20 minas of copper (kt c/k 254.30 (21 minas in kt c/k 263.24);
Rent of 1 wagon (from Turhumit? to Wahšušana): 14 minas of copper (kt 87/k 436.23; by courtesy of K. Hecker);
Rent of 1 wagon: 2 minas (of tin) (kt n/k 80.17-18; by courtesy of S. Bayram);
Rent of unspecified number of wagons: 19 1/2 shekels of silver (KTS 2, 46.12-13: [12] 1/3 *ma-na* LÁ 1/2 GÍN [13] *ig-ri e-re-qá-tim*); Paid for 1 wagon: 6 1/2 shekels of silver (kt b/k 665.11-14: [11] ... 6 1/2 GÍN KB [12] *a-na e-ri-qí-im* [13] *[š]a ú-nu-ut-kà* [14] *[na-áš]-ú áš-qúl*);
Paid for 3 wagons + their *parsum*, 39 1/2 minas of poor copper (TC 3, 5.21-24: [21] 40 LÁ? 1/2 *ma-na* URUDU *lá-mu-nam* [22] 3 *e-re-qá-tum$_8$* «*ú*» *qá-dum* [23] *pá*-ar-sí-ší-na* [24] *it-ba-lá*, '3 wagons together with their *parsum* cost 39 1/2 (or 40 1/2 ?) minas of poor copper' (collation by K. Reiter);
(Paid for unspecified number of wagons loaded with wood?: 4 1/2 shekels of silver, CCT 1, 15a.10-11);
Price of 1 wagon: 1/2 mina of tin (kt e/k 46.16).

Promise (to pay) their hire over there, and let Šamaš-tappā'ī go along with the tin.'[216] CCT 2, 40a is a letter to Iddi(n)-Abum and Kulumāya from Puzur-Aššur: 'The men from Kanish are carrying to you 3 sacks weighing 5 talents of refined copper, under my seal. Give them 6 shekels of silver as their hire and instruct the servants to provide them with bread. If you have not yet transported my tin, give it to the men from Kanish; they can carry it with a *s.* Give them 5 shekels of silver in tin as their hire, and let Šamaš-tappā'ī go along with them. Copper is in short supply here.'[217]

Copper was transported by means of *sikkum* in RA 58, Goudchaux 2 and TC 3, 10. The first is letter from Puzur-Aššur to Ilī-wēdāku: 'My dear brother, take care to send 20 talents of fine copper to me. The man from Ninašša cannot obtain a *s.* so (bring it) with donkeys or let the servants carry 1 or 2 talents each hither; copper is in short supply here!'[218] TC 3, 10 is a letter to Puzur-Aššur (in Wašhania) from Ilī-wēdāku: 'You wrote about copper to me as follows: "The man from Ninašša should *bring* 20 talents of copper by *s.* as far as Wašhania to me." The man from Ninašša has checked it, but it is not the right moment to make a *s.* (...) It is the wrong moment for a *s.* to be obtained.'[219]

It occurs also in the letter TTC 28, which Adu sent to Amur-ilī: 'Spend some silver over there by either transporting the tin successively with a *s.*, or by loading (it on) donkeys, and enter Purušhattum with the first party.'[220] The word in line 13 is usually identified as *sisā'um*, and interpreted as 'horse'. This reading offers some problems, because 'horse' occurs in Old Assyrian as *sīsī'um* in the Anatolian title *rabi sīsī'ē*.[221] Besides, the use of horses for transport is until now unattested in Old Assyrian texts. It is better to assume that the ancient scribe forgot to incise a few small horizontal wedges and instead of the intended KI wrote the sign SÁ, which is clearly on C. Michel's handcopy of this text in RA 80, p.139.[222]

In these texts, the word occurs with the following prepositions and verbs:
to carry (*našā'um*) in/by means of (*ina*) a *sikkum*, ATHE 66 and TTC 28;

---

216 ATHE 66.4-17: [4] ... AN.NA-*ki* [5] *ša bi*$_4$*-lá-tim ag-ra-ma* [6] *a-na Ni-na-ša-a li-ší-ú-nim* [7] *a-ša-me-ma a-li-ki* [8] *a-na Wa-áš-ha-ni-a lá ú-šu-ru* [9] *šu-ma Wa-ah-šu-ša-na-i-im* [10] *ù Kà-ni-ší-im ša ma-mì-tám* [11] *ub-lu-ni-ni ṭup-pí* [12] *ik-ta-áš-dam i-na* [13] *sí-ki-im dí-na-ma* [14] *li-ší-ú-nim ig-ri-šu-nu* [15] *a-ma-kam-ma qí-bi-a* [16] *ù Š. . iš-tí* [17] AN.NA *li-li-kam.*

217 CCT 2, 40a.4-20: [4] 3 *i-lá-tum* 5 GÚ [5] URUDU *ma-sí-am šu-qúl-ta-ší-na* [6] *ku-nu-ki-a Kà-ni-ší-ú* [7] *na-áš-ú-ni-ku-nu-tí* [8] 6 GÍN KB *ig-ri-šu-nu* [9] *dí-na-šu-nu-tí ù ṣú-ha-ri* [10] *ša-hi-za-ma ak-lam* [11] *lu-ša-ki-lu-šu-nu* [12] *šu-ma* AN.NA-*ki a-dí-ni* [13] *lá ta-bu-kà-nim* [14] *a-na Kà-ni-ší-e dí-na* [15] *iš-tí sí-ki-im li-ší-ú-nim* [16] 5 GÍN KB *ig-ri* [17] AN.NA *dí-na-šu-nu-tí* [18] *ù Š. iš-tí-šu-nu* [19] *li-tal-kam* URUDU [20] *a-na-kam-ma wa-qar*. Line 15 was collated by K.R. Veenhof.

218 RA 58, Goudchaux 2.16-23: [16] *a-hi a-ta i-hi-id-ma* [17] 20 GÚ URUDU SIG$_5$ *šé-bi*$_4$*-lam* [18] *Ni-na-ša-i-um sí-kam* [19] *lá e-pá-ša-am-ma i-na* [20] ANŠE.HI.A *lu ṣú-ha-ru* [21] 1 GÚ.TA *ù* 2 GÚ [22] *li-iz-bi*$_4$*-lu-nim* [23] URUDU *a-na-kam i-qí-ir.*

219 TC 3, 10.3-9, 28-29: [3] *a-dí* URUDU *ta-áš-pu-ra-am* [4] *um-ma a-ta-ma* 20 GÚ URUDU [5] *i-sí-ki-im Ni-na-ša-i-/um* [6] *a-dí Wa-áš-ha-ni-a* [7] *le-pu-ša-am Ni-na-ša-/i-um* [8] *sà-ni-iq lá na-ṭù-ma* [9] *sí-kam lá i-pá-áš* (...) [28] ... *sí-ku-um* [29] *lá na-ṭù-ma lá i-né-pá-áš.*

220 TTC 28.11-19: [11] *a-ma-kam* KB [12] 1 GÍN *gu*$_5$*-mu-ur-ma* [13] *lu i-na sí-ki*[!](SÁ)*-im* [14] AN.NA *i-ta-ší* [15] *ú-lá e-ma-re-e* [16] *sé-er-da-ma* [17] *i-pá-nim-ma* [18] *a-Pu-ru-uš-ha-*[*t*]*im* [19] *e-ru-ub.*

221 See Veenhof 1989, p.521 n.16.

222 The non-existing form *sisā'um* and in AHw also *sisi'um* can be deleted from the dictionaries: CAD S, pp.171a b) and 329b b) whole OA section; AHw, p.1051b from '(Zinn) - 396[150];'.

to carry together with (*išti*) a *sikkum*, CCT 2, 40a;
to make (*epāšum*) a *sikkum*, RA 58, Goudchaux 2.
The combination of *i*(*n*)-*sikkim* (line 5) with *lēpušam* (line 7) appears to result from a contamination of two expressions: *ina sikkim našā'um* and *sikkam epāšum*.

TC 3, 10.4-7 was interpreted by the CAD E, p.208a within the context of a metallurgical treatment of copper, and translated: 'The ruler of Ninašše should smelt twenty talents of copper in the *sīkum*.' P. Garelli (1963, pp.224-225), apparently following the CAD, considered *sikkum* a technical term for copper melting (TC 3, 10 is 'l'unique texte relatif à la fonte du cuivre'). In his view, the Assyrians would have turned to the ruler of a town when they wanted to have copper refined. In the commentary on RA 58, Goudchaux 2 and TC 3, 10 in RA 58, pp.134-135, Garelli again assumed that a metallurgical treatment was meant, adding that the 'man from Ninašša' might also have been a coppersmith. His reconstruction of events underlying both texts differs from the one given here in § 4.5.1 (Ah-šalim) and § 4.5.3. CAD S, which appeared in 1984, refrained from giving a translation for *sikkum* (CAD S, p.255b, *sikku* B: 'meaning unknown'). K.R. Veenhof (1989, p.521 n.16) proposed to link the word with *sikkum* known from Mari texts, and to consider it a kind of bag.

In all texts, *sikkum* appears as an alternative to transport with donkeys or porters, which makes an metallurgical interpretation impossible. It could be 'made' by a man, perhaps the ruler of the town. Goods were carried with or by means of it. A *sikkum* could then be interpreted as a kind of bag, or as a wagon. The same word is also attested in documents from the time of the Third Dynasty of Ur. These texts have been analysed by W. Heimpel who concluded that *sikkum* refers there to a state-run service, which supported royal messengers on their travels by providing them with coaches, mules, and ferries.[223] This institution seems to have disappeared after Ur III. In the Old Assyrian texts quoted above, *sikkum* might refer to a kind of transport system, under which convoys equipped with wagons and draft animals were organized, which shipped goods from one town to another. The 'man from Ninašša' in RA 58, Goudchaux 2 and TC 3, 10, is likely to be the ruler of that town, who had the means of organizing such a convoy to Wašhania, the next large town on the road to Kanish.

3.4.4 *Storage*

Merchandise was usually stored within the precinct of the houses belonging to the traders. The *huršum* especially is named in this context. The word means kitchen, but as the kitchen was the place where supplies were kept, *huršum* came to mean storehouse also. It occurred in the title borne by an important official in the Anatolian palace hierarchy, the 'head of the (royal) storehouses', *rabi huršātim*. His activities are poorly documented. In Assyrian documents, he appears only in the context of accepting textiles as a gift, which in itself is an indication of his importance.[224] Two texts found on the mound and written by

---

223 Heimpel 1994, p.31.

224 CCT 3, 28b.10-14: [10] 22 TÚG *a-na ší-mì-im* [11] É.GAL-*lúm* . *il₅-qé* [12] 1 TÚG *a-na* GAL *hu-ur-ša-tim* [13] *A-ta-ta* [14] *i-ší*, 'The palace took 22 textiles to buy at a discount price; Atata made a present of

the palace administration mention a *rabi huršātim* called Halkiašu. According to kt g/t 36, eight of his servants were put at the disposal of the *rabi simmiltim*, while in g/t 42+z/t 11 he is listed as having 15 persons in his service.[225] The palace possessed its own storehouses, yet they are rarely mentioned, as for instance in kt n/k 1318: 'He (= the ruler) maltreated us, (and) brought our merchandise to his palace and just locked (it) away ("sealed") in a storehouse.'[226] The Assyrian administration office, the *bēt kārim*, probably had its own storage facilities, but they are not mentioned (see § 5.4 for the term *suhuppum*).

The *huršum* was not a single room, but rather a section of the house, that could be sealed and was sometimes divided into several smaller strongrooms, also called *huršum*, or *maṣṣartum*, 'the safe'. The division of the storehouse into separate compartments is apparent in VS 26, 53: 'Guard the house there; 8 talents and 50 minas of tin, under seal, are in the *huršum*, in the foremost strongroom; 230 textiles of current quality, 120 fine textiles and 13 black donkeys, all this under seal I left behind with you in the house.'[227] Whereas the tin is stated to be stored in the *huršum* of the house, it remains unclear whether the 350 textiles were stored there as well, or (partly) in other rooms, perhaps even in the courtyard, like the donkeys. References to copper stored in a *huršum* are rare.[228] Nevertheless, we may assume that for example the large quantity of copper mentioned in kt u/k 4 was stored in the *huršum* of the owner's house. This text mentions 4,000 (minas) of Kunanamit copper, and 5,500 (minas) of *šikkum* copper, in all 4,750 kg, which might have been packed in circa 158 bags.[229] The packing of copper in straw, mentioned in kt n/k 276, may not have been exceptional. The pied-à-terre (*bēt wabrim*) where a trader stayed when visiting another town, usually translated 'inn', was equipped with storage facilities, as in

---

1 textile to the head of the storehouses'; kt 91/k 108.1-3: [1] 5 TÚG.HI SIG$_5$ *a-na* [2] GAL *hu-ur-ša-tim* [3] *a-dí-in*, 'I gave 5 fine textiles to the head of the storehouses' (by courtesy of K.R. Veenhof); and VS 26, 146.5-8: [5] *a-ha-ma* 1 TÚG *ku-sí-tám* [6] *i-na* É [d]IM [7] GAL *hu-ur-ša-tim* [8] *il*$_5$*-qé*, 'Furthermore, the head of the storehouses received 1 *kusītum*-garment in the temple of the Stormgod.'

225 Kt g/t 36.5-7: [5] 8 *a-wi-lu a-nu-tum* [6] *ur-ki Ha-al-ki-a-šu* [7] GAL *hu-ur-ša-tim*; kt g/t 42+z/t 11.34-36: [34] ... 15 *qá-qá-da-t*[*um*] [35] *ur-ki Ha-al-ki-a-šu* [36] GAL *hu-ur-ša-tim*.

226 kt n/k 1318.9-12: [9] *e-mu-qá-tám e-pu-uš-ni-a-tí* [10] *ú-nu-ut-ni . a-na* [11] É.GAL-*lì-šu . ú-šé-li-ma* [12] *i-hu-ur-ší-im . ik-ta-na-/ak*.

227 VS 26, 53.3-18: [3] ... *a-ma-kam* [4] *a-šar* . É-*tim ša-ṣí-ri* [5] 8 GÚ 50 *ma-na* [6] AN.NA *ku-nu-ku-ú* [7] *i-na hu-ur-ší-im* [8] *i-na ma-ṣa-ar-tim* [9] *ma-ah*[!]*-ri-im* [10] 2 *me-at* 30 TÚG.HI.A [11] *ša qá-tim* [12] 1 *me-at* 20 TÚG.HI.A [13] SIG$_5$-*tum* [14] 13 ANŠE.HI.A [15] *ṣa-lá-me-ú mì-ma* [16] *a-nim . i-na* [17] É-*tim ku-nu-ke-e* [18] *e-zi-ba-ki-im*. The reference in kt 91/k 346.1-4 that '4 1/2 talents of fine copper (and) 1 talent 20 minas of poor copper' were removed from Amur-Aššur's *maṣṣartum* means that documents recording claims worth these amounts of copper were taken out. See for the 'safe' as place where tablets were stored, Veenhof 1986, p.12 with n.43.

228 Kt a/k 339.17-18: [17] ⌜2 GÚ⌝ 5 *ma-na* URUDU *ú* 40 [+x] *ma-na* URUDU [18] *i-na hu-ur-ší-kà* . [*li-i*]*š-qú-lam*, 'Let him pay me in your storehouse 2 talents and 5 minas of copper and also 40 [+x] minas of copper'; kt e/k 67.3-9: [3] *ta-áš-pu-ra-am . um-ma* [4] *a-ta-ma* . 1 *me-at* URUDU [5] *ša a-dí-na-ku-ni . ù* 50 *ma-na* [6] *ša hu-ur-ší-im . a-na Šu-Be-lim* [7] *dí-in hu-ur-ša-am* [8] *Ni-mar-Ištar ù Puzur*$_2$*-A-šur* [9] *ip-té-ú-ma*, 'You wrote me as follows: "Give the 100 minas of copper that I gave you and the 50 minas that are in the storehouse, to Šū-Bēlim." Nimar-Ištar and Puzur-Aššur opened the storehouse.'

229 Kt u/k 4.12, 18-21: [12] *a-na* É-*tí-kà . né-ru-ma* (...) [18] 4 *li-mì-im* URUDU *Ku-na-na-ma-i-/u*[*m*] [19] *i-na* É *be-tí-kà na-dí-i* [20] *i-na* 5 *li-mì-im* 5 *me-at* URUDU *ší-ka*[*m*] [21] *ša té-zi-bu-ú-ni* (...), 'We entered your house, (...) 4,000 (minas) of copper from Kunanamit are stored in your house; out of the 5,500 (minas) of *šikkum* copper which you left behind (...).'

ATHE 29: 'We seized 8 talents of fine copper (and) 4 minas 45 shekels of silver, and sealed it with Šumabia's and my seal; it lies in my inn.'[230]

3.5 *Copper and bronze artefacts*

3.5.1 *The coppersmith*

Artefacts used as household utensils, implements, or weapons, were produced and repaired by the coppersmith in his workshop. The excavations in the *kārum* area at Kültepe have revealed the ruins of a number of workshops, both from level II,[231] and from level Ib.[232] Some clustering of level Ib workshops can be observed in the northern central part of the *kārum*. In level II the south-eastern part of the *kārum* area, where the workshop discovered in 1982 was situated, was the area where industrial activities were conducted. It did not produce archives or single tablets, presumably because the inhabitants were artisans who did not act as merchants and therefore did not write on a large scale. A number of texts contain references to smiths (*nappāhum*) living and working in Kanish. They often bear Anatolian names, such as Happuašu (CCT 5, 20a.2), Kadudu (TC 1, 87.19), or Šuharpia (kt m/k 5.38). But Assyrian smiths were also present in Kanish, for instance Aššur-mālik (ATHE 14.12) and Anina son of Puzur-Ana (kt n/k 85.7-9). It is uncertain whether the Anatolian palace possessed its own workshops where copper was refined and cast, and if so, whether these were located only on the mound or in an 'industrial area'.[233] Coppersmiths, like those living in the *kārum* of Kanish, may have been private entrepreneurs, whom the palace could have ordered to refine an amount of poor copper or to manufacture a particular object; the functionary called 'head of the coppersmiths' (*rabi nappāhī*, Kurāya in TC 3, 254.1) could have been the controller. In kt g/t 36, two coppersmiths from Kanish were among the workforce (*aštapīrum*) of the *rabi simmiltim*, the highest functionary in the city-state after the ruler.[234] The workshops, like the rest of the *kārum*, were destroyed and the scarcity of valuable objects like pieces of copper or bronze suggests that the rooms were evacuated or plundered before they burnt down. The room which functioned as workshop in these buildings contained a furnace.[235] Traces of metalworking include, apart from the furnace remains, the terracotta lower halves of pot-bellows.[236] The Akkadian word for pot-bellows is *nappahātum*. It does not occur in Old

---

230 ATHE 29.15-20: [15] ... 8 GÚ URUDU [16] SIG_5 . 4 2/3 *ma-na* 5 GÍN [17] KB *ni-iṣ-ba-at-ma* [18] *i-na ku-nu-ki Šu-ma-bi₄-a* [19] *ú ku-nu-ki-a ni-ik-nu-uk-ma* [20] É *be-et ub-ri-a . na-dí.*

231 1982: Özgüç 1986, pp.43-44, 48-50, Pl. 84 and 85, Plan 6.

232 1954 in squares Z-aa/26-27: Özgüç 1955b; 1959 in squares aa-bb/19-20: Özgüç 1986, p.41, Pl.77-78; in square L/14: Özgüç 1986, p.42, Pl. 81, 9-10, Pl. 82, 1-2; in squares bb-cc/19-20: Özgüç 1986, p.42, Pl. 83, 1-2; in square c/8, Özgüç 1986, p.42.

233 In Acemhöyük, traces of metallurgical activity were found on the mound (Çukur and Kunç, p.33). Below the 'kitchens' of the Hatipler palace, 'in the earlier level 4, a large residential complex was attached to a copper workshop, with an oven, a crucible, and stone pounders' (*AJA* 99 [1995] p.219).

234 See Veenhof 1989, p.524.

235 Özgüç 1986, Pl. 82, 1-2.

236 Kt r/k 197, Özgüç 1986, p.41, Pl. 95, 4 and Fig. 46; Özgüç 1986, p.42, Pl. 83, 2; of different shape is Özgüç 1955b, Fig. 25, height c. 24 cm.

Assyrian texts, but once in a text from Mari, where the expenditure of oil for oiling (the leather upper part of) a pot-bellows is recorded.[237] Blow-pipes,[238] tuyères,[239] and crucibles[240] were found together with a variety of moulds and stone tools.[241] In level Ib, some workshops were equipped with trough-shaped tables.[242]

The division of labour between the foundryman and the coppersmith was as yet unknown during the Old Assyrian period. The findings in the Kanishite workshops suggest that a *nappāhum* worked not only copper and bronze, but other metals as well, and could even have a section in his workshop where non-metals were worked. The coppersmith was able to refine silver and cast it into a shape that suited the mercantile demands. Likewise, he melted copper, refined it when necessary, and produced bronze. A few texts mention the possibility of buying ready-made metal objects, like kettles, from a coppersmith (see the citations from AnOr 6, 15 and TMH 17a in Appendix 3 under *šannum*), so he had always an opportunity to use copper. This made him a potential buyer of copper. It is against this background that a passage from AKT 2, 26 has to be understood: 'Huzala, Uṣur-ša-Ištar's servant, is bringing 10 minas of fine Taritar copper (in a package) sealed by me to Tarīš-mātum. She must not by herself sell the copper to the metalworkers without experts (being present), lest they deprive her of a couple of shekels of silver. (Instead you must) take the copper and sell it for silver. (...) let her buy barley and make malt and "beer-bread".'[243] The silver obtained by selling the copper was to be used for buying barley, with which the lady could make ingredients for the brewing of beer.

One went to a metalworker to have a metal object made or repaired. Usually the copper or tin required was brought along. The finds in private houses of moulds for axes (kt i/k 183; kt t/k 75), for a dagger (kt j/k 132), for a chisel (kt 81/k -), for files (kt k/k 169; kt t/k 73; kt 82/k 97), and for rings (kt i/k 185), indicate that in some cases also the mould for casting the implement belonged to the customer. This will have applied mostly to craftsmen, who could have moulds made according to their specific design. Files could for instance have been used by local house builders;[244] but apparently Assyrian 'merchants' saw the advantage of possessing such moulds as well.

---

237 ARM 23, 481+ARM 25, 41; cuneiform copy by C. Michel in MARI 6, p.459, [4] 1/2 SÌLA Ì.GIŠ [5] *a-na pa-š⌈a-aš* [kuš]*n⌉a-ap-pa-ha-tim* [6] *i-nu-ma* 1 GÚ URUDU [7] *ša-pí-il-ti* URUDU [8] *ša* MU.TÚM *Ia-ri-im-Li-im* [9] *im-sú-ú ù ip-tu*, 'Half a litre of oil for the oiling of the pot-bellows at the time they refined and ... 1 talent of copper, the remainder of the copper from Yarim-Līm's delivery.'

238 Özgüç 1986, Pl. 86, 7.

239 Özgüç 1986, p.43.

240 Özgüç 1955b, Fig. 26, D. c. 13.5 cm; and 27, L. c. 8.5 cm; Özgüç 1986, Pl. 82, 1; kt 82/k 267 is small, L. 4.2 cm, D. 4 cm, Özgüç 1986, p.50.

241 Özgüç 1986, p.43.

242 Özgüç 1986, Pl. 77, 1 and 78, 2, L. c. 190 cm, W. c. 70 c., height from paved floor to rim c. 35 cm. Photo of a second one on Pl. 78, 3, L. c. 140 cm, W. c. 80 cm.

243 AKT 2, 26.4-14, 16-18: [4] 10 *ma-na* URUDU SIG$_5$ *ša Ta-ri-ta-ar* [5] *ku-nu-ki-a a-na Ta-ri-iš-ma-tim* [6] *Hu-za-lá ṣú-ha-ru-um* [7] *ša Ú-ṣur-ša-Ištar na-ší-i* [8] URUDU *i-na lá mu-dí-ú-tim* [9] *ší-it-ma a-na na-pá-hi-i* [10] *e ta-dí-in-ma* KB [11] 1 GÍN *ù* 2 GÍN [12] *e ú-ṣa-hu-ru-ší* [13] URUDU [*le*]*-qé-a-šu-ma a-na* [14] KB [*dí*]*-na-šu-ma* (...) [16] ... ŠE-*am* [17] *lu ta-áš-a-ma bu-uq-lam* [18] *ù ba-pì-ra-am lu*[!] *té-pu-uš*.

244 A certain Perua, qualified as *itinnum*, occurs as witness in kt a/k 801/b reverse: [4] IGI *Pè-ru-wa* [5] *i-ti-nim*.

A recent study of metal artefacts from Iraq revealed which technical skills were mastered by Mesopotamian metalworkers about four centuries before the Old Assyrian Period.[245] Copper was melted in a crucible, and by adding tin to the melt, bronze was made. The minimum percentage of tin required for the intentional alloying is disputed.[246] A true bronze would contain circa 10% tin. Bronze is malleable with a tin content of up to 6%. From 6-15% tin the bronze is suitable for casting. From Mari a number of texts have been published giving the ratio between copper and tin for making bronze. This depended on the purpose of the object. A frequent ratio was 1 part of tin to 8 parts of copper, ideally resulting in a bronze with 11.1% tin, but different ratios occur as well.[247] The ratio of 1:8 is attested in one Old Assyrian text: 'The smith received 4 minas of copper and 1/2 mina of tin (to make an object with).'[248] The object, e.g. dishes, bowls, vases, and even larger vessels, could be shaped by hammering a (round) sheet of copper of high purity or a plate of bronze. With special hammers and anvils it was hammered on the inside (sinking), or on the outside (raising). Pieces of hammered sheet could be joined by rivetting, by soldering and brazing, and by cutting teeth that fitted into gaps on the opposite side. By further hammering and polishing, the joints could be made nearly invisible.

There is no evidence for the tinning of the interior of copper or bronze vessels used in the preparation of food.[249] Apart from the fact that such a thin layer would be very difficult to identify because of patina or the generally poor state of preservation of the artefact, people seem to have been unaware of the health risks posed by copper, and thus never thought of a precaution to protect themselves.

Implements and jewellery could be cast, whereby the molten metal was poured into a mould. It is thought that many vessels were cast in a mould as well and afterwards were hammered into final shape and polished. These operations make it impossible to judge by the vessels' exterior whether it had been cast and subsequently hammered, or only hammered. Other criteria are then necessary, like the thickness of the side or the structure of the dendrites.[250] For castings using moulding sand or the lost-wax process, each cast demanded the creation of a new clay mould. Reusable moulds were made of stone or clay. Both open moulds and bivalve moulds occur. The open moulds were frequently carved into the sides of a cubic or rectangular block of stone. Such a block offered a variety of moulds, sometimes more than one on the same side. The advantage of having several moulds in the same block instead of in a number of separate moulds outweighed the disadvantage of the block's greater weight. Implements and weapons, as well as ingots, were mainly cast in open or bivalve moulds. The number of moulds related to the size of the workshop. The level II workshop excavated in 1982 consisted of two rooms measuring 7x3 and 6x5 m;

---

245 Müller-Karpe 1993, esp. pp.264-268.

246 Müller-Karpe 1989, p.183; J.D. Muhly in *AJA* 97, p.240 n.7: 'From a metallurgical point of view there is little noticeable difference in a copper with less than 2% tin.'

247 1:8 in ARM 25, 690 (12.8%); 693 and 695; 1:6 in ARM 25, 694 (13%); 1:6 2/3 in ARM 25, 691 (12.8%).

248 CCT 1, 37b.12-14: [12] ... 4 *ma-na* URUDU [13] *ù* 1/2 *ma-na* AN.NA *na-pá-hu-um* [14] *il*$_5$*-qé*.

249 Tylecote 1976, p.78.

250 See Müller-Karpe 1993, p.267.

these yielded about 50 moulds.[251] The less well preserved workshop from level Ib discovered in 1954 consisted of three rooms (room 1 measured 2.70x2 m). Room 1 and 2 yielded 14 moulds. The largest mould, in which long thin spikes were cast, was of clay and placed at the foot of a wall.[252] A number of these reusable moulds found in Kültepe in workshops and private houses have been published. Those used in the manufacture of copper or bronze implements (jewellery not included) have been listed in Appendix 2.

Going to a workshop as a customer required some commercial insight and some knowledge of copper working. This is illustrated by the passage from AKT 2, 26 cited above. Due to their largely commercial character, the Old Assyrian texts contain little information about the manufacture of metal objects. If they do, the verb 'to make' (*epāšum*) is used without providing further details, as in TC 3, 101.[253] The plan to have a new kettle made forms the subject of a letter (TC 3, 97) which Buzāzu sent to his wife Lamašša in Assur, and of which a copy survived in Kanish. Details are given about the repair of the old kettle: 'Kashanuel is bringing you 10 minas of fine copper to make the kettle. Keep the copper in store until I come. When I come, I will bring an additional quantity of copper along and then I will have the kettle made. Let Anina give 1 mina of copper, and let them *solder* the bottom of the 20 minas kettle; keep it (= the kettle) until I come!'[254]

The metal for the kettle's manufacture was furnished by Buzāzu himself, which made it cheaper than a bought one. In view of line 14, the desired kettle may be regarded as a replacement of an old one weighing 20 minas (10 kg); 10 minas of copper of high purity was sent ahead of Buzāzu, who was to bring the remaining amount of copper (10 minas?) himself. The high purity of the copper was a necessity if the metal was to be hammered into a kettle. If the kettle was to be of bronze, some minas of tin were needed as well.

Lines 11b-16 describe that another kettle, apparently worn and the one to be substituted, was to be repaired and afterwards kept (imperative plural = L. and A.) till Buzāzu's arrival. The meaning of the words in line 13 is secured by a similar passage in I 429: [47] 2/3 *ma-na a-na i-ša-ad zi-ri* [48] *a*[!]*-na*[!] *na-pá-hi* (erasure), '2/3 mina (of copper was given) for (the repair of) the bottom of a cauldron to the smith.' The repair for instance of a crack in the bottom of a kettle is performed by soldering (solder's melting point >600° C) or brazing (<400° C), whereby two pieces of the same material are joined by a fusible alloy. This technique was known in Mesopotamia from at least the middle of the third millennium B.C. onwards.[255] The amounts of copper in line 11 (c. 480 gr) and in I 429.47 (c. 320 gr) served as metal to be soldered onto the copper kettles; part of these amounts may

---

251 According to the report in *AJA* 87 (1983), p.430.

252 Özgüç 1955b, Fig. 4.

253 TC 3, 101.23-24: [23] *pá-at-ri ša ú-ša-hi-zu* [24] *šé-pì-iš*, 'Have the daggers made concerning which I gave instructions'. The verb *patāqum*, 'to pour', or derivatives of it are not yet attested in Old Assyrian. A.L. Oppenheim read in BIN 6, 28.16 'a unique reference for *erûm pitqum*, "refined copper",' (*JNES* 5 (1946), p.278). The word concerned is rather URUDU *bi-it-qá-/t*[*um*], 'copper losses'.

254 TC 3, 97.3-16: [3] 10 *ma-na* URUDU SIG_5 [4] *a-na ša-ni-im* [5] *e-pá-ší-im Kà-as-ha-nu-/el* [6] *na-áš-a-ki-im* . URUDU [7] *a-dí a-lá-kà-ni* [8] *li-bi-ší i-na* [9] *a-lá-ki-a* . *mu-lá-e-šu* [10] *ú-ba-lam-ma* . *ša-nam* [11] *ú-šé-pá-áš* . 1 *ma-na* URUDU [12] *A-ni-na li-dí-in-ma* [13] *i-ša-ad* . *ša-nim* [14] *ša* 20 *ma-na* . *ta-ba-kà-tám* [15] *li-dí-ú-ma* . *a-dí* [16] *a-lá-ki-a* *kà-i-lá-ší.*

255 Müller-Karpe 1993, p.266.

also have been used to pay the smiths. The word in line 14 is read with W. von Soden (AHw, p.1295a) as *tabākattam*, an adverb of *tabākum*, 'to pour' (usually said of liquids). The meaning of the expression *tabākattam nadā'um* would then be 'to solder'.

A word written *ta-bu-e-im* in the genitive occurs in TC 3, 163, where lines 7-9 run: [7] 1/3 GÍN KB *a-na* [8] URUDU *ta-bu-e-im* [9] *i-na Ra-za-ma*. P. Garelli interpreted the word following URUDU as an adjective (1963, p.83 n.3 'pour du cuivre ...'). In his study on the Old Assyrian itineraries, Kh. Nashef read *ša-bu-e-im* and derived it from *šabbu'um*, 'um das Kupfer in Razamā zu bezahlen' (1987, p.47). The first sign of this word is TA rather than ŠA, which makes Nashef's reading impossible. Besides, costs incurred for paying a debt are rarely mentioned, unless one assumes that the silver was paid instead of a tiny amount of copper (less than 1 mina). Garelli's interpretation is possible but unlikely since it would mean that an Assyrian merchant had bought less than 1 mina of copper in Razama, a journey of two days from Assur, whereas in the same text copper is used to pay for expenses. There remains the possibility that the word is an infinitive of a verb in the D-stem. The small amount of silver would then have been paid 'for *t*-ing copper'. The verb might refer to a metallurgical or mechanical treatment and the 1/3 shekel of silver might have been paid to a smith, as in CCT 1, 26b.8-9, where 3 1/2 shekels of tin were paid to a *napppāhum* without however specifying what his work had been (text discussed in Nashef 1987, p.21). In ATHE 28.8, it is said that 1 shekel of silver was paid to a smith and for firewood, apparently for refining the amount of silver mentioned earlier in that text. If the tin in CCT 1, 26b was bought in Assur at a rate of 12:1, its value in silver would have been less than 1/3 shekel. For this price, a smith had perhaps repaired or manufactured a tool for the Assyrians. The TC 3 text lists expenses made on a journey to the west, probably from Assur, and the copper carried along would have been a limited quantity used for making payments en route. The amount paid in TC 3, 163.7 was then probably to a smith, but the question remains of what he did to the copper. The repair of an object seems excluded as only the general word 'copper' is used, while refining is called *masā'um*.

The verb used is different from that (listed under *ṭebû* D) which occurs in the meaning 'to hide' in CCT 5, 8b.25 and kt 89/k 206.8. In the first text, which has been edited by C. Michel (1991a part II, pp.80-82 as text 54), it is used when silver should be withheld: 'I seized the ladies and we said: "Alas, the silver of my father will be lost!" They replied: "Do not bring us to the *kārum*!" and they promised me (to pay) 5 minas of silver. Later on, P., Š., and his brother, the sons of M., advised the ladies: "Hide the silver! Are you not women? He is a man, and they will seize him for his father's debt."'[256] In kt 89/k 206, pieces of *aši'um*-mineral must be hidden (*ta-be-e-ma*). I tentatively rendered this form '*take out*' in *BiOr* 49, col.797. In Old Assyrian the verb *ṭebû* D is used in the meaning 'to get rid of something' (AHw, p.1383: 'etwa beseitigen, aus der Welt schaffen'), but this translation does not fit in with the passage from TC 3, 163. It appears to signify some mechanical treatment of copper (objects) performed by a smith, which resulted in copper qualified as

---

256 CCT 5, 8b.13-27: [13] *ù a-wi-lá-tim* [14] *aṣ-ba-at-ma um-ma* [15] *né-nu-ma . ú-zu-um* [16] KB-*áp a-bi*$_4$*-a . i-ha-li-/iq* [17] *um-ma ší-na-ma* [18] *a-na kà-ri-im lá ta-ra-/dí-ni-a-tí* [19] 5 *ma-na* KB [20] *iq-bi*$_4$*-a-nim wa-*<*ar*>*-ki-tá-ma* [21] *P. ù Š.* [22] *ù a-hu-šu* DUMU-*er-ú M.* [23] *a-na a-wi-lá-tim im-li-ku-ma* [24] *um-ma šu-nu-ma* KB [25] *ta-bi*$_4$*-a . lá sí-ni-ša-tí-ni* [26] *šu-ut za-kà-ar a-hu-bu-ul* [27] *a-bi*$_4$*-šu i-ṣa-bu-tù-šu.*

tabbu'um. This adjective occurs in kt n/k 524: [11] 2/3 *ma-na* KB 20 *ma-na hu-ša-e e-li-a-tim* [12] 4 *ma-na ta-bu-ú-tim a-na* GN [13] PN *ú-bi*$_4$*-il*$_5$*-šum*, 'PN brought him in GN 40 shekels of silver, 20 minas of scrap copper in the shape of *e*.-s (and) 4 minas of *t*. (scrap)' (by courtesy of C. Günbattı). The plural of this adjective is masculine, like *hušā'ū*, and since the 4 minas of *t*. are mentioned after scrap in a specific shape (*elītum*), I am inclined to consider this amount as scrap copper resulting from a mechanical treatment called *tabbu'um*.

In KTH 39 we read: [4] ... 67 *ma-na* / URUDU [5] *ur-kà-num* [6] *i-ša* URUDU x (TA without the final vertical wedge)*-be-e* [7] É *kà-ri-im* [8] *E.* [9] *il*$_5$*-qé*, 'later on, E. took at the *kārum*-office 67 minas from that of the ... copper.' If we read *ta-be-e* it would be a form of the verb *tabb'um*, but then in the G-stem. J. Lewy interpreted it as 'from the promised (*qá-be-e*) copper,' but neither CAD Q nor AHw list this expression.

3.5.2 *Textual and archaeological evidence of copper and bronze artefacts*

The Old Assyrian cuneiform texts and metal finds excavated at Kültepe provide information on the use of objects consisting of copper or bronze. Part of the excavated copper and bronze artefacts were analysed in Germany (Stuttgart) by means of the quantitative spectral analysis method. The results were published in Esin 1969. The section from U. Esin's book containing analyses of the Kültepe artefacts has been used here to determine whether an object can be considered to consist of copper or of bronze.[257]

Bronze vessels were expensive items and in many cases pottery was used instead of metal. As an alternative to bronze, lead was used for some pan-like vessels in level Ib,[258] and it is likely that more artefacts were made of this metal in spite of its rarity in level II texts. Two inventory texts are the main sources of information about the bronze utensils possessed by a wealthy Assyrian merchant's household, kt c/k 1517 and kt h/k 87.[259] The translation of these texts is difficult because the exact meaning of many words remains uncertain. The first lists 76 bronze items and reads (kt c/k 1517):

[1] 1 *ri-bu-um* . 1 *uš-hi-um*	'1 *r*.-jar, 1 lamp,
[2] 1 GÍR *ap-šu-hu-um*	1 *a*.-knife,
[3] 2 *ṣú-ur-ṣú-pá-an*	2 *ṣ*.-vessels,
[4] 2 *it-qú-ra-tum*	2 ladles,
10 *sà-pu-ú* [5] *ša-ah-tù-tum*	10 *s*.-bowls without metal band,
17 *sà-pu-ú* [6] *ša mu-sà-ri*	17 *s*.-bowls with metal band,
[7] 9 *ma-ak-ra-kà-tum*	9 *m*.-s,
[8] 6 *hu-ru-pá-tum*	6 *h*.-dishes,
[9] 6 *ša-ha-tum*	6 *š*.-vessels,
[10] 10! *ma-áš-qá-lá-tum*	10 measuring cups,

257 Printing-errors in the Typological Catalogue lead one to suspect that the analyses may contain some as well.

258 Özgüç 1986, p.73.

259 K. Balkan quoted from kt c/k 1517 ('Text B') lines 3-6, 9-10, 14, 16-18 and from kt h/k 87 ('Text C') lines 1-2, 4-5, 7-11, 13, '15-16, 18-19, 21-24' in his review of ICK 2 (Balkan 1965, p.160).

^{11}4 *kà-sá-tum*	4 cups,
^{12}2 *li-bu* .	2 *l*.-cups,
1 *kà-sú-um* ^{13}*ša Ší-mu-ri-im*	1 cup of Šimurrum-type,
^{14}3 *zu-ur-ša-tum*	3 *z*.-cups,
^{15}1 *mu-ša-lu-um*	1 mirror;
^{16}ŠUNIGIN 1 GÚ 17 *ma-na* ^{17}ZABAR	in total 1 talent 17 minas of bronze (objects)
^{18}*ša A-šur-e-mu-qí*	belonging to Aššur-emūqī.'

Kt h/k 87.1-28, 32-33, lists 93 items and a number of other household pieces:

^{1}10 *ša-áp-lá-tum ša tí-ra-*⌜*ni*⌝	'10 grooved stands,
^{2}1 *ša-pí-il*$_5$*-tum ša-šu-ru-um*	1 stand, (for) a sieve (?),
^{3}2 *uš-hi-ú ša bi-ṣí-ni*	2 lamps,
^{4}1 *ša-pì-il*$_5$*-tum ša sà-pé-e*	1 stand for *s*.-bowls,
^{5}2 *ṣú-ur-ṣú-pá-tum*	2 *ṣ*.-vessels,
^{6}3 *sú-pá-nu* . *ša Kà-ni-iš*^{ki}	3 *s*.-bowls of Kanish-type,
^{7}*maš-qal-tum ša* 2 SÌLA	a measuring cup of 2 sila,
^{8}*maš-qal-tum ša* 1 SÌLA	a measuring cup of 1 sila,
^{9}9 *ha-bu-ra-a-tum*	9 *h*.-vessels,
^{10} ŠÀ.BA *sà-pu-um ša na-aṣ-bi*$_4$*-tim*	one thereof is a *s*.-bowl with handle,
^{11}18 *ša-ha-tum*	18 *š*.-vessels,
^{12}4 *hu-ub-ku-ú ra-bi*$_4$*-ú-tum*	4 large and
^{13}4 *hu-ub-ku-ú ṣa-hu-ru-tum*	4 small *h*.-vessels,
^{14}6 *sà-pu-ú ša mu-sà-ri*	6 *s*.-bowls with metal band,
^{15}5 *ku-na-ki-ú*	5 *k*.-s,
^{16}2 *zu-ur-ša-tum*	2 *z*.-cups,
^{17}5 *hu-tù-lá-tum*	5 *h*.-s,
^{18}2 *áš-hu-lu* 2 *mu-lá-/tum*	2 *a*.-containers, 2 *m*.-s,
^{19}3 *sà-pu-ú* ^{20}*ša-ah-tù-tum*	3 *s*.-bowls without metal band,
^{21}1 *a-ga-nu-um* . *ša qá-tim*	1 *a*.-hand bowl
^{22}1 *it-qú-ru-um ša qá-tim*	(with) 1 spoon;
^{23}ŠUNIGIN 1 GÚ 40 *ma-na* ZABAR	in total 1 talent 40 minas of bronze (objects).
^{24}14 GÚ URUDU *ša ṣí-ib-tim*	14 talents of interest-bearing copper,
^{25}14 *pá-šu-ru* 7 *ú-ru-za-na-tum*	14 tables, 7 *u*.-tables,
^{26}6 *qá-áb-li-a-tum* .	6 *q*.-containers,
3 *zi-ra-tum* ^{27}*ša* 30 *ma-na*.TA	3 cauldrons of 30 minas each,
mì-ší-it ^{28}*zi-ri-im ša hu-ur-ší-a*	the stock of cauldrons in my kitchen,
(...)	(...)
^{32}*iš-tù A-a me-ta-at-ni*	all this is with Šat-Aššur
^{33}*mì-ma a-nim iš-tí Ša-at-A-šur*	since Aya died.'

The amounts of bronze are considerable in both texts, 77 minas (c. 38.5 kg) and 100 minas (c. 50 kg), but might well be normal quantities of bronze used in a merchant's household.[260] Apart from these texts, details are scarce on what other complete households actually comprised. That of Hunia in Assur possessed at least 40 minas in bronze objects according to kt n/k 212: 'I (= H.'s sister) handed over to her (= H.'s wife) before your *m.*-representatives her belongings and bronze objects weighing 40 minas from your kitchen and she took them out (of the house).'[261]

Tableware and other metal objects were exchanged between family members living in Kanish and Assur. In CCT 3, 20, Pūšu-kēn sent *supannum*-bowls and ladles to his wife Lamassī in Assur: '[1-4] Speak to Pūšu-kēn, thus says Lamassī: Kulumāya is bringing 9 and Idī-Suen is bringing 3 textiles to you. [5-6a] Dān-Aššur brought me 2 bronze *s.*-bowls; [6b-8a] Ea-šar brought me 1 bronze *s.*-bowl and 1 bronze ladle; [8b-9a] Iddi(n)-Suen brought me 2 bronze ladles; [9b-10] Ahu-waqar brought me 1 bronze ladle; [11] Ikuppia brought me 1 silver *s.*-bowl; [12] a silver *s.*-bowl was also brought by the son of Ibisua. [13-14] They brought me cups, ladles, and ... (, items not of bronze or silver, as well - J.G.D.)' (lines 13-14 are not a summary of the items mentioned in the previous lines).[262] In CCT 4, 20a, he received tableware and implements: '[1] Speak to Muza and Pūšu-kēn. [2-3] Speak to Muza, thus says Lamašša (= wife of Buzāzu, Pūšu-kēn's son): [4] 3 bronze *a.*-s, [5] three *š.*-knives, [6] three forks, [7] two bronze ladles, [8] 2 *s.*-bowls *mamiru* [9] 3 *h.*-axes, 1 axe? [10-11a] 4 *a.*-s and 1 cauldron weighing 20 minas, [11b-14] all this I gave to you. My dear sister, entrust these items now to Pūšu-kēn!'[263] Like many other texts, these letters do not mention the logic behind the actions described in them. In the first, Lamassī is sending textiles that must be sold in Anatolia. The bronze and silver objects which she receives are not merchandise, but seem to be intended for use in her own household. Also the second text is about bronze/copper objects, presumably for a household. The shipment of a few objects to Assur can be compared to the sending of small amounts of refined copper to persons in Assur or Anatolia; if the donkey was not fully loaded, the transporter could bring some additional weight along in the shape of metal.

The bronze objects listed in the inventory texts were largely kitchen utensils, like dishes, bowls, cups, and cooking vessels. Not listed as 'bronze object' were kettles

---

260 The passage from RA 81, 55.14': 1 GÚ *sí-pá-ri*, probably refers to bronze objects from a household as well.

261 Kt n/k 212.11-14: [11] ... *ú-nu-sà* [12] *ù sí-pá-ri* 40 *ma-na . ša* [13] *hu-ur-ší-kà* IGI *mu-zi-zi-kà* [14] *as-ni-qá-ší-ma . tù-uš-té-ṣí* (by courtesy of S. Bayram).

262 CCT 3, 20.1-14: [1] *a-na Pu-šu-ke-en*$_6$ *qí-bi-ma* [2] *um-ma Lá-ma-sí-ma* 9 TÚG.HI.A [3] *Ku-lu-ma-a na-áš-a-kum* 3 TÚG.HI.A [4] *I-dí-Sú-in na-áš-a-kum* [5] 2 *sú-pá-ni ša* ZABAR *Dan-na-A-šùr* [6] *ub-lam .* 1 *sú-pá-nam ša* ZABAR [7] *ú it-qú-ra-am ša* ZABAR *I-a-šar* [8] *ub-lam* 2 *it-qú-ra-tim ša* ZABAR [9] *I-dí-Sú-in ub-lam .* 1 *it-qú-ra-am* [10] *ša* ZABAR *A-hu-wa-qar ub-lam* [11] 1 *sú-pá-nam ša* KB *I-ku-pí-a ub-lam* [12] *sú-pá-nam ša* KB DUMU *I-bi-sú-a ub-/lam* [13] *sà-ma-lá-tim it-qú-ra-tim* [14] *ú sí-ba-<ra>?-tim ub-lu-nim.*

263 CCT 4, 20a.1-14: [1] *a-na Mu-za ú Pu-šu-ke-en*$_6$ [2] *a-na Mu-za qí-bi-ma* [3] *um-ma Lá-ma-ša-ma* [4] 3 *a-lu-na-tum ša sí-pá-ri* [5] *a-ša-lá-ší-šu šu-ga-ri-a-ú* [6] *a-ša-lá-ší-šu ma-az-lu-/gu*$_5$ [7] *ší-ta it-qú-ra-tum . ša sí-pá-/ri* [8] 2 *sú-pá-nu . ma-mì-ri* [9] 3 *ha-ṣí-nu .* 1 *kà-lá-pu* [10] 4 *ar-za-lu . ù* 1 *zi-ru* [11] *ša* 20 *ma-na mì-ma* [12] *a-nim a-dí-na-ki-im* [13] [*a*]*-ha-tí . a-tí-i . a-ma* [14] *Pu-šu-ke-en*$_6$ *pí-qí-sú-/nu* (see for *mamiru* Bilgiç 1954, p.68).

(*šannum*) and cauldrons (*zirum*), although from other texts it is known they were sometimes made of bronze. Whether of copper or bronze, they were apparently distinguished from the other bronze kitchen utensils, like in CCT 5, 14b, 4 ... *lu zi-ru-um* 5 *lu ú-tù-up-tù-um* 6 *lu sí-pá-ru*, 'be it the cauldron, furniture, or bronze household objects'. Every household had one or more of these items. The word for cauldron, here rendered as *zirum*, has to be distinguished from a type of cloth, which was written in cuneiform with the same signs, but was preceded by the determinative for 'cloth' to prevent confusion.[264] J. Lewy explained *zirum* as 'tent' (1950, p.21 n.2), but the CAD Ṣ, p.213b (*ṣīru* B) did not translate it. In kt h/k 87, three cauldrons are mentioned; it is unclear whether they were intended as merchandise in this particular case. Pottery imitations of metal cauldrons and jars were found in level II and Ib.[265] The size of a kettle and cauldron was indicated by weight and not by capacity. A kettle weighing 19 minas (9.5 kg) is known; for cauldrons the evidence is richer: 15 minas (7.5 kg), 20 minas (10 kg), 30 minas (15 kg), 40 minas (20 kg), and even 57 1/3 minas (c. 29 kg) are given as weight of a *zirum*. These cooking pots must have been heavy to handle.[266] Although the epigraphic material testifies to a widespread use of copper and bronze objects in daily life, only a limited range of such artefacts was found during excavations. The reason for this is that the majority of the excavated metal artefacts are mortuary evidence, coming from tombs and graves, which mainly contained items of jewellery, weaponry, and household utensils like bowls, dishes, jars, and cauldrons. The available textual and archaeological evidence is itemized below. In general, it is impossible to identify individual words with excavated objects. Some copper or bronze items known to have been used in the kitchen, because they were found during excavations, cannot be identified in the texts, because of the translation difficulties mentioned above. For instance the word for large copper or bronze frying pans is unknown, although every kitchen must have possessed one or more. One frying pan was found in a tomb in Kanish, two others are also mortuary evidence, excavated in the Old Assyrian level in Assur.[267]

In a schooltext published by K. Hecker, copper and bronze are listed together with some objects made of them: kt t/k 76+79 V: 4 URUDU DIŠ 5 ZABAR 6 GAL ZABAR 7 URUDU *ša ša-du-i-šu* 8 URUDU *zi-ra-tum* 9 URUDU *e-li-t*[*um*], 'copper, bronze, a bronze cup, native copper, copper cauldrons, a copper *elītum*.'[268]

Metal household objects also functioned as 'money'. Copper or bronze had an intrinsic value which was high enough to make them acceptable as pledge for creditors. They were, moreover, widely available in Assyrian households. A person selling straw recorded the

264 KTS 2, 17.20': 1 TÚG *zi-ru-um*. It occurs also in texts from Mari (plural *zi-ra-tum*), see ARM 18, p.133 and ARM 23, p.169.

265 Özgüç 1986, p.74.

266 'Was der Zeitgenosse nicht überlieferte ist die Mühseligkeit, die das Kochen am offenen Herdfeuer mit sich brachte, die unglaubliche Schwere der Arbeit der Hausfrauen. So sprechen schon die Gewichte der gußeisernen Töpfe für sich: mit bis zu 20 kg Eigengewicht - gefüllt bis zu 50 kg Gewicht - stellte das Hantieren mit ihnen selbst schon Schwerstarbeit dar.' H. Kaiser and H. Ottenjann in the guide to the Museumsdorf Cloppenburg [1985], p.154 on rural life around 1800 AD in Lower Saxony, Germany.

267 Ass. 20504ac and Ass. 21410f. P. Calmeyer discussed these objects, which he preferred to call *Omphalos-Becken*, and presented the rather curious interpretation that they were used for panning nuggets of placer tin from a stream (1977, pp.90-93).

268 Hecker 1993, p.287.

items he held as pledge in TC 2, 61: [1] 10 *ku-lu-pì-nu* [2] *šu-ga-ri-a-e* [3] *a-ša-pá-ar-tim* [4] *ú-kà-lu* 10 *ku-lu-pì-nu* [5] *ar-za-lá-am* [6] *ú-kà-al* [7] 10 *ku-lu-pì-nu* [8] *ma-sí-a-am* [9] *ú-kà-al* [10] 5 *ku-lu-pì-nu* [11] *sú-e-en* [12] *ú-kà-al*, '10 bundles: I(!) am holding as pledge *š*.-s; 10 bundles: I am holding as pledge an *arzallum*; 10 bundles: I am holding as pledge refined (copper); 5 bundles: I am holding as pledge 2 *su'um*-(discs of metal).' These were all copper or bronze objects, but this sketchy list does not even specify the quantity of items serving as a pledge in lines 2 and 8. The word in line 8 is in the copy clearly *masiam*, 'refined (copper)', and not *kussiam*, 'saddle', as read in CAD K, pp.528 and 593a 6. Often a ladle was given as a pledge, sometimes together with a kettle, as in CCT 4, 19b: 'Redeem our father's *q*.-container, the cauldron, and the ladle, and send them to me in Wahšušana.'[269] Lamašša reports in BIN 4, 90 that a ladle was given as a pledge for foodstuff.[270] A kettle and a *kulium* had been given to an Anatolian in kt n/k 172: 'We released from the house of the Anatolian creditor a kettle and a *kulium* weighing 23 minas.'[271] In Assur, bronze household utensils were seized by the city authorities as a pledge in kt n/k 204: 'The *bērū* (= functionaries) and Iddi(n)-Aššur took out of the storehouse 29 1/2 minas of bronze (objects) from the bronze (objects) which Akaddia had left behind and they weighed it out to the *limmum*-eponym. The remaining bronze (objects) were deposited as pledges.'[272]

3.6 *On weights*

3.6.1 *The Old Assyrian weight-stones*

Weight-stones, usually made of haematite, less frequently of another stone or of lead, have been discovered in Kültepe and were published by T. Özgüç (1986, pp.78-81). The published stones are presented in Appendix 5, and the average weight of the Old Assyrian shekel is 8.257368 gr. Those representing fractions of a shekel, especially of 2/3 and less, show major deviations from the mean weight computed for 1 shekel, and must have been inaccurate when small quantities of silver or gold were weighed. Also weights of 1 shekel and more show at times important deviations. All this demonstrates how difficult it must have been for the merchants and others working with such weights to adhere to the 'standard'. Based on this and on textual evidence, it is possible to determine which weights were represented by a stone. Two of these weighing 1.1 gram could be interpreted as 22 1/2 še weights deviating only +5.1% and +11.4% instead of the -20.1% if they would have been 30 še weights. But the amount of 22 1/2 še was probably not established by a special weight, but rather by using a 15 and a 7 1/2 še weight. Amounts like 100 and 125 še

---

269 CCT 4, 19b.7-12: [7] ... *qá-áb-li-tám* [8] *ša a-bi-ni zi-ra-am* [9] *ù it-qú-ra-am* [10] *pu-ṭu-ur-ma* [11] *a-na Wa-ah-šu-ša-na* [12] *a-ṣé-ri-a šé-bi-lam.*

270 BIN 4, 90.17-19: [17] ... 1 *it-qú-ru-um* [18] *a-na ki-ší-ni a-na ša-pá-ar-tim* [19] *i-ni-dí-ma.*

271 Kt n/k 172.15-18: [15] ... 23 *ma-na* [16] *ša-nu-um ù ku-li-um* [17] *šu-qú-ul-ta-šu-nu* [18] É *nu-a-im nu-šé-ṣí* (by courtesy of S. Bayram).

272 Kt n/k 204.15-22: [15] ... *i-na* ZABAR [16] *ša A-kà-dí-a . té-zi-bu-ni* [17] 30 LÁ 1/2 *ma-na* ZABAR [18] É *ku-nu-ki . bi-ru . ú-šé-ṣí-ú-nim-ma* [19] *a-na li-mì-im* [20] *bi-ru ù I-dí-A-šùr iš-qú-lu* [21] *ší-tí .* ZABAR *a-na ša-áp-ra-tim* [22] *i-ta-du-ú* (by courtesy of S. Bayram).

suggest that weights of 10 and 5 še also existed. This makes it probable that the following weights were used (compared with the evidence from Mari texts; 1 še (1/180 shekel) = 0.04587426 gr; 1 shekel ('g'; 1/60 mina) = 8.257368 gr, and 1 mina ('m') = 495.44208 gr):

	Old Assyrian	Mari[273]
(5 še)		ARM 25, 226.3
(10 še)		
15 še (= 1/12 g)	weight kt f/k 144	ARM 25, 684.4
30 še (= 1/6 g)	weight kt 82/k 138	
45 še (= 1/4 g)	CCT 1, 2.20 (1 1/4); weight kt 76/k 183	ARM 25, 279.4
60 še (= 1/3 g)	weight kt 84/k 96	ARM 25, 191.4
90 še (= 1/2 g)	TC 3, 213.34; weight kt o/k 155	ARM 25, 688.4
120 še (= 2/3 g)	weight kt 76/k 179	
135 še (= 3/4 g)	weight kt 85/k 14	
1 g	BIN 6, 242.4; weight kt 84/k 80	ARM 25, 366 tr.1
1 1/4 g	weight kt r/k 125	
1 1/2 g	weight kt i/k 60	
1 2/3 g?	weight kt r/k 124	
2 g	weight kt 85/k 80	ARM 25, 260.4
2 1/2 g		ARM 25, 282.6
3 g	weight kt 74/t 21	ARM 25, 283.2
5 g	weight kt b/k 472	ARM 25, 282.6
6 g	weight kt j/k 152 (lead)	
7 g	weight kt j/k 153 (lead)	
8 g	weight kt b/k 317	
10 g (= 1/6 m)	kt c/k 254.20; weight kt p/k 172	ARM 25, 282.6
12 g	weight kt 77/k 131	
15 g (= 1/4 m)	weight kt i/k 58	
20 g (= 1/3 m)	weight kt p/k 165	ARM 25, 279.2
30 g (= 1/2 m)	BIN 6, 237.3; weight kt l/k 184	ARM 25, 366.2
50 g (= 5/6 m)	weight kt 82/k 180 (lead)	
1 m	TC 3, 171.13; weight kt f/k 103	ARM 25, 191.2
(1 m 15 g?)	weight kt e/k 142 (lead)	
2 m	weight kt e/k 137 (lead)	ARM 25, 282.3

273 See Joannès 1989, p.140.

3 m		ARM 25, 172.2 (lead); 282.3
5 m		ARM 25, 282.2; 196.3 (lead)
10 m	kt c/k 254.21; weight kt u/k 108	ARM 25, 196.3 (lead)

The biggest weight in use was that of 10 minas, which means that Tahašaili, when he weighed out the 14 talents of *šikkum* copper, had to weigh this amount 84 times (see below).

3.6.2 *Weight corrections*

The weighing procedures in Old Babylonian Mari have been studied by F. Joannès (1989). The expression *ana pān x parik* occurs in texts from Mari when weights for round figures were used, and the scales had to be balanced by adding the necessary weight to the scale with the object to be weighed, usually a fraction of a shekel or of a mina (Joannès 1989, pp.133-135). Sometimes the terminology was mixed up: *ana pān x parik* means on the scale opposite to the scale with *x*, in for instance ARM 25, 282, commodity: 9 minas 42 1/2 shekels of silver weighed by means of a 5 mina, a 3 mina, and a 2 mina weight; *parik* on scale with silver: 17 1/2 shekels.[274] *Itti x parik,* 'balanced by means of *x*', means in ARM 25, 424 that a weight was placed in the same scale as *x*.[275] But in other texts *ana pān x parik* refers to an addition on the scale with *x*, as in ARM 25, 279.[276]

In ARM 25, 477, 684, and 688, the addition of fractions of a shekel is surprising because the amounts of gold weighed are all round figures in shekels. Joannès suspected that in such cases a deviation caused by the balance was corrected. These additions are not similar to those found in some Old Assyrian texts, where it is stated that a small amount is 'brought back' to a bigger weight, usually with silver, twice in the case of copper, because there it does not appear to result from the balance. The phrase 'a 10 shekel weight brought back (= added) on top of each 10 mina weight', in kt c/k 254.5 and kt c/k 535.5, 8, followed by the expression *kaspū waddū*, '(its) silver (weight) established,' occurs in the case of consignments 2 (5 talents of fine copper) and 4 (6 talents of fine and 14 talents of *šikkum* copper) mentioned in the letters by Tahašaili who cooperated with Alāhum son of Sukuhum (see § 4.3). The pertinent lines from kt c/k 535 read: [3] 6 GÚ URUDU SIG_5 10 GÍN-*tum* [4] *a-ṣé-er* 10 *ma-na-e* [5] *ta-ú-ur-ma* KB *ú-du* [6] *a-ha-ma* 14 GÚ URUDU *ší-kam* [7] [10 GÍ]N-*tum ta-ú-ur* [8] [KB] *ú-du ku-nu-ki-a* [9] [*A-lá*]-*hu-um ù Ú-li-a* [10] *ub-lu-ni-kum*, '6 talents of fine copper, 10 shekels brought back to (each) 10 mina weight, (its) silver (weight) established, and furthermore 14 talents of *šikkum* copper, 10 shekels

---

274 ARM 25, 282.1-7: [1] 9 2/3 *ma-na* 2 1/2 GÍN KB [2] *i-na* NA_4 5 *ma-na* [3] NA_4 3 *ma-na* 2 *ma-na* NÍG.ŠU LUGAL [4] *a-na pa-an* NA_4.HI.A *a-hu-ne-e* [5] NÍG.ŠU LUGAL [6] NA_4 10 GÍN 5 GÍN 2 1/2 GÍN *ša ma-hi-ri* [7] *pa-ri-ik.*

275 ARM 25, 424.1- reverse 2: [1] 1 *ma-na* 1/3 GÍN KB A.RÁ.1.KAM [2] *it-ti ma-ni-im* 1/3! GÍN *pa-ri-ik* [3] 1/2 *ma-na* IGI.6.GÁL KB A.RÁ.2.KAM [4] 1/2 *ma-na pa-ri-ik* [5] ŠU.NÍGIN 1 1/2 *ma-na* 1/2 GÍN KB ... [2'] *ša i-na* NA_4.HI.A (reverse) [1] *Bur-qa-an* [2] *iš-ša-aq-lu.*

276 ARM 25, 279.1-4: [1] 1/3 *ma-na* 2 2/3 GÍN [15] ŠE KÙ.[ ] [2] *i-na* NA_4 1/3 *ma-na ù* NA_4 [3 GÍN] [3] NÍG.ŠU LUGAL *a-na pa-an* KÙ.[ ] [4] IGI.4.GÁL *pa-ri-ik.*

brought back to (each) 10 mina weight, (its) silver (weight) established, Alāhum and (W)alia brought you under my seal.' In consignment 4, *kaspū waddū* is omitted in the parallel passages in kt c/k 263, and also in the lines concerning *šikkum* copper in kt c/k 254.8 and kt c/k 671.27. In contrast with the other consignments recorded in his letters, here Tahašaili is himself handing over an amount of copper and that to persons who are no sons of Alāhum, but acted as transporter and (retail-)seller for the firm. The fourth consignment was sealed by Tahašaili, and this probably happened with the second consignment as well; there the witnesses are mentioned in whose presence the hand-over took place. The terminology '... *aṣṣēr* ... *ta''ur(-ma) kaspū waddū*' has been discussed by K.R. Veenhof,[277] who considered the first part to refer to some kind of 'increase', which, in the documentation available at the time (consisting of documents recording debts in silver), the debtor would have had to add when he returned the silver. The second part of this expression refers to the result of the action preceding it, stating that the amount of silver was established by that means. Veenhof's interpretation was followed by B.L. Rosen in his treatment of BIN 6, 238.[278] The passages from the kt c/k texts, however, do not concern debts, but specify amounts of copper transferred to another person, which makes an interpretation 'increase' unlikely. The fact that Tahašaili probably used his private weight-stones in weighing out the copper, suggests that the 10 mina weight was deficient, and that as compensation each time it was used, a 10 shekel weight was added to ensure that the intended weight of 10 minas was reached. The second part of the expression, '(the amount of) silver (was) established,' is understandable when silver is weighed, but not in the case of copper. Nevertheless, also in the kt c/k texts it belongs to the expression as a whole. Its meaning, as in the case of silver, is that the correct amount was established by the manipulation of the weights described. The object was in most cases silver, which made *kaspū waddū* a kind of standard expression, that could be used in other instances as well, and was in this case transferred to copper. The following texts contain new evidence for this and related expressions:

1) I 608.1-10: ^{1} 8 *ma-na* (erasure) «GÍN» KB ^{2} *i-na* 1 *ma-na-i-a* ^{3} *ta-ki-tám A-šur-ṣú-lu-li* ^{4} *is-ni-qam* 7 *ma-na* 10 GÍN ^{5} *A-mur-Ištar ù Ú-zu-a* ^{6} *is-ni-qú* 10 *ma-na* KB ^{7} *A-mur-*^{d}UTU *is-ni-iq* ^{8} 11 *ma-na* LÁ 4 GÍN ^{9} *Pu-šu-ke-en*$_{6}$ ^{10} *is-ni-iq* (by courtesy of K. Hecker);

2) kt a/k 459/b.5-9: ^{5} *i-na* 1 *ma-na-im* ^{6} *ša* ^{d}IM-*ṣú-lu-li* ^{7} 2/3 GÍN.TA *du-na-ú* ^{8} *na-ad-ú-ma i-ta-ki-sú* ^{9} *i-*SÀ.BA *ú-ta-ar*;

3) kt a/k 516/b.10-14: ^{10} *i-na* 1 *ma-na-im* ^{11} *ša A-šur-ni-šu il*$_{5}$*-qé-ú* ^{12} *du-na-ú ú-lá na-du-ú* ^{13} KB *ta-ki-tám* ^{14} *na-ší-e*;

4) kt c/k 1500.46-49a: ^{46} ... KB ^{47} *i-ma-na-im . ša A-šur-ni-im-ri* ^{48} *il*$_{5}$*-qé* 1/2 GÍN.TA ^{49} *a-ṣé-er ma-na-im ta-ur*, 'The silver has been received, weighed with Aššur-nimri's (= one of the witnesses) 1 mina weight; 1/2 shekel has been brought back to (each) mina';

[277] Veenhof 1972, pp.57-68.

[278] Rosen 1977, pp.176-178.

5) kt n/k 157.1-3a, 9-12: [1] 20 *ma-na* KB *ṣa-ru-pá-am* [2] *ù* 1 *ma-na* KÙ.GI [3] *pá-ša-lam* (...) [9] KB *i-na* 1 *ma-na-im* [10] *ša En-na-Sú-en*$_6$ [11] 1/2 GÍN *lá ta-ú-ur* [12] *i-ta-ki-is na-ší* (by courtesy of S. Bayram);
6) kt n/k 220.1, 7-10, 16-18: [1] 3 *ma-na* KB (...) [7] *i-na* 1 *ma-na-i-a* [8] KB *wa-du . ba-lúm* [9] *du-na-e . il*$_5$*-qé* [10] 1 *ma-na* KB (...) [16] *i-na* 1 *ma-na-i-a* [17] *ba-lúm du-na-e* [18] KB *wa-du il*$_5$*-qé* (by courtesy of S. Bayram);
7) kt n/k 492.17-23: [17] KB ... [18] *ša* 15 1/2 *ma-na* KB . 1/4 GÍN.TA [19] *a-na* 1 *ma-na-im ša* É *kà-ri-im* [20] *nu-ta-er-ma a-na A-ta-ta* [21] *ni-iš-qú-ul-ma ṭup-pá-am* [22] *ša E-na-A-šùr ša* [23] 15 1/2 *ma-na* KB *nu-šé-ṣí-am* (by courtesy of S. Bayram);
8) kt 91/k 140.58, 60 (debts): [58] 1/3 *ma-na* 1 GÍN *A-lá-hu-um lá áp-sí-iq* (...) [60] 1/3 *ma-na A-šur-ṣú-lu-li ša tí-ki-ta lá i-šu* (by courtesy of K.R. Veenhof);
9) kt 91/k 518/b.3'-4': [3'] KB *i-na ab-nim ni-*⌜*a*⌝*-[im]* [4'] *ta-ki-tám i-ša-qú-lu* (by courtesy of K.R. Veenhof);
10) kt 91/k 523.1-7: [1] 2/3 *ma-na* KB *ṣa-ru-pá-am* [2] *i-na ma-na-im* [3] *ša A-mur-*^{d}UTU *ba-lúm du-na-e* [4] *i-ta-ki-is i-ṣé-er* [5] *Ir-ma-A-šùr I-ri-ší-im* [6] *ù Ili*$_5$*-me-dá-ku* [7] *Ú-ṣur-ša-A-šùr i-šu* (by courtesy of K.R. Veenhof);
11) TC 1, 26.1-13: [1] *um-ma Ša-lim-a-hu-um-ma* [2] *a-na Pu-šu-ke-en*$_6$ [3] *ù* ^{d}NIN.ŠUBUR*-ba-ni qí-bi*$_4$*-ma* [4] *a-na* ^{d}NIN.ŠUBUR*-ba-ni qí-bi-ma* [5] *ta-áš-pu-ra-am um-ma a-ta-/ma* [6] *té-er-ta-kà a-na ṣé-er* [7] *ša ki-ma ku-a-tí li-li-kam-ma* [8] AN.NA *ù* TÚG.HI.A *ki-ma* [9] *i-qí-pu-ni ša* KB [10] 1 GÚ *li-dí-nu-nim ù i-ta-ki-/it* [11] *a-ni-a-tim* 10 *ma-na*.TA [12] KB*-pí li-li-kà-ku-ma* [13] *a-na té-er-tí-a i-hi-id*, 'Thus says Šalim-ahum, speak to Pūšu-kēn and Ilabrat-bānī, speak to Ilabrat-bānī: you wrote me as follows: "Send a message to your representatives to give me tin and textiles, on the conditions one usually entrusts (these goods to a retailer), worth in silver about 1 talent; then from the *t.* of these (goods), silver acquired by me will come to you in quantities of 10 minas at a time; please heed my message!" (Šalim-ahum then details who is keeping his tin and textiles in store, and urges Ilabrat-bānī to agree with these persons on the conditions on which he will sell the merchandise)';
12) TC 3, 171.12-16: [12] ... 16 *ma-na* KB [13] *i-1 ma-na-im ša kà-/ri-im* [14] *ta-ki-tám a-na* [15] *Ú-ṣur-ša-Ištar* [16] *ni-dí-in*;
13) UF 7, p. 327, no. 5.6-8a: [6] 2 *ma-na* KB 1 1/4 GÍN.TA [7] *a-na ṣé-er* 1 *ma-na-im* [8] *ta-ú-ur*. In line 15 the debt-note is referred to as 'the tablet of 2 minas of silver (and its interest)';
14) VS 26, 134: [1] 14 GÚ 30 *ma-na* [2] *i-na áb-na-tí-ni* [3] *ta-ki-tám* [4] *ú-sà-ni-qám*, 'He transferred to me 14 talents 30 minas of (copper or wool), weighed with our weight-stones, a(s) *t.*'[279]

The exact meaning of the terms *ta''ur*, 'it has been brought back (on the scale)' and *dunnā'ū nad'ū*, 'an increase was laid on (the scale)' is not always clear. It appears that a weight

279 See the remarks on this text made by K.R. Veenhof in VS 26, p.28. The meaning 'to transfer' for *sanāqum* D is also attested in CCT 2, 29.6-8: [6] 10 GÚ 55 *ma-na* URUDU SIG$_5$ [7] *a-na A-šur-ma-lik a-ma-lá* [8] *té-er-tí-kà ú-sà-ni-iq*, 'In accordance with your instructions I transferred to Aššur-mālik 10 talents 55 minas of fine copper.'

deficiency caused by the stone and not by the balance, called *bitqat abnim* in EL 278.6, was restored by the addition of a small weight on one of the scales. Such a correction resulted in the correct or true amount, called *takittum*, and this word may have been used as well for the more or less correct weight-stone in text 8. The suggestion made by J.-M. Durand (MARI 5, p.610) that *ēṣamma* in KTS 6.21 might refer to an inexact weight-stone, is unnecessary, as it simply means 'at least'. The 'corrections' were of course performed as far as the balances could determine the 'exact' weight. In his study on ancient Mesopotamian weight metrology, M.A. Powell wrote that 'Mesopotamian weights tolerated an inaccuracy of about 3% of the mass of the object being weighed, which accords closely with the range of accuracy indicated for ancient balances.'[280] If a weight had a deficiency which had been noticed by the Assyrians, the correct amount was established by several means. Nearly all these texts mention that the amount was weighed by means of a 1 mina weight, and often add a form of the verb *ta"urum*:

ta"ur can be equated with *dunnā'ū ulā nad'ū kaspum takittam naši* in 3;

lā ta"ur ittakkis naši (5) can be equated with *bālum dunnā'ē kaspum waddu* (6) = *bālum dunnā'ē ittakkis* (10) = *dunnā'ū nad'ū-ma ittakkisu iqqerbīšu uta"ar* (2). What exactly the *dunnā'ū* are that 'have been cut off' is unclear; *naši* in 5 is elliptic for *takittam naši*, and refers to the amount.[281]

Text 8 is difficult. The first part might mean 'I have not cut off,' which is then synonymous with the form of *nakāsum* attested in 2 and 10.[282] The last 4 words in line 60 might refer to the debtor, 'who does not possess a calibrated weight-stone (*takittum*).' *Takittum* means 'correct amount' in 3, 9, and perhaps also in 11, and 'as the correct amount' in 1, 12, and (with Veenhof) 14.

Inaccuracies created great problems when tiny amounts of silver or gold were weighed. By mentioning which (defective) mina weight had been used, and by using the same weight when e.g. the debt was repaid, the negative effect could be neutralized. The restoration of the weights' deficiency by manipulating one of the scales with a weight until the precision was deemed functional enough, must have been the result of calibration of private weight stones against a standard weight, perhaps in the *kārum*-office. The weight could be used if the deviation was not too big, and when the owner of the weights carefully implemented the correction every time he used it. Not all inaccuracies would have been noticed, and this may account for the fact that only a number of weight-stones were corrected. Even the *kārum*-office used weights that needed some correction. The deviations are 0.4% (1/4g), 0.8% (1/2g), 1.1% (2/3g), 1.7% (1g and 10g), and 2.1% (1 1/4g) of the '(standard) 1 mina weight'. These percentages are similar to those from Mari texts where *parik* is used to correct a weight: 0.3% (ARM 25, 366), 0.6% (213), 0.8% (477), 1% (366), 1.5% (213), 1.7% (688), 2.6% (684). The only 1 mina weight from Kültepe published so far, kt f/k

---

280 Powell 1979, p.83.

281 In BIN 6, 237, the last word in line 5 (4-5: K[B] *ú-du-ú . na-ší*) is probably also short for *takittam naši*. Veenhof (1972, p.58) and Rosen (1977, p.175) interpreted *n.* as a personal name (V.: of the creditor, R.: of the debtor), which would then be attested only in this text. The debtor is rather *Ku-ur-ku-ra-n[im]*; read in line 8, in spite of Rosen's objections: *ah-bu-ul*, 'I lent K. (the silver).'

282 The form *e-ba-zi-ku-ni* in CCT 3, 5a.16, listed in AHw, p.838 under *pasāqum*, probably is a scribal mistake for *e-zi-ba-ku-ni*, from the verb *ezābum*.

103, has a deviation of 4.1% (see Appendix 5), which would seem rather big. Compared with the deviations computed for 1 shekel weights, the corrections mentioned in the Old Assyrian and Mari texts are not uncommon.

Deviations of weights:

text	amount weighed	weight	addition
kt n/k 492	15m 30g silver	1 mina weight of the *kārum*-office	1/4g (1 1/4??)
BIN 6, 238	1m 20g silver	1 mina weight of the *kārum*	1/2g
kt c/k 1500	3m silver	1 mina weight of Aššur-nimri	1/2g
kt n/k 157	20m silver	1 mina weight of Enna-Suen	1/2g
TC 3, 213	4m silver	1 mina weight of the *kārum*	1/2g
VS 26, 119.35	1m silver	1 mina weight	1/2g
kt a/k 459	9m 11g silver	1 mina weight of Adad-ṣulūlī	2/3g
BIN 6, 242	10m silver	1 mina weight of Šū-Ištar	1g
CCT 1, 2	42m 1g; 40m silver	1 mina weight of the *kārum*-office	1 1/4g
UF 7, 5	2m silver	1 mina weight	1 1/4g
kt c/k 254	5t copper	10 mina weight (of Tahašaili?)	10g
kt c/k 254	6t; 14t copper	10 mina weight (of Tahašaili?)	10g

3.6.3 *The Anatolian weight* aban mātim

Beside the Assyrian-Babylonian weight system, there existed a local mina weight, called *aban mātim*, literally 'weight stone of the land'. In this expression, the Assyrians meant by 'the land' Anatolia and *aban mātim* is therefore rendered here 'the Anatolian mina'.[283] In the evidence available it was used once for silver and in all other cases for weighing amounts of copper, probably *šikkum*. The expression is attested in:

CCT 2, 24.21-29: [21] ... 20 TÚG-*tù-kà ša qá-/tim* [22] *i-na* É.GAL-*lim i-li-qí-*⌜*ú*⌝ [23] ⌜40⌝ *ma-na*.TA URUDU *i-na a-ba-an* [24] *ma-tim ša ší-ni-pè-e ša* TÚG-*tí-kà* [25] 4 *me-at* 87 *ma-na* URUDU [26] É *kà-ri-im lá-áp-ta-tí* [27] *i-wa-ṣa* URUDU *ta-lá-qé ša šál-ša-at* [28] TÚG-*tí-kà* 9 GÍN.TA 1 *ma-na* 1/6 GÍN [29] KB *a-dí* 10 u_4-*me* . *ta-lá-qé*, 'Your 20 textiles of current quality have been bought in the palace; (at a price of) 40 minas of copper each, according to the *a.m.*, you are credited in the *kārum*-office for 487 minas of copper for two-third of your texiles, you can receive it when the copper becomes available. For one-third of your textiles, at 9 shekels (of silver) each, you can collect within 10 days 1 mina 1/6 shekel of silver.'

HUCA 27, p.13 n.58.6'-10': [6'] ... 2 GÍN [7'] KB *ù* 40 *ma-na* [8'] URUDU *iš-tí A-šur-i-dí* [9'] 3 GÚ URUDU *i-na a-ba-an* [10'] [*m*]*a-tim*, 'Aššur-iddi(n) owes 2 shekels of silver and 40 minas of copper; 3 talents of copper (weighed) by means of the *a.m.*'

I 520.18-20: [18] 40 *ma-na* URUDU *a-ba-an ma-tim* [19] É *kà-ri-im ni-il*$_5$-*qé* KB-*áp-šu* [20] 6 1/2 GÍN, 'We received 40 minas of copper, according to the *a.m.*, in the *kārum* office, its value in silver was 6 1/2 shekels' (by courtesy of K. Hecker).

---

283 See Veenhof 1972, pp.54-57 with previous literature.

Ka 275.4-6: ^{4} 81 *ma-n[a* URUDU] ^{5} *a-ba-an m[a-tim]* ^{6} *ša*$^{?}$*-mì-ma.*

kt c/k 481.8-19 (// kt c/k 459.7-12): ^{8} 5 *ma-na* LÁ 3 GÍN ^{9} AN.NA *i-na* ^{10} *áb-na-tí-ni* ^{11} *i-na li-bi*$_{4}$ *Ha-dí-ni* . 28 *ma-na* URUDU ^{13} *ší-ku-um a-ba-an ma-tim* ^{14} *Ì-lí-áš-ra-ni* ^{15} *ni-ša-al-ma* ^{16} *šu-ma um-ma šu-ut-ma a-na nu-a-im* ^{17} *Ha-dí-ni* . *iš-qú-ul* ^{18} *i-na* 1/3 *ma-na* 2 GÍN ^{19} KB *i-ṣa-he-er*, 'Hadi(a)ni owes 5 minas minus 3 shekels of tin, according to our weight-stones. We will demand 28 (or 29?) minas of *šikkum* copper, according to the Anatolian weight-stone, from Ilī-ašranni; if he says: "Hadi(a)ni paid it to the Anatolian," he will deduct it from the 23 shekels of silver (that Hadiani owes from the proceeds of tin, lines 1-4).'

kt u/k 3.1-24: ^{1} *a-na Šu-Ištar qí-bí-ma um-ma A-šùr-*GAL*-ma* ^{2} *i-na* 16 TÚG *ša ik-ri-bi-kà* ^{3} *a-na* 2 TÚG 2 *me-at* URUDU *a-ba-an ma-tim* ^{4} *Tù-ra-mì-il* . *i-dí-na-am a-na* 2 TÚG ^{5} 1 *me-at* 80 *ma-na* URUDU *a-ba-an ma-tim* ^{6} *Šu-Hu-bur i-dí-na-am a-na* 2$^{!}$ TÚG ^{7} 1 *me-at* 80 *ma-na* URUDU *a-ba-an ma-tim* ^{8} *A-šùr-ba-ni i-dí-na-am* 2 *me-at* 80 *ma-na* ^{9} URUDU *a-ba-an ma-tim ša* 3$^{?}$ TÚG *A-la-hu-um* ^{10} *i-dí-na-am ší-tí ṣú-ba-tí* 3 TÚG *ma-aṭ-ú-tí-šu-nu* ^{11} *e-zi-ba-ma* 16 GÍN.TA *ta-ad-nu* ^{12} 1 TÚG *ša qá-tim ša i-a-tí i-dí-na-ni a-na* ^{13} 10 GÍN KB *ta-dí-in a-na* 2 TÚG 2/3 *ma-na* / KB ^{14} *Lu-lu i-dí-na-am* ŠUNIGIN URUDU *ša i-ta-dí-nu* ^{15} 8 *me-at* 40 *ma-na a-ba-an ma-tim a-na* ^{16} *áb-na-tí-ni i-tù-ur-ma* 7 *me-at* 60 *ma-/na* ^{17} 12 *ma-na*.TA *a-na hi-tim* KB *i-na áb-na-tí-/ni* ^{18} *ù* URUDU *i-na áb-na-tí-ni* ŠUNIGIN KB-*šu* ^{19} 3 *ma-na* 10 GÍN KB *lu* 2/3 *ma-na* KB ^{20} *ša* 2 TÚG *ša Lu-lu i-dí-na-ni lu ší-im* ^{21} 1 TÚG *ša qá-tim ša i-a-tí ta-dí-na-ni* 10 GÍN KB ^{22} *ší-im-šu lu* 3 TÚG *ma-ṭí-ú-tí-šu-nu ša i-a-tí e-zi-/ba-ni* ^{23} 16 GÍN.TA KB-*áp-šu* 5/6 *ma-na* LÁ 2 GÍN ^{24} ŠUNIGIN KB-*pì-kà* 4 2/3 *ma-na* 8 GÍN KB, '^{1} Speak to Šū-Ištar, thus says Aššur-rabi: ^{2} out of the 16 textiles belonging to your *ikribū*-capital, $^{3-4a}$ Tūram-il gave me 200 (minas of) copper according to the *a.m.* for 2 textiles; $^{4b-6a}$ Šū-Hubur gave me 180 minas of copper according to the *a.m.* for 2 textiles; $^{6b-8a}$ Aššur-bānī gave me 180 minas of copper according to the *a.m.* for 2 textiles; $^{8b-10a}$ Alāhum gave me 280 minas of copper according to the *a.m.* for 3 textiles. $^{10b-11}$ The remainder of the textiles, their 3 textiles of bad quality, he left with me and they were sold at 16 shekels (of silver) each. $^{12-14a}$ The textile of normal quality which he gave me, was sold for 10 shekels of silver. Lulu gave me 40 shekels of silver for 2 textiles. $^{14b-16}$ They paid me successively a total of 840 minas of copper according to the *a.m.*; (this quantity) amounted to 760 minas expressed in our own weight-stones. $^{17-19a}$ (Exchanged for silver beads) at 12 minas (of copper for each) bead, both silver and copper weighed with our weight-stones, resulted in a total of 3 minas 10 shekels of silver. $^{19b-24}$ The 40 shekels of silver that Lulu gave me for the 2 textiles, and the proceeds of the textile of normal quality you gave me (and) of which the price was 10 shekels of silver, as well as their 3 textiles of bad quality which he left behind with me: at 16 shekels each, the value in silver was 48 shekels, the total of your silver is 4 minas 48 shekels' (photo in Özgüç 1986, Pl. 60, 1a-b; collation by K.R. Veenhof).

KTS 7a.20-22: ^{20} ⌜6⌝ *sú-a-tim* 8 GÚ LÁ 10 *ma-na* ^{21} *a-na* 4 *me-at* 40 *ma-na* ^{22} *a-ba-an ma-tim ta-ad-na*, '6 *sû*-stones weighing 7 talents 50 minas have been sold for 440 minas (of copper), according to the *a.m.*'

CCT 2, 24 has been discussed by J. Lewy (1950, p.418 n.289) and K.R. Veenhof (1972, pp.56-57). In this text, a palace paid copper weighed with the Anatolian mina to an Assyrian merchant, which he could collect in the *kārum*-office. The sender of the letter gives the amount of copper to be collected by its owner from the *kārum*-office as 487 minas, while the amount expected is 40 x 13 1/3 = 533 1/3 Anatolian minas. One-third of the proceeds will be paid out in silver: 6 2/3 x 9 = 1 mina (see also below, p.172 and n.561), but the amount written in line 28 looks like 1 *ma-na* 10 1/4 GÍN. Lewy (1956, p.38 n.132) proposed to emend it to 1 *ma-na* 10 <LÁ> 1/4 GÍN, '1 mina 9 3/4 shekels'. Veenhof accepted this emendation and computed that at a price of 9 shekels each, not 6 2/3 but 7 3/4 textiles were therefore taken as 'one-third', while the remaining 12 1/4 x 40 make 490 minas of copper in line 25. He concluded that 'one-third' and 'two-thirds' were used only as approximations and that this text did not record a conversion from one weight standard into another, as assumed by Lewy. However, when one reads '1/6! GÍN', the trader would receive only 1/6 shekel of silver too much, and the difference between the amounts of copper can be explained by the use of two different weight standards. It is therefore uncertain whether this text can be used to calculate the difference between the Assyrian and the Anatolian mina, which would have been c. 10% if conversion was implied and 487 Assyrian minas equalled 533 1/3 Anatolian minas. Nevertheless, the difference between both weight standards was in fact c. 10%. This appears from text kt u/k 3, which does mention a conversion: 840 Anatolian minas are 760 Assyrian minas. With an Assyrian mina of c. 495 grams, the Anatolian mina is then c. 448 grams. One can estimate that the Assyrian mina was around 10% heavier than its Anatolian namesake, which gives a weight of c. 450 grams for the Anatolian mina. This links the Anatolian mina weight during the Old Assyrian period with the so-called 'mina of Carchemish' of about 470 gr.[284] It can be assumed that the Assyrians used some standard ratio when dealing with the Anatolian mina, because in kt u/k 3, the amounts of copper are given in Anatolian minas, although the people concerned are all Assyrians. The single reference to an amount of silver expressed in *a.m.* has an Assyrian mina 20% heavier instead of 10%. The text is VS 26, 72: [12] *a-na* 8 1/3 GÍN KB [13] 10 GÍN KB *i-na a-ba-an ma-/tim / ku-nu-kà-kà*, '(Send me at the first opportunity) 10 shekels of silver according to the *a.m.* under your seal, for 8 1/3 shekels of silver.' K.R. Veenhof (VS 26, p.24) placed *ina aban mātim* in line 12 and this may be correct. The *a.m.* refers in any case to the 10 shekels.

Some weight-stones from Kültepe and one from Boghazköy appear to be based on the *aban mātim* or Anatolian mina weight. Kültepe weights (see above) which differ strongly if 1 mina = c. 495 grams, but less if based on the Anatolian mina of c. 450 grams, are:

kt 75/k 20, a 15 shekel weight, would differ 4.2% instead of 12%;

kt n/k 244, a 10 mina weight, would differ 6.6% instead of 14.2%.

The contemporary lead disc found in Boghazköy weighing 115.7 gr. fits well into the Anatolian mina-system, as a quarter of one mina it differs only about 2.8%.

The copper that was weighed with the Anatolian mina is only specified in kt c/k 481 as *šikkum*, but when one considers the equivalents of the copper in the other texts, this quality

284 See M.A. Powell in *RlA* 7, p.516 and A. Archi, *Eblaitica* 1, p.47.

must be understood there as well. In CCT 2, 24, one textile yields 40 minas *a.m.* or 9 shekels of silver; that is 266 2/3:1. The equivalent in I 520 is 369 1/4:1. In kt u/k 3, the price per textile varies from 100 to 90 minas *a.m.*; the textiles sold by Lulu yield 20 shekels of silver per piece, which means that the equivalent is circa 300:1. These ratios between copper and silver are only understandable if the copper is of poor quality, e.g. *šikkum*. It may have been copper that was mined in the vicinity of the town and was still unrefined. The fact that this copper was used as a means of payment by the local palace is an indication of its involvement in and probably control over local copper production. It is unclear whether this metal was handed over as an ingot of a standard size.

CHAPTER 4
THE TRADE

4.1 *The archival approach*

Individual documents, letters, debt-notes, so-called *Sammelmemoranda* (often large tablets containing the text of a number of debt-notes, see Veenhof 1985), bullae and seals, and anonymous notes, provide many useful details on several aspects of the Assyrian trade on and in Anatolia. To exploit fully the informative wealth of a text, it should be studied within the context of related documents. These can be texts written for identical purposes on similar occasions, like the types of letter discussed by M.T. Larsen (1967), which were exchanged between colleagues in Assur and Kanish reporting on the transport of silver and gold to Assur, and on the way these metals were spent there on merchandise and transport, and who was guiding the donkeys with the merchandise to Kanish. They may also be documents originating from the archive of the same limited group of people, like a family or firm, or even of a single trader. The ideal would be the analysis of a complete merchant's archive, covering several years and documenting a variety of activities, which would provide a wealth of information and answer many questions about facts which would otherwise remain isolated and obscure. Such a study, in spite of the excavation of many merchant's archives in Kanish, with at times 1,000, often still unpublished, clay tablets, remains an ideal. The ancient trader did not preserve every document indefinitely; certain types, especially debt-notes, will have been discarded or destroyed after repayment of the debt recorded in them. Letters and various other documents were sent to different places and the sender did not always keep a copy. Parts of archives were transferred to the capital Assur in view of lawsuits or upon the death of an Assyrian merchant to facilitate investigations of claims and counter claims put forward by his creditors and family. Unfortunately, very few of these archaeologically excavated archives have been published.

Analysing the business of an Old Assyrian merchant on the basis of his archive is not easy. One has to establish the context he was working in: his family ties, business relations, his status and rank among colleagues, the period during which and the area in which he was active, his methods and preferences, and if possible, something of his character. The numerous homonyms and the fact that the envelopes of the majority of business letters are missing, hamper identification. These envelopes often contained the patronymics and usually the seal of the sender. Another confusing factor was that personal names in letters were often without patronymics, especially those of people not of importance in the family or firm, and those with common names. Occasionally the scribe provides some help by adding a patronymic, by designating a person as 'son of PN', or by mentioning that an individual belonged to a particular family/firm (*ša bēt* PN, 'of PN's firm'; PN *niāum*, 'our PN'), or that he was in the service of a particular trader.

Another source of uncertainty is that letters frequently refer to persons not by their name but by their title, profession, or function. It is often difficult to discover who is the person designated as 'the/our *tamkārum*', since he could be the head of a firm, a creditor, a commission agent, or a fellow merchant who owned merchandise or money, but wished to remain anonymous. Similar problems arise when someone is designated simply as *kaṣṣārum*, 'transporter', 'donkey driver', or as *ša kīma jāti*, 'my representative', and in the latter case it must be asked what the difference is between this designation and *šazzuztī*, 'my stand-in', or *muzzizum*. Some persons were called *ummiānum*, 'investor, principal', or *mēr ummiānim*, 'son of/belonging to the category of *ummiānum*', without their names being mentioned. Designations like *abī*, 'my father', or *bēlī*, 'my lord', are ambiguous, since they may refer both to family and to business relations, and are frequently employed when urgent requests are made. They do however offer some insight into the hierarchical relationship.

Business relations, often unspecified, could be of various kinds. They could rest on formal arrangements or contracts such as that with a *kaṣṣārum*, engaged by a so-called *be'ūlātum*-contract (Veenhof 1994, pp.183-184), with an *ummiānum* or fellow merchant who had invested gold in a trader's *naruqqum*-capital, or a *tappā'um*, 'partner', with whom a partnership-agreement (*tappā'uttum*) was concluded. But in other cases the formal relation is unclear or absent. The letters contain many references to Assyrians travelling, selling, and buying, but it is often difficult to establish whether they did so in the service of others or for their own profit, as member of a firm or as independent commission agent.

The uncertainty about the sequence of the year eponyms in combination with frequently insufficient data on particular traders (letters were not dated, for instance), makes it difficult to assess exactly how many years a trader was active and during which phase of the *kārum* level II period. Certain merchants, in between or near the end of their career, might return to Assur (the 'City'), and reading letters one has to be very careful in finding out whether these were written from Assur from where the trader directed his affairs, or whether he was still in Anatolia. Among letters found in Kültepe are archive copies of letters sent to Assur and to other places in Anatolia, as well as letters received from Assur. Inside Anatolia a distinction can be made between traders based firmly in one city, and others, usually younger members of a firm, who travelled frequently and stayed elsewhere. Finding out what 'here' and 'there' in a letter actually means is at times as difficult as it is important for understanding the business described.

A thorough analysis of an archive of some size and importance and a careful study of the often difficult letters is necessary for establishing the identity of the main persons and of the particulars of the firm's business. In a thematic study, however, which cuts across archives and investigates a particular aspect of the trade, such time-consuming analyses of a firm's complete business have to be the exception. In the following sections the analyses of a number of interesting archives are presented, but it must be stressed that from some not all the surviving texts were accessible. The focus in these analyses is of course on the role which copper played in the archive holder's business. After presenting the main participants of the archive, often those involved in the trade in copper, an outline is given on how this

- (Amur-ilī) ----
  - Šū-Laban
    - Imdilum (4.4)
      - Amur-ilī (4.4)
      - Puzur-Ištar (4.4)
      - Ištar-bāštī (4.4)
    - Ennum-Bēlum (4.4)
    - Puzur-Aššur (4.4)
    - Aššur-ṣulūlī
  - Aššur-imittī (4.6)
    - Uṣur-ša-Ištar (4.6)
    - Iddi(n)-Aššur (4.6)
    - Hunia (4.6)
    - Ikuppia (4.6)
  - Amur-Ištar (4.4; 4.6)
    - Suen-rē'ī (4.6)

- (Išar-kitt-Aššur)
  - Puzur-Aššur (4.5)

- (Suea)
  - Pūšu-kēn (4.5)
    - Buzāzu (4.2; 4.5)
    - Suea (4.5)

- (Kuskusum/Šū-Anim)
  - Adad-ṣulūlī (4.2; 4.5)
  - Šumma-libbi-Aššur (4.2)

- (Sukuhum)
  - Alāhum (4.3)
    - Aššur-bēl-šadue (4.3)
    - Aššur-mālik (4.3)
    - Šū-Aššur (4.3)
    - Aššur-ṭāb ? (4.3)

- (Šuli)
  - Aššur-iddi(n) (4.3)
    - Aššur-nādā (4.2)
      - Iddi(n)-Ištar (4.2; 4.3)
    - (Aššur-taklāku (4.2)

Genealogical relationship of the archive holders from §§ 4.2-4.6

metal was obtained, sold, and where this happened. For this purpose it is necessary to include basic information from letters and other documents, although this may seem at times anecdotal, and to present only part of the evidence.

4.2 *Adad-ṣulūlī son of Kuskusum/Šū-Anim*

4.2.1 *Participants*

In 1948, during the first season of excavations in the *kārum* area of Kültepe, a building was excavated in square F-G/9-10. In room 1 and 3 nearly 800 tablets, (fragments of) envelopes, and bullae were discovered, which reportedly belonged to Adad-ṣulūlī.[285] These, together with nearly 500 other texts found during the same season, were labelled 'kt a/k'. As the exact discovery site of these texts is at this moment uncertain and as it is not possible to infer from the excavation number whether a text was actually discovered in this house, one can consider instead the persons named in these documents and try to discern groups of documents centered around a key individual among the kt a/k texts. Some of these groups are:

- around Adad-ṣulūlī; in witnessed documents the persons Puzur-Šamaš son of Šū-Nunu and Aššur-dūrī son of Puzur-Ištar frequently act as witnesses.
- a small group of texts related to Iddin-Ištar son of Aššur-nādā, the same person as Adad-ṣulūlī's partner (see below).
- a number of letters addressed to Asānum.[286] Till now only one patronymic of Asānum is known, Ummasa (CCT 6, 14.33, 55, 64, 71), which makes it likely that this person is meant. Asānum is never mentioned in the same text as Adad-ṣulūlī.
- an Aššur-iddin is creditor in a number of texts.[287] Two important documents deal with the affairs of the deceased Aššur-taklāku son of Aššur-iddin, kt a/k 394 and 534. An Aššur-taklāku concluded *be'ūlātum*-contracts with transporters in several texts.[288]

The name Adad-ṣulūlī is so far attested with four patronymics, of which the son of Amur-Aššur occurs only in AKT 3, 93.7, and the son of Enna-Aššur only in kt n/k 661.3. But both Adad-ṣulūlī son of Kuskusum and Adad-ṣulūlī son of Šū-Anim are presently known from kt a/k texts:

1) Adad-ṣulūlī son of Kuskusum occurs three times. In AKT 1, 50 (= kt a/k 452) he is the creditor (witnesses: Akua son of Puzur-Aššur, Uṣur-ša-Aššur son of Butāya);[289] in AKT

---

285 Özgüç 1949, pp.143-144.

286 Kt a/k 301; 332; 345; 406; 411; 441; 526; 530; 536; 547; 548; 579; 583; 584; 913; 937; 955.

287 AKT 1, 52 (= kt a/k 569); kt a/k 477; 602/b; 604; 608.

288 Kt a/k 393: Išma-Adad son of Kurub-Ištar receives 25 shekels of refined silver; no date;
kt a/k 473: Adad-bānī receives 26 shekels of refined silver for 3 years; date: month III of the year following that of Amur-Aššur;
kt a/k 492: Šū-Kūbim had received 20 minas of refined copper; date: month V year-eponym Iddi(n)-Abum.

289 AKT 1, 50/a.4-7: [4] 1 1/2 GÚ URUDU SIG5 *i-ṣé-er* [5] *I-ri-ší-im* DUMU *A-mur-*dUTU [6] dIM-*ṣú-lu-li* DUMU *Ku-ús-ku-sí-im* [7] *i-šu.*

1, 61 (= kt a/k 421) a witness together with Šū-Ištar son of Eradi;[290] in kt a/k 299 he collects silver from Dān-Aššur son of Šalim-ahum.[291] Outside the kt a/k texts he occurs in BIN 6, 250.13 as witness to a consignment of silver to Pūšu-kēn, and as employer of a servant in RA 58, Sch. 8.[292]

Grouped around this person are Erišum son of Amur-Šamaš who is a debtor in AKT 1, 50 and a business partner in kt a/k 481. Dān-Aššur son of Šalim-ahum is connected with Pūšu-kēn in I 560 and I 652; in Pa. 2 with Iddi(n)-Ištar.[293] If BIN 6, 250 means that Adad-ṣulūlī son of Kuskusum is involved in the affairs of Pūšu-kēn, it is likely that he is also the person who possesses a house (his pied-à-terre, *bēt wabrim*?) in Purušhattum mentioned in connection with Pūšu-kēn's son Buzāzu (kt a/k 403 // 513 with Aššur-dūrī son of Puzur-Ištar as a witness; kt a/k 499/b).[294]

2) Adad-ṣulūlī son of Šū-Anim occurs outside the kt a/k texts as a witness in KTS 48c and TC 3, 257, both involving Uṣur-ša-Aššur (son of Akua? then the same as in kt a/k 294 and 410) and Suea (the son of Pūšu-kēn? no other patronymic attested for Suea). It is unclear whether in KTS 48c he is the same person as the Adad-ṣulūlī mentioned in line 11. With the patronymic he is attested in the kt a/k texts only in kt a/k 440. This text quotes two documents recording the repayment of a debt and in both cases the debtor is Adad-ṣulūlī, in line 19 identified as the son of Šū-Anim. Kt a/k 440: ^{1} KIŠIB *I-ku-pì-a* DUMU *A-šùr-ba-ni* ^{2} KIŠIB *Puzur*$_4$*-A-šur* DUMU *Šu-Be-lim* ^{3} KIŠIB DÙG-*ṣí-lá-A-šur* DUMU *Puzur*$_2$-^{d}MUŠ ^{4} KIŠIB *Ma-num-ba-lúm-A-šur* DUMU *A-šur-ma-/lik* ^{5} 11 *ma-na* KB *ṣa-ru-pá-am* ^{6} *ša i-ṣé-er* ^{d}IM-ANDUL *A-šur-ma-/lik* ^{7} DUMU *A-šur-ba-ni i-šu-ú ki-ma* ^{8} *A-šur-ma-lik* KB *Ma-num-ba-lúm-A-/šur* ^{9} *ša-bu* . *ṭup-pu-um ša* 11 *ma-/na* ^{10} KB *ṣa-ru-pì-im ša hu-bu-ul* ^{11} ^{d}IM-ANDUL . *ša e-li-a-ni sá-ar* ^{12} *šu-ma a-na* ^{d}IM-ANDUL *A-šur-ma-lik* ^{13} *i-tù-wa-ar* . *Ma-num-ba-lúm-A-šur* ^{14} *ú-ba-áb-šu* KIŠIB *Šu-Ištar* ^{15} DUMU *A-šur-ba-ni* . KIŠIB *Ili*$_5$*-we-da-ku* ^{16} DUMU *I-da-a* KIŠIB *Da-da-a* ^{17} [DUMU] *A-šur-i-mì-tí* 4 1/2 *ma na* KB ^{18} *ṣa-ru-pá-am ša Šu-Ištar* ^{19} DUMU *A-šur-ba-ni* . *a-na* ^{d}IM-[ANDUL] ^{20} DUMU *Šu-A-nim a-na e-bu-ṭí-e* ^{21} *i-dí-nu-šu-ni* . *Šu-Ištar*

---

290 AKT 1, 61/a.2-3: ^{2} KIŠIB [^{d}IM]-*ṣú-lu-li* ^{3} [DUMU *Ku-ús-ku-s*]*í-im*.

291 Kt a/k 299.4-9: ^{4} ... KIŠIB ^{5} ^{d}IM-*ṣú-lu-li* ^{6} DUMU *Ku-ús-ku-sí-im* ^{7} 1 1/2 *ma-na* 5 GÍN KB *ṣa-ru-pá-am* ^{8} *i-na* KB-*áp* ^{9} *Dan-A-šur* DUMU *Šál-lim-a-hi-im* (*ilqe*).

292 RA 58, Sch. 8.5-9: ^{5} *a-Za-al-pá* ^{6} *a-na pá-ni* ^{7} *ṣú-ha-ri* ^{8} *ša* ^{d}IM-ANDUL ^{9} DUMU *Ku-ús-ku-sí-im*, '(Silver and sundries) I sent to Zalpa, to the servant(s) of Adad-ṣulūlī son of Kuskusum.'

293 Pa. 2 records a financial arrangement between Aššur-nādā and Iddi(n)-Ištar, witnessed by Erišum, Adad-ṣulūlī, and Pilah-Aššur. Same group of people?

294 Kt a/k 403.1-12: ^{1} *i-na Pu-ru-uš-ha-tim* ^{2} *a-na* É ^{d}IM-*ṣú-lu-li* ^{3} *né-ru-ub-ma um-ma* ^{4} *Bu-za-zu-ma a-na* ^{5} ^{d}IM-*ṣú-lu-li-ma* ^{6} 4 *ma-na* KB *En-nam-A-/šur* ^{7} DUMU *Ú-zu-a ší-im* ^{8} *pè-er-dí ù a-mu-tim* ^{9} *ha-bu-lam me-eh-ra-tim* ^{10} *dí-na-ma* KB-*áp-kà* ^{11} *En-nam-A-šur* ^{12} *li-dí-na-kum*; 'In Purušhattum we entered Adad-ṣulūlī's house, and Buzāzu said to Adad-ṣulūlī: "Ennam-Aššur son of Uzua owes me 4 minas of silver, the proceeds of *perdum*-animals and *amūtum*-mineral; give me its equivalent and let Ennam-Aššur (then) give your silver (back) to you."'

Kt a/k 499/b.1-9: ^{1} *kà-ru-um Wa-ah-šu-ša-na* ^{2} *ṣa-he-er* GAL *dí-nam* ^{3} *i-dí-ma Bu-za-zu* 3 ^{4} *a-na* ^{d}IM-*ṣú-lu-li* ^{5} *i-ṣa-ba-at-ma i-ša-al-/šu-ma* ^{6} *e-pá-al-šu-ma* ^{7} *ú* ^{d}IM-*ṣú-lu-li* ^{8} *Bu-za-zu i-ša-al-ma* ^{9} *Bu-za-zu e-pá-al-šu-ma*, 'The plenary assembly of the *kārum* of Wahšušana gave a verdict: Buzāzu can seize 3 people (as witness) against Adad-ṣulūlī and interrogate him and he (= A-ṣ.) will answer him; also Adad-ṣulūlī will interrogate Buzāzu, and Buzāzu will answer him.'

KB *ša-bu* [22] *ṭup-pu-um ša ku-nu-uk* [d]IM-ANDUL [23] *ša e-li-a-ni sà-ar* [24] *um-ma* [d]IM-ANDUL-*ma a-na* [d]IM-SIG$_5$-*ma* [25] 2 *ṭup-pè . ša ku-nu-ki-šu-nu* [26] *ha-ru-mu-tim . a-dí-na-kum* [27] *i-na Kà-ni-ìš . ṭup-pè-e* [28] *ša hu-bu-li-a li-dí-nu-ni-ku-nu-/tí-ma* [29] *ṭup-pè-e . ša ku-nu-ki-šu-nu* [30] *wa-šé-er . ú ṭup-pì-a ša tù-šé-ṣa-/ni-ni* [31] *a-ta ù A-šùr-na-da* [32] *ṭup-pì-a . pí-tí-a-ma i-li-bi*$_4$ *ṭup-pì-a* [33] *ha-ru-mu-tim šu-uk-na* [34] *A-šur-na-da li-ik-nu-uk-/šu-nu* [35] *a-na a-wa-tim* [36] *a-ni-a-tim kà-ru-um* [37] *Wa-ah-šu-ša-na i-dí-ni-a-tí-ma* [38] IGI GÍR *ša A-šur ší-bu-tí-ni* [39] *ni-dí-in* IGI *Puzur*$_4$-*Ištar* [40] DUMU *A-šur-ma-lik* [41] IGI *A-šur-du-ri* DUMU *Puzur*$_4$-*Ištar* [42] *Puzur*$_4$-[d]UTU DUMU *Šu-Nu-nu tap-pá-i-ni*, '[1-4] Seal of Ikuppia son of Aššur-bānī, seal of Puzur-Aššur son of Šū-Bēlim, seal of Ṭāb-ṣill-Aššur son of Puzur-Nirah, seal of Mannum-bālum-Aššur son of Aššur-mālik. [5-14a] The 11 minas of refined silver that Aššur-mālik son of Aššur-bānī had as claim on Adad-ṣulūlī - instead of Aššur-mālik, Mannum-bālum-Aššur has been satisfied with the silver. A tablet about 11 minas of refined silver being the debt of Adad-ṣulūlī which will emerge, is invalid. If Aššur-mālik makes claims on Adad-ṣulūlī, Mannum-bālum-Aššur will clear him. [14b-17a] Seal of Šū-Ištar son of Aššur-bānī, seal of Ilī-wēdāku son of Idāya, seal of Dadāya son of Aššur-imittī. [17b-23] The 4 1/2 minas of refined silver which Šū-Ištar son of Aššur-bānī had given to Adad-ṣulūlī son of Šū-Anim as *ebuṭṭum*-loan, Šū-Ištar has been satisfied with the silver. A tablet bearing the seal impression of Adad-ṣulūlī that will emerge, is invalid. [24-33] Thus said Adad-ṣulūlī to Adad-damiq: "I hereby give you the two quittances with their seal impressions. Let them give to you (pl.) in Kanish my debt-notes and hand over the tablets bearing their seal impressions. And as for my tablets that you will obtain, you and Aššur-nādā must gain access to my tablets and place (them) among my sealed quittances. Aššur-nādā must seal them." (protocol before the *kārum* of Wahšušana).'

In the group of people around Adad-ṣulūlī son of Šū-Anim on the basis of kt a/k 440, the first person belonging to his close business partners is Adad-damiq. This is a rare name; only Pilah-Ištar is attested as a patronymic (TC 1, 67.4). But in kt a/k 478/b, an Adad-damiq occurs in combination with Adad-ṣulūlī.[295] This also links Hududu, the writer of this letter, with Adad-ṣulūlī son of Šū-Anim. In this letter also Šamaš-ennam occurs (a rare name, only Nūr-Suen is known as a patronymic in kt a/k 456.3), who is known as a transporter of Adad-ṣulūlī from kt a/k 486 and 556 (in kt a/k 456 the anonymous creditor, *tamkārum*, could well be Adad-ṣulūlī). Qatātum is also named in Hududu's letter, and this name is attested in other texts in which Adad-ṣulūlī is involved as well. In kt a/k 440, Adad-damiq, together with Aššur-nādā, must place tablets in Adad-ṣulūlī's archive, which Aššur-nādā will seal, which implies that he was a close and trusted partner of Adad-ṣulūlī. In fact, he and Ennam-[DN] are mentioned together with his representative (= Iddi(n)-Ištar?) by Adad-ṣulūlī in kt a/k 333.[296]

---

295 Kt a/k 478/b is a letter from Hududu to Adad-ṣulūlī. Lines 25-29: [25] *um-ma šu-nu-ma a-dí* [d]IM-*ṣú-lu-/li* [26] *ù* [d]IM-SIG$_5$ *ší-ma-tim* [27] *ša A-ta-a . tù-uš-ba-al-ki-ta* [28] *qá-qá-da-tí-ni ù be-el ší-ma-tim* [29] *i-šé-e-ú*, 'They said: "Until you, Adad-ṣulūlī and Adad-damiq, convey Atāya's will, they will sue ourselves and the heirs."'

296 Kt a/k 333.1-2, 23-26: [1] [d]IM-*ṣú-lu-li a-n[a]* [2] *A-šùr-i-dí iṣ-ba-at-[ni-a-tí-ma]* (...) [23] KB *a-na ša ki-ma i-a-tí* [24] *A-šur-na-da ù En-nam-[ ]* [25] *dí-in um-ma A-š[ur-i-dí-ma]* [26] *lá-dí-in* (...), 'Adad-ṣulūlī seized us against Aššur-iddi(n) (...) "Give the silver to my representative, Aššur-nādā and Ennam-[DN]." Aššur-iddi(n) replied: "I will give it."' In view of kt a/k 403, quoted above, it seems possible that

As the excavation reports do not give the registration numbers of texts found in a specific room, the question arises which of the two persons named Adad-ṣulūlī was the archiveholder of building F-G/9-10, and is meant in the many kt a/k texts that deal with the copper trade. An important part of kt a/k belongs to the son of Šū-Anim. But both seem to have been connected at times with Pūšu-kēn and his family. Instead of assuming that archives of two different persons called Adad-ṣulūlī were found, one can assume that they are the same person: Adad-ṣulūlī's father Šū-Anim occasionally would have been called by the informal name Kuskusum.[297] That two other sons of Kuskusum, Atāya and Šumma-libb-Aššur, are never identified as sons of Šū-Anim, could be due to the small number of texts in which they occur with patronymics.

The documentation about the business of *Adad-ṣulūlī* is rather one-sided since I have selected mainly texts relating to copper. It is doubtful whether he belonged to a wealthy and influential family like Uṣur-ša-Ištar, Imdilum, or Puzur-Aššur did, because he never held the presumably prestigious post of week eponym, *hamuštum*, as these persons did, who can be considered as well-to-do businessmen. Adad-ṣulūlī appears to have been one or more steps behind them in wealth and importance. There are as yet no caravan documents published which give insight into the amount of goods imported by him. Although his financial situation is poorly recorded in the documentation available, Adad-ṣulūlī was wealthy enough to invest 2 minas of gold in the *naruqqum*-capital of Puzur-Aššur,[298] and could employ several men as transporter (*kaṣṣārum*).[299] He was married and his wife lived in Assur.[300] His sister Hududu sent him textiles from Assur which he sold in Anatolia, and the amounts vary from a few when shipped with other people,[301] to a complete donkey-load which her brother took with him when he left Assur.[302] Adad-ṣulūlī maintained business contacts, be it as their junior, with well-known and important traders like Pūšu-kēn and his

---

Ennam-Aššur son of Uzua is the second person. Kt a/k 508 is a letter from Aššur-nādā and Adad-damiq to Adad-ṣulūlī.

297 See for Old Babylonian cases of different names for the same individual, M. Stol, *SEL* 8 (1991), p.210. In Old Assyrian the evidence is meagre; there are two examples of an alias: Pūšu-kēn's daughter Ahaha, the '*kubabtum*-priestess', is often called *Waqqurtum*, 'Precious One'. The other is the year eponym Šū-Suen son of Šū-Bēlim, who appears to be the same person as the year eponym Šū-Suen son of Babalum.

298 Kt a/k 494/b: [1] 2 *ma-na* KÙ.KI [2] *ša* DAM.GÀR-*ru-um* [3] *i-na na-ru-uq* [4] *Puzur$_4$-A-šur wa-du-ú* [5] KÙ.KI *ša* ^{d}IM-*ṣú-lu-li* [6] ^{d}IM-*ṣú-lu-li* [7] *tám-kà-ru-um* [8] IGI *Dan-A-šur* [9] IGI *Puzur$_4$*-^{d}UTU, '(As for) the 2 minas of gold for which "the merchant" has been marked (as owner) in Puzur-Aššur's *naruqqum*-capital, the gold is Adad-ṣulūlī's, Adad-ṣulūlī is the merchant. Witnessed by Dān-Aššur and Puzur-Šamaš.' See for the institution of *naruqqum*-investments Larsen 1977.

299 Il(ī-)anum (kt a/k 264), Šamaš-ennam (kt a/k 486; in kt a/k 556 sent to Assur and later to Turhumit), Qatātum son of Azuza (kt a/k 433; 478.14), and Šū-Anim son of Azuza (AKT 1, 56).

300 Kt a/k 478/b (written by his sister Hududu), 38-41: [38] ... *a-ni-ma um-ma* [39] *a-ša-at-kà-ma a-na a-hi-[k]i* [40] *ta-áš-ta-pí-ri-ma a-hu-ki mar-ṣa-tim* [41] *iš-ta-na-pá-<ra>-am*, 'Now, your wife said (to me): "You have often written to your brother, but your brother sends me everytime only harsh words (in reply)."'

301 Kt a/k 478/b.1-4: [1] *um-ma Hu-du-du-ma a-na* ^{d}IM-*ṣú-/lu-li* [2] *qí-bi-ma ṣú-ba-ti iš-té-na* [3] *ù ša-na ú-na-ša-ar-ma* [4] *uš-té-né-ba-lá-kum*, 'Thus Hududu, say to Adad-ṣulūlī: I will set aside a few textiles and send (them) to you everytime (a consignment leaves).'

302 Kt a/k 532.3, 9: [3] 30 *ku-ta-nu* 1 ANŠE *ù ú-nu-sú* (...) [9] *i-nu-mì tù-úṣ-ú-ni a-dí-na-/kum*, 'I gave you 30 *kutānum*-textiles and 1 donkey (...) when you left.'

family,[303] with Pūšu-kēn's partner Puzur-Aššur son of Išar-kitt-Aššur, and with the sons of Aššur-iddin son of Šuli: Aššur-nādā, Aššur-taklāku, and with the son of Aššur-nādā, Iddin-Ištar, for whom see below. It is perhaps because of this relationship that Puzur-Aššur son of Išar-kitt-Aššur was one of the arbitrators of the settlements quoted in CCT 1, 48, which wil be discussed below. Adad-ṣulūlī (son of Kuskusum) was also a contemporary of Ikuppia son of Aššur-imittī (see § 4.6); both sealed text kt a/k 299.[304]

A conflict arose from a transaction with copper involving on the one hand *Šumma-libb-Aššur* as creditor and his representative Enna-Suen son of Ilī-Anum,[305] and Adad-ṣulūlī and *Aššur-damiq* as debtors on the other. This conflict is documented by several texts. The first is kt a/k 497/b, followed by AKT 1, 74, and ends with a settlement by arbitration recorded in CCT 1, 48 (published and discussed as EL 268). Kt a/k 497/b: ^{1} *kà-ru-um Tur$_4$-hu-mì-it* ^{2} *ṣa-he-er* . GAL . *dí-nam* ^{3} *i-dí-in-ma* . *iš-tù* ^{4} *ha-mu-uš-tim ša* ^{5} *A-šùr-mu-ta-bi-il$_5$* ^{6} [ITU].KAM . *hu-bu-ur* . DINGIR ^{7} *ša-pá-tám* . *i-li-ik* ^{8} *li-mu-um I-na-a* ^{9} DUMU [*A*]-*mu-ra-a* . 1 GÚ 15 [*ma-na*.TA] ^{10} [URUDU] SIG$_5$. ^{d}IM-ANDUL ^{11} [*ù*] *A-šùr*-SIG$_5$ ^{12} *i-na-dí-ú-ma* ^{13} *En-na-Sú-in* ^{14} DUMU *Ì-lí-*⌜*nim*⌝ ^{15} *ú-ša-bu-ú ší-bi* ^{16} ^{d}IM-ANDUL *i-ša-kà-šu-um* ^{17} *ša ki-ma* 7 1/2 GÚ 2/3 *ma-na* ^{18} URUDU SIG$_5$. *ki-ma Šu-ma-li-/bi$_4$-*[*A-šur*] ^{19} *ša-bu-ú* ^{20} *ú-ra-ma* . ^{d}IM-ANDUL-[*ma*] ^{21} *ší-bi-šu a-na A-šùr*-S[IG$_5$] ^{22} *i-za-kà-ar-ma* . *a-l*[*i*] ^{23} *i-ma-gu$_5$-ru-ni* ^{24} *e-ta-wu-ú*, 'The plenary session of the *kārum* of Turhumit gave as (its) verdict: from the week of Aššur-muttabbil, month VI half elapsed, year-eponym Innāya son of Amurāya, onwards, Adad-ṣulūlī and Aššur-damiq will [each] deposit 1 talent 15 [minas] of fine [copper] and satisfy with it Enna-Suen son of Ilī-Anum (*Ilēnum*?). Adad-ṣulūlī will engage witnesses against him to testify that (Enna-Suen) has been satisfied instead of Šumma-libb-Aššur with 7 talents 30 2/3 minas of fine copper. The next day, Adad-ṣulūlī will name his witnesses to Aššur-damiq and they will negotiate in order to arrive at an agreement.' AKT 1, 74: ^{1} *kà-ru-um* ^{2} *Tur$_4$-hu-mì-it* ^{3} *ṣa-he-er* . GAL ^{4} *dí-nam* . *i-dí-in-ma* ^{5} *a-ma-lá* . *dí-in* ^{6} *kà-ri-im* ^{7} *pá-ni-im* ^{8} *ší-bi-šu* ^{9} ^{d}IM-*ṣú-lu-li* ^{10} *a-na A-šùr*-SIG$_5$ ^{11} *i-za-kà-ar* ^{12} *Tù-ra-mì-il$_5$* ^{13} *ú-lá ú-da*, 'The plenary session of the *kārum* of Turhumit gave as (its) verdict: in accordance with the previous *kārum* verdict Adad-ṣulūlī will name his witnesses to Aššur-damiq. He does not have to identify Tūram-il (as witness).' CCT 1, 48 reads: ^{1} *Šu-ma-li-bi-A-šur ù* ^{d}IM-/ANDUL ^{2} *iṣ-bu-tù-ni-a-tí-ma* ^{3} *a-wa-tí-šu-nu* . *nu-ga-me-er-/ma* ^{4} 1 5/6 *ma-na* KB *ša Šu-ma-li-bi-/A-šùr* ^{5} *a-na* ^{d}IM-ANDUL . *ha-bu-lu* ^{6} KB ^{d}IM-ANDUL *ša-bu* ^{7} *ṭup-pu-šu* . *ha-ar-mu-um ša ki-ma* ^{8} ^{d}IM-ANDUL *a-na Šu-ma-li-bi-A-/šùr* ^{9} *i-du-nu-ma i-du-ak* ^{10} 1 *ma-na* 1 GÍN KB *ša A-šur*-SIG$_5$ ^{11} *i-ṣé-er Da-da-a* DUMU ^{12} *Tí-li-a* . ^{d}IM-ANDUL *il$_5$-qé-ú* ^{13} *Šu-ma-li-bi-A-šur* KB *il$_5$-qé* ^{14} *šu-ma A-šur*-SIG$_5$ *a-na* 1 *ma-na* 1 / GÍN ^{15} KB *a-na* ^{d}IM-ANDUL ^{16} *i-tù-ar Šu-ma-li-ba-A-šùr* ^{17} KB *ú-ta-ar um-ma* ^{d}IM-ANDUL-*ma* ^{18} 7 1/2 GÚ 2/3 *ma-na* URUDU-*kà* ^{19} *ma-sí-am En-na-Sú-in* ^{20} DUMU *Ì-lí-a-lim* . *il$_5$-*

303 In AnOr 6, 21, he owed 5 minas of silver as an *ebuṭṭu*-loan to Pūšu-kēn; witnesses were Puzur-Aššur son of Išar-kitt-Aššur and Puzur-Šamaš son of Šū-Nunu. Date: month V, year-eponym Kubia.

304 Kt a/k 299.1-6: ^{1} KIŠIB *La-ba-na-da* ^{2} DUMU *I-dí-Sú-in* ^{3} KIŠIB *I-ku-pí-a* DUMU ^{4} *A-šur-i-mì-tí* KIŠIB ^{5} ^{d}IM-*ṣú-lu-li* ^{6} DUMU *Ku-ús-ku-sí-im.*

305 The signs A-LIM in the name *Ì-lí-A-LIM* may have to be read A-num$_x$, see *N.A.B.U.* 1991, p.7 n.9. Cf. V. Donbaz, *N.A.B.U.* 1993, p.5 n.6.

qé um-ma [21] *Šu-ma-li-ba-A-šur-ma* 5 GÚ / LÁ 5 *ma-na* [22] URUDU *il*$_5$*-qé* ITU.KAM *ša ke-na-/tim* [23] *li-mu-um Šu-Sú-in* [24] DUMU *Ba-ba-lim a-na* ITU.3.KAM *ṭup-pá-am* [25] *ša ší-bi-šu* . dIM-ANDUL *ša* 7 1/2 GÚ / 2/3 *ma-na* [26] *i-lá-qé-ma a-na Šu-ma-li-/bi-A-šur* [27] *i-da-šu* . *šu-ma ṭup-pá-am* [28] *ša ší-bi-šu* . dIM-ANDUL *lá il*$_5$*-té-qé* [29] 3 1/2 GÚ 5 2/3 *ma-na* URUDU (erasure) [30] URUDU *ma-sí-am hu-bu-ul* dIM-/ANDUL [31] *a-na a-wa-tim a-ni-a-/tim* [32] *kà-ru-um Wa-ah-šu-ša-na* [33] *ṣa-he-er* GAL *i-dí-ni-a-tí-ma* [34] IGI GÍR *ša A-šur ší-bu-tí-ni* [35] *ni-dí-in* IGI *A-hu-qar* [36] IGI *Ú-ṣur-ša-A-šur* [37] *Puzur*$_4$*-A-šur* [38] DUMU *I-šar-ki-it-A-šur tap-pá-i-/ni*, '[1-3] Šumma-libb-Aššur and Adad-ṣulūlī seized us and we brought their dispute to an end: [4-9] (as for) the 1 5/6 minas of silver which Šumma-libb-Aššur owed to Adad-ṣulūlī, Adad-ṣulūlī has been satisfied with the silver; Adad-ṣulūlī's representatives will give his certified debt-note to Šumma-libb-Aššur and it will be destroyed. [10-17a] (As for) the 1 mina 1 shekel of silver which Aššur-damiq had as claim on Dadāya son of Tilia, and (which) Adad-ṣulūlī had collected, Šumma-libb-Aššur has (now) received the silver. If Aššur-damiq claims the 1 mina 1 shekel of silver from Adad-ṣulūlī, Šumma-libb-Aššur will return the silver. [17b-22a] Adad-ṣulūlī declared: "Enna-Suen son of Ilī-Anum took 7 talents 30 2/3 minas, your refined copper." Šumma-libb-Aššur replied: "He took (only) 4 talents 55 minas of copper." [22b-30] (Date:) month III, year-eponym Šū-Suen son of Babalum. Within 3 months Adad-ṣulūlī will obtain his document certified by witnesses stating that (Enna-Suen took) 7 talents 30 2/3 minas, and give it to Šumma-libb-Aššur. Should Adad-ṣulūlī prove unable to obtain his document certified by witnesses, 2! talents 35 2/3 minas of copper, refined copper, will be considered to be Adad-ṣulūlī's debt. [31-38] The plenary session of the *kārum* of Wahšušana gave us (as witness) for this case, and we gave our testimony in front of Assur's dagger. Witnesses: Ahu-(wa)qar (and) Uṣur-ša-Aššur; Puzur-Aššur son of Išar-kitt-Aššur was our partner.'

In this text, three cases between Šumma-libb-Aššur and Adad-ṣulūlī are settled by arbitration:

(1) Šumma-libb-Aššur paid 1 5/6 mina of silver which he owed to Adad-ṣulūlī;

(2) Šumma-libb-Aššur had received from Adad-ṣulūlī 1 mina 1 shekel of silver which the latter had received from Dadāya son of Tilia, who owed this amount to Aššur-damiq; Šumma-libb-Aššur will repay this silver in case Aššur-damiq claims it from Adad-ṣulūlī;

(3) Adad-ṣulūlī declares that Enna-Suen son of Ilī-Anum received 7 talents 30 2/3 minas of refined copper due to Šumma-libb-Aššur, but according to Šumma-libb-Aššur Enna-Suen received only 4 talents 55 minas. The *kārum* of Wahšušana rules that within 3 months, Adad-ṣulūlī has to produce a document recording those who witnessed the transfer of 7 talents 30 2/3 minas, and hand this document over to Šumma-libb-Aššur. If Adad-ṣulūlī cannot obtain such document, the difference of 2! talents 35 2/3 minas will be considered his debt.

Little is known about the background of the amount of copper around which this dispute arose. One can surmise that Adad-ṣulūlī and Aššur-damiq had sold in Turhumit merchandise on behalf of Šumma-libb-Aššur, and that the proceeds thereof were 7 talents 30 2/3 minas of refined copper. The history of the conflict may have been as follows: Enna-Suen, collecting the copper for Šumma-libb-Aššur, had received 4 talents 55 minas. Enna-

Suen or Šumma-libb-Aššur filed a complaint against Adad-ṣulūlī and Aššur-damiq with the *kārum* of Turhumit for the remaining 2 talents 35 2/3 minas. The *kārum* decreed that Adad-ṣulūlī and Aššur-damiq had to pay a total of 2 talents 30 minas to Enna-Suen, after which he would be fully satisfied on behalf of Šumma-libb-Aššur (kt a/k 497/b). The reason for the difference of 5 2/3 minas remains obscure. Adad-ṣulūlī could now declare that Enna-Suen had received the full amount. It seems that the dispute arose when Adad-ṣulūlī was incapable of producing evidence for making these payments. The solution was that he would prove by means of witnesses that he had made the payments; in case he failed to produce this evidence, the difference would be considered his debt, and he and Aššur-damiq would have had to pay 3 talents 35 2/3 minas of copper. This amount also creates a problem; either it must be read 2[!] talents, or it is correct, and might contain a fine of 1 talent; but the 5 2/3 minas are included here, contrary to kt a/k 497/b.

This small group of texts demonstrates that Šumma-libb-Aššur, Adad-ṣulūlī, and Aššur-damiq were working together. This also appears from Adad-ṣulūlī's archive, where Aššur-damiq and Šumma-libbi-Aššur occur together in a letter sent by Hududu to Adad-ṣulūlī (kt a/k 478.44-45). Their relationship to Adad-ṣulūlī is unclear. A Šumma-libb-Aššur son of Kuskusum is known as a witness in documents concerning Pūšu-kēn (CCT 5, 40b; I 471; I 606). If he is the same person as his homonym from CCT 1, 48, he would be Adad-ṣulūlī's brother.

Adad-ṣulūlī acted as agent for *Puzur-Aššur*'s firm; this made him buy, transport, and sell copper on his behalf (see for the texts § 4.5). In BIN 4, 36, where he calls him 'brother and lord', Adad-ṣulūlī transports 4 donkeys loaded with textiles belonging to Puzur-Aššur to Turhumit, which he wished to exchange for copper. Together with Ikuppia he transports copper belonging to Puzur-Aššur in CCT 3, 49b, and in KTS 7a, he and Buzāzu sell Puzur-Aššur's antimony and *sû*-stones for copper.

Adāya (not the same as Adad-ṣulūlī's brother Atāya) is one of Adad-ṣulūlī's agents. There are some texts in which his task is to sell textiles; in kt a/k 514 he reports that out of 41 *kutānum*-textiles which Adad-ṣulūlī had entrusted to him, he and Adad-ṣulūlī's representatives gave 28 to Iddi(n)-Ištar (who sold them to the palace).[306] The remaining 13 textiles were sold and as Alāhum and Iddi(n)-Ištar write to Adad-ṣulūlī in kt a/k 391, Adāya was responsible for paying the proceeds of these textiles.[307] Adāya gave 12 to a trustworthy

---

306 Kt a/k 514.1-18: [1] *a-na* ^{d}IM-*ṣú-lu-li* [2] *qí-bi$_4$-ma um-ma A-da-a-ma* [3] *i-na* 41 TÚG *ku-ta-ni* [4] *ša ta-áp-qí-da-ni* [5] 28 TÚG *ku-ta-ni a-na-ku* [6] *ù ša ki-ma ku-a-tí* [7] *a-na I-dí-Ištar ni-dí-/in* [8] 12 TÚG *ku-ta-nu iš-tí-a* [9] *ší-im-šu-nu iš-tí* [10] *tám-kà-ri-im* [11] *lá š[a] ta-pá-ri-/du* [12] *ú ku-ta-nam* [13] *a-na* 11 *ma-na* URUDU SIG$_5$ [14] *a-na Tù-i-mì-im* [15] DUMU *Zu-gi-li* [16] *A-lá-hu-um i-dí-in* [17] *ha-ra-šu a-mì-ša-am* [18] *ṣa-ba-sú-ma* URUDU *ša-áš-qí-/il$_5$-šu*, 'Speak to Adad-ṣulūlī, thus (says) Adāya: out of the 41 *kutānum*-textiles that you entrusted to me, I and your representatives gave 28 to Iddi(n)-Ištar; 12 *kutānum*-textiles are still with me (= I owe their price to you). Their proceeds are with a retailer about whom you do not have to worry. Alāhum gave a *k.*-textile for 11 minas of fine copper to Tu'imum son of Zugili; he is on his way to you, seize him and make him pay the copper to you!'

307 Kt a/k 391.1-10, 16-19: [1] *um-ma A-lá-hu-um-ma* [2] *I-dí-Ištar ù A-da-a-/ma* [3] *a-na* ^{d}IM-ANDUL [4] *qí-bi-ma . um-ma I-dí-Ištar-/ma* [5] *ba-áb-šu . ša* URUDU SIG$_5$ [6] *li-bi-ší-ma . me-eh-ra-at* [7] TÚG.HI-*tí-kà . ša i-na* [8] É.GAL-*lim . il$_5$-qé-ú-ni* [9] *i-ra-mì-ni-a .* URUDU SIG$_5$ [10] *a-na-dí-ma . ù-šé-ba-lá-kum* (...) [16] ... 30 LÁ 2 TÚG.HI-*tí* [17] *a-na* É.GAL-*lim a-dí-in* [18] *ší-im .* 13 TÚG.HI-*tí* [19] *iš-tí . A-da-a* (...), 'Thus (say) Alāhum, Iddi(n)-Ištar, and Adāya, speak to Adad-ṣulūlī, thus speaks Iddi(n)-Ištar: should fine

tamkārum who was to pay the proceeds to him, and 1 textile was sold for 11 minas of fine copper to Tu'imum son of Zugili by Alāhum. He owes (the proceeds of) 12 textiles to Adad-ṣulūlī. It seems that the proceeds of these textiles often reached Adad-ṣulūlī in the form of copper (kt a/k 401).[308] At times, proceeds belonging to Adad-ṣulūlī were lent to others; in kt a/k 480, Adāya writes to Kikidānum that the 21 1/2 minas of fine copper which he had given him, belong to Adad-ṣulūlī.[309]

Erišum son of Amur-Šamaš is also an agent or partner. He owes Adad-ṣulūlī 1 1/2 talent of fine copper which he has to pay within one year (AKT 1, 50). In kt a/k 300, Adad-ṣulūlī turns to Puzur-Šamaš, who acts as guarantor for Erišum, to ensure payment of 15 talents 52 minas of poor copper which Erišum owed him as part of the 'balance payments of Turhumit'.[310] In kt a/k 481 (// kt a/k 512), *Ladia*, to whom Adad-ṣulūlī owes copper as proceeds from (the sale of) his textiles, is allowed to take the copper from the shares of Adad-ṣulūlī and of Erišum as soon as it leaves the palace.[311]

Aššur-malik and Adad-ṣulūlī act together for Puzur-Aššur according to two letters they wrote to Pūšu-kēn; in RA 81, 1, they ask him to send Puzur-Aššur 7 minas of silver as proceeds of some 7 talents of copper. In TC 2, 18, they report that Ea-šar paid in copper the final instalment of the debt he owed to Puzur-Aššur, and promised to bring the copper to Purušhattum (to sell it there; see for these texts § 4.5).

Hanāya is a partner of Adad-ṣulūlī. He takes silver from Adad-ṣulūlī's money as compensation for textiles Adad-ṣulūlī bought from him and for which he would have paid in copper. The goods which Adad-ṣulūlī had given as pledge (for this payment?) were released to a third person (kt a/k 265). In kt a/k 474, Adad-ṣulūlī is told by Uzubiškum, Šuli, and

---

copper appear on the market, then I will secure fine copper from my own resources as counter value for your textiles which they bought in the palace, and send it to you. (...) I gave 28 textiles to the palace; the proceeds of 13 textiles are due by Adāya.' Does Iddi(n)-Ištar secure the copper from his share in the copper which becomes available?

308 Kt a/k 401.1-9: ^{1} *a-na* ^{d}IM-*ṣú-lu-li* ^{2} *q[í-bi]-ma um-ma A-da-a-ma* ^{3} *i-[na* 1]1 TÚG.HI.A ^{4} *ša tám-kà-ri-im* ^{5} 6 TÚG.HI.A . *ta-ad-nu* ^{6} 5 TÚG.HI.A . *i-ba-ší-ú* ^{7} *a-na* URUDU-*im* . *i-tù-ru-ma* ^{8} *ù iš-tí té-er-tí-kà-/ma* ^{9} *nu-šé-ba-lá-kum*, 'Speak to Adad-ṣulūlī, thus (says) Adāya: of the 11 textiles of the retailer 6 have been sold, 5 are still on hand. When they are converted into copper and we will send you (the copper) together with your own goods.'

309 Kt a/k 480.1-11: ^{1} *a-na* ^{d}IM-*ṣú-lu-li* ^{2} *ù Ki-ki-da-nim qí-bi$_4$-ma* ^{3} *um-ma A-da-a-ma* ^{4} 21 1/2 *ma-na* URUDU SIG$_5$ ^{5} *ša* ^{d}IM-*ṣú-lu-li* ^{6} *a-dí-na-kum-ma* ^{7} *i-na ṭup-pì-im* ^{8} DAM.GÀR *ú-dí* ^{9} ^{d}IM-*ṣú-lu-li* ^{10} DAM.GÀR *a-hi a-ta* ^{11} URUDU *dí-šu-ma*, 'Speak to Adad-ṣulūlī and Kikidānum, thus (says) Adāya: I gave you 21 1/2 minas of fine copper belonging to Adad-ṣulūlī, but on the bond "merchant" is marked (as the owner). Adad-ṣulūlī is the merchant! My dear brother, give him the copper.'

310 Kt a/k 300.1-11: ^{1} ^{d}IM-*ṣú-lu-li* ^{2} *E-ri-ša-am* DUMU *A-mur*-^{d}UTU ^{3} *a-na Puzur$_4$*-^{d}UTU ^{4} *a-na qá-ta-tim a-na* ^{5} 15 GÚ 52 *ma-na* URUDU ^{6} *lá-mu-nim ša ni-ip-lá-/tim* ^{7} *ša Tur$_4$-hu-mì-it* ^{8} *ša* 1 *ma-na-um* ^{9} 3 GÍN.TA *i-lu-ku* ^{10} *a-na Puzur$_4$*-^{d}UTU ^{11} *i-dí-šu* (...), 'Adad-ṣulūlī (= creditor) made Puzur-Šamaš assume guarantee for Erišum son of Amur-Šamaš (= debtor), for the 15 talents 52 minas of poor copper of the balance payments of Turhumit, on which per mina 3 shekels accrue (as interest).'

311 Kt a/k 481.3-12: ^{3} *um-ma Lá-dí-a-ma* ^{4} *a-na* ^{d}IM-*ṣú-lu-li-ma* ^{5} *ší-im* TÚG.HI-*tí-a* URUDU SIG$_5$ *dí-nam* ^{6} *um-ma* ^{d}IM-*ṣú-lu-li-ma* ^{7} *i-na a-wa-tí-ni* ^{8} *i-na wa-ṣa* URUDU ^{9} *ša* É.GAL-*lim i-na* ^{10} *qá-tí-a* . *ù qá-tí* ^{11} *I-ri-šum lá-qá-am* ^{12} *qá-bi$_4$-a-tí* (...), 'Ladia said to Adad-ṣulūlī: "Give me fine copper as proceeds for my textiles!" Adad-ṣulūlī replied: "According to our agreement you are entitled to take (it) from my share and that of Erišum when the copper leaves the palace."'

Aššur-lamassī that the settlement of accounts in the *kārum*-office in their place of residence revealed that he has to make a balance payment of 12 talents of copper to Hanāya.[312]

Iddi(n)-Ištar son of Aššur-nādā is another partner of Adad-ṣulūlī. His patronymic is derived from a comparison between kt a/k 294 (with the patronymic on the case) and kt a/k 405, both texts mentioning Aduda son of the king (= Šarru-kēn I). His father, Aššur-nādā son of Aššur-iddin, is witness in AKT 1, 69. Aššur-iddin son of Šuli, his grandfather, lived in Assur at this time.[313] Iddi(n)-Ištar's uncle, Aššur-taklāku, appears to have died early. Aššur-iddin's sons are often mentioned together with sons of Alāhum son of Sukuhum (see for the commercial relationship between Aššur-iddin and Alāhum § 4.3). In kt a/k 410, Aššur-ṭāb son of Alāhum collects shares of copper for Aššur-nādā and Aššur-iddin. As witness in a document where Adad-ṣulūlī figures as creditor, Iddi(n)-Ištar is not identified on the case with his father's name, but remarkably enough with that of his grandfather.[314] In AKT 1, 65, he has transferred 2 1/2 talents of fine copper to Puzur-Šamaš, and this identifies him as the Iddi(n)-Ištar cooperating with Adad-ṣulūlī. He writes together with Alāhum and Adāya a letter to Adad-ṣulūlī in which he announces his decision not to wait till the palace has paid the copper it is due, but instead to ship him out of his own resources fine copper equal to the proceeds of the textiles which the palace bought (kt a/k 391, see above, n.307). He further reports that he sold the 28 textiles, mentioned in kt a/k 514 (see above, n.306), to the palace (the same as those for which the palace would pay with fine copper?), and reminds Adad-ṣulūlī that the proceeds of 13 textiles are still owed by Adāya (kt a/k 391). A debt-note was written for 6 talents of fine copper which he owed to Adad-ṣulūlī, probably for merchandise he was to sell for him (kt a/k 436).[315] In TC 3, 5, Ikuppia reports to Puzur-Aššur that he asked Adad-ṣulūlī's representative Iddi(n)-Ištar in vain for copper as proceeds of 5 *kutānum*-textiles; this text also shows that Iddi(n)-Ištar was active as the agent of Puzur-Aššur.[316] The same Iddi(n)-Ištar son of Aššur-nādā (by then already 3 years dead!) appears in CTMMA 1, 84, as the retailer of an Ennum-Aššur, to whom he owed 5 talents of silver when he died. He possessed a 'house' in Turhumit, the town famous for its

---

312 Kt a/k 474.4-11: ^{4} ... *a-na-kam* ^{5} É *kà-ri-im* NÍG.KA-*sí* ^{6} *i-sí-ú-ma* 7 *me-at* ^{7} 20 *ma-na* URUDU . *a-na* ^{8} *Ha-na-a* . *ta-na-pá-al* ^{9} *šu-up-ra-ma* . *a-na* ^{10} *ša ki-ma ku-a-tí* ^{11} *a-na-kam* . URUDU . *li-iš-qú-lu*, 'Here, the accounts have been settled in the *kārum*-office, and as a result you have to make a balance payment of 720 minas of copper to Hanāya. Write to your representatives that they pay the copper here.'

313 AKT 1, 40 is dated to the year when Aššur-iddin son of Šuli was eponym in Assur; Adad-ṣulūlī was then in Anatolia.

314 AKT 1, 42/a.2: KIŠIB *I-dí-Ištar* DUMU *A-šùr-i-dí*.

315 Kt a/k 436/a.1'-6': $^{1'}$ [KIŠIB *I-dí-Ištar*] $^{2'}$ DUMU *A-šur-na-da* (...) $^{4'}$ ... 6 GÚ URUDU SIG_5 $^{5'}$ [*i*]-*ṣé-er I-dí-Ištar* . ^{d}IM-*ṣú-lu-li* $^{6'}$ [*i*]-*šu*.

316 TC 3, 5.1-17: ^{1} *a-na* Puzur$_4$-*A-šur* ^{2} *qí-bi*$_4$*-ma um-ma I-ku-pì-a-ma* ^{3} *ta-áš-pu-ra-am um-ma* ^{4} *a-ta-ma ší-im* 5 TÚG *ku-ta-/ni* ^{5} *ša ki-ma* ^{d}IM-*ṣú-lu-/li* ^{6} URUDU *li-dí-nu-ni-kum* ^{7} *a-na-kam I-dí-Ištar e-ri-/iš-ma* ^{8} *um-ma I-dí-Ištar-ma* ^{9} *mì-ma* URUDU-*šu ša* ^{10} ^{d}IM-*ṣú-lu-li* ^{11} *lá ú-kà-al* . *a-ma-kam* ^{12} ^{d}IM-*ṣú-lu-li* ^{13} *e-ri-iš* . *ší-im* ^{14} KB . *da-nim I-dí-Ištar* ^{15} *e-ri-iš-ma um-ma* ^{16} *šu-ut-ma* URUDU *a-qá-tí-a* ^{17} *la ma-qí-it* (...), 'Speak to Puzur-Aššur, thus says Iūppia: you wrote me as follows: "Let Adad-ṣulūlī's representatives give you copper as proceeds of 5 *kutānum*-textiles." I demanded (it) here from Iddi(n)-Ištar, but he said: "I do not possess any copper belonging to Adad-ṣulūlī." Ask Adad-ṣulūlī over there! I also demanded the proceeds of the "strong silver" from Iddi(n)-Ištar, but he said: "I did not obtain copper."'

copper market, and it was in this city where, according to BIN 4, 64, he met with Ilī-Alum, who worked for his father Aššur-nādā and for Alāhum.

Šū-Bēlim is also a partner of Adad-ṣulūlī. He sells a consignment of copper belonging to Adad-ṣulūlī for silver; the donkeys which were used to transport the metal were sold for copper and out of that amount he took what was due to him (kt a/k 259).[317]

Balṭua seized in Šalatuar 5 1/2 talents of refined copper from Adad-ṣulūlī's transporter (*kaṣṣārum*) Šamaš-ennam. He seems to have thought that he was entitled to it, but Adad-ṣulūlī turned to the *kārum* of Wahšušana for help (kt a/k 489, followed by kt a/k 486 // 509).[318]

Iddi(n)-Suen son of Alāhum assumes the obligation to buy 10 talents of refined copper from Adad-ṣulūlī in Purušhattum for the high price of 12 minas of silver (kt a/k 252 // a/k 422/b; case is kt a/k 432/a).[319] Adad-ṣulūlī occurs as his representative in TC 3, 272, where he receives silver from Mannum-bālum-Aššur in Turhumit.

A Puzur-Ištar collected for a certain *Iddi(n)-Aššur* 4 talents of poor copper from Adad-ṣulūlī in kt a/k 625.[320]

Aššur-ṭāb son of Ennam-Anum received merchandise on profitable conditions from Adad-ṣulūlī to lessen his financial difficulties. He received 20 minas of tin and had to pay after two months 3 talents 20 minas of broken fine copper to Adad-ṣulūlī in Turhumit (AKT 1, 66).

---

317 Kt a/k 259.1-13: ^{1} *a-na* ^{d}IM-ANDUL ^{2} *qí-bi-ma um-ma* ^{3} *Šu-Be-lúm-ma* URUDU-*a-kà* ^{4} 1 *ma-na* 12 1/2 GÍN.TA ^{5} *a-na* KB *ta-dí-in* ^{6} ANŠE.HI.A *a-na* ^{7} 62 *ma-na* URUDU *ta-dí-in* ^{8} *a-ma-lá té-er-tí-kà* ^{9} URUDU-*i al-qé* ^{10} 2 *ma-na* URUDU DIRI ^{11} *a-dí i-zi-zu-<ni>* ^{12} *a-na* ŠE-*im ú tí-ib-nim* ^{13} *áš-qúl* (...), 'Speak to Adad-ṣulūlī, thus (says) Šū-Bēlim: your copper has been sold at a rate of 72 1/2 shekels of copper per shekel of silver, the donkeys (sold) for 62 minas of copper. In accordance with your instructions I took my (share of the) copper. I paid 2 minas of excellent copper for barley and straw as long as (the donkeys) were standing.'

318 Kt a/k 489/b.1-10: ^{1} *kà-ru-um Wa-ah-šu-ša-na* ^{2} *ṣa-he-er* GAL ^{3} *dí-nam i-dí-ma* ^{4} ITU.KAM *té-i-na-tim* ^{5} *li-mu-um Šu-Sú-in* ^{6} DUMU *Ba-bi-lim šu-ma* ^{7} URUDU *ša* ^{d}IM-ANDUL ^{8} *Ba-al-ṭù-a* ^{9} *i-na Ša-lá-tù-a*r ^{10} *iṣ-ba-at*, 'The plenary session of the *kārum* of Wahšušana passed this verdict: date: month X, year-eponym Šū-Suen son of Babulum; if Balṭua has taken copper belonging to Adad-ṣulūlī in Šalatuar (Adad-ṣulūlī will present his terms and they will produce their witnesses).'

Kt a/k 486/b.1-17: ^{1} ^{d}IM-*ṣú-lu-li* ^{2} *ù Ba-al-ṭù-a i-na* ^{3} *mì-ig-ra-tí-šu-nu* ^{4} *iṣ-bu-tù-ni-a-tí-ma* ^{5} *um-ma* ^{d}IM-ANDUL-*ma* ^{6} *a-na Ba-al-ṭù-a-ma* ^{7} 5 1/2 GÚ URUDU *ma-sí-am* ^{8} *i-na Ša-lá-tù-ar* ^{9} *iš-tí* ^{d}UTU-*e-nam* ^{10} *kà-ṣa-ri-a* ^{11} *ta-ak-lú-ú* ^{12} *šu-ma* URUDU *a-na* ^{13} URUDU-*i-a i-tù-a*r ^{14} 1/2 *ma-na*.TA ù ^{15} 1 GÍN.TA KB *a-ma-lá* ^{16} *dí-in kà-ri-im* ^{17} *ta-ša-qá-lam*, 'In mutual agreement Adad-ṣulūlī and Balṭua seized us, and Adad-ṣulūlī said to Balṭua: "You withheld in Šalatuar 5 1/2 talents of refined copper from my transporter Šamaš-ennam. If it is proved that this copper is mine, you will pay me (silver for my copper) at a rate of 30 shekels of copper per shekel of silver, in accordance with the *kārum* verdict."'

319 Kt a/k 252.1-11: ^{1} 10 GÚ URUDU *ma-sí-am* ^{2} ^{d}IM-*ṣú-lu-li* ^{3} *i-na Pu-ru-uš-ha-/tim* ^{4} *a-na I-dí-Sú-in* ^{5} *i-da-an-ma* ^{6} *i-na re-eš*$_{15}$ ^{7} URUDU-*i-šu* 12 *ma-na* ^{8} KB *ṣa-ru-pá-am* ^{9} *I-dí-Sú-in* ^{10} *a-na* ^{d}IM-*ṣú-lu-li* ^{11} *i-ša-qal*, 'Adad-ṣulūlī will give 10 talents of refined copper to Iddi(n)-Suen in Purušhattum, and at the very moment (he accepts) his copper, Iddi(n)-Suen will pay Adad-ṣulūlī 12 minas of refined silver.'

320 Kt a/k 625.1-7: ^{1} *a-na I-dí-A-šur qí-bi*$_4$*-ma* ^{2} [*um*]-*ma Puzur*$_4$*-Ištar-ma* ^{3} [*ša*]-*zu-úz-ta-kà* ^{4} ⸢*a*⸣-*zi-iz-ma* 4 GÚ URUDU *lá-mu-num* ^{5} *ša* ^{d}IM-*ṣú*-ANDUL *i*$^{!}$*-hi-ib-lá-ku-ni* ^{6} *al-qé* URUDU *a-ṣí-tí-kà* ^{7} *ša-qú-ul*, 'Speak to Iddi(n)-Aššur, thus Puzur-Ištar: I acted as your representative and collected the 4 talents of poor copper that Adad-ṣulūlī owed you; the copper has been paid for various expenses of yours.'

4.2.2 *Procurement and sale of copper*

There are several contracts for the delivery of copper among the kt a/k texts from Adad-ṣulūlī's archive. They were written when Adad-ṣulūlī gave merchandise to someone in order to have it sold for copper. It is possible that some of this merchandise belonged not to Adad-ṣulūlī himself, but to another merchant who had entrusted it to him for the same purpose, in which case Adad-ṣulūlī used a third person to make the desired exchange. The texts are:

kt a/k 416: Ikuppī-Aššur son of Puzur-Ištar has to pay 43 minas of refined copper within 15 weeks (month XI, year of Enna-Suen son of Šū-Aššur);
kt a/k 436: Iddi(n)-Ištar has to pay 6 talents of fine copper within 3 months (month X, year of Innāya);
kt a/k 461: Imlikāya (as guarantor), Kaza, and Nuhšātum have to pay 53 minas of copper from Ninašša within 1 year (month V, year Aššur-imittī);[321]
kt a/k 621: Adāya of the firm (house) of Idī-A[ ] has to pay 30 minas of fine copper within 100 weeks (month X, year of Enna-Suen);
AKT 1, 41: Ilī-mīšar son of Kulumum has to pay 2 talents 13 1/2 minas of *šikkum* copper (month I, year following on that of Innāya);
AKT 1, 42: Ṣill-Adad son of Naplis has to pay 40 minas of refined copper in Wahšušana within 3 months (month I, year following on that of Aššur-nādā);
AKT 1, 50: Irišum son of Amur-Šamaš has to pay 1 1/2 talents of fine copper within one year (month VI, year not given);
AKT 1, 66: Aššur-ṭāb son of Ennam-Anum receives 20 minas of tin and has to pay 3 talents 20 minas of broken fine copper in Turhumit within 2 months (no date is given).[322]

321 Kt a/k 461.1-21: ^{1} 53 *ma-na* URUDU *ša* ^{2} *Ni*$^{!}$*-na-ša-a-a* ^{3} *i-ṣé-er Im-lik-a-a* ^{4} *Ga-za-a ù Nu-ùh-ša-tim* ^{5} ^{d}IM-*ṣú-lu-li* ^{6} *i-šu iš-tù* ITU.KAM ^{7} *áb-ša-ra-ni* ^{8} *li-mu-um* ^{9} *A-šur-i-mì-tí* ^{10} *a-na ša-na-at* ^{11} *i-ša-qal* URUDU *i-qá-qá-ad* ^{12} *šál-mì-šu-*<*nu*> . *ra-ki-is šu-ma lá ìš-qú-ul* ^{13} *ki-ma a-wa-at kà-ri-im* ^{14} *ṣí-ib-tám ú-ṣa-áb* ^{15} URUDU *a-num iš-tù* ^{16} 30 *ma-na* URUDU *ša-aq-/lu* ^{17} *a-šu-mì Ga-za-a* ^{18} *ù* 53 *ma-na* URUDU ^{19} *a-šu-mì Ga-za-ma* ^{20} *Im-lik-a-a* ^{21} *qá-ta-tum*, 'Adad-ṣulūlī has 53 minas of copper from Ninašša as a claim on Imlikāya, Gaza, and Nuhšātum. (Gaza) will pay within one year from month V, year eponym Aššur-imittī. The obligation to pay back the copper is upon whoever among them is solvent. If he does not pay, he will add interest according to the rate decreed by the *kārum*. This copper is (the amount) after 30 minas had been paid; on behalf of Gaza; and the 53 minas of copper are also on behalf of Gaza; Imlikāya is guarantor.'

322 AKT 1, 66.1-16: ^{1} *i-nu-mì A-šur-*DÙG *iš-tù* ^{2} *a-lim*^{ki} *e-li-a-ni* ^{3} 1 GÚ AN.NA *ku-nu-ki* ^{4} *ša a-lim*^{ki} ^{d}IM-*ṣú-lu-li* ^{5} IGI *A-šur-*DÙG *i-ša-kà-ma* ^{6} *a-šál-ší-šu i-ma-hu-ṣú-ma* ^{7} 20 *ma-na* AN.NA *A-šur-*DÙG ^{8} *i-lá-qé-ma* 3 GÚ 20 *ma-na* ^{9} URUDU SIG$_5$ *ša-bu-ra-am* ^{10} *iš-tù u$_4$-mì-im* ^{11} *ša* AN.NA *i-lá-qé-ú* ^{12} *a-na* ITU.2.KAM ^{13} *i-na Tur$_4$-hu-mì-it* ^{14} *A-šur-*DÙG *a-na* ^{15} ^{d}IM-*ṣú-lu-li* ^{16} *i-ša-qal*, 'When Aššur-ṭāb turns up from a trip to the City, Adad-ṣulūlī will put before him 1 talent of tin, still bearing the City's seals, and they will divide it into three parts and Aššur-ṭāb will take 20 minas of tin; after 2 months from the day when he received the tin, he will pay to Adad-ṣulūlī in Turhumit 3 talents 20 minas of broken fine copper.'

Not every text seems to be a regular delivery contract where the seller wants to receive his copper within a reasonable period of time. In cases where payment is due within 1 year or 100 weeks (2 years?), the debt is rather a soft loan. The purpose of AKT 1, 66, probably is to help the debtor, using the standard exchange ratios for tin (6:1) and copper (60:1) to silver, whereby the amounts of tin and copper both have a value of 3 1/3 minas of silver. To make a profit, the debtor had to sell the tin for more than 10:1 in copper, which is possible as the exchange ratios between tin and copper show in Appendix 1. Tin also occurs in kt a/k 849,[323] where Adad-ṣulūlī receives from a *limmum*-representative of the *kārum* of Wahšušana a package sealed by Man-māhir which contained 1 talent of tin. It was seized by the *kārum* as a means of forcing Man-māhir to make a balance payment (*nipiltum*) in copper to Adad-ṣulūlī. Adad-ṣulūlī promises to return the tin to the *kārum* as soon as Man-māhir pays him 17 talents 47 minas of copper with its interest; in the meantime he keeps the tin as security. The quality of the copper is not mentioned, and the security has not necessarily exactly the same value. But the loss of the tin will persuade Man-māhir to pay the copper, even if it is less. The value of the tin would equal that of the copper if one assumes a rate of 120:1 in silver and the total amount including interest would add up to some 20 talents; the talent of tin equals the value of the copper converted into silver, namely 10 minas. Adad-ṣulūlī, however, cannot sell the tin immediately, he has to wait till Man-māhir pays him the copper, which would increase with interest in time.

In kt a/k 512, Adad-ṣulūlī sold textiles for Ladia for fine copper.[324] He apparently sold textiles for copper for Hanāya as well, who used an exchange ratio of 180:1 when taking silver as payment instead of the copper he was to receive from Adad-ṣulūlī (kt a/k 265).[325] If Adad-ṣulūlī had been able to exchange the copper at a better rate, e.g. 90:1, he would

---

323 Kt a/k 849.1-22: [1] ⸢1⸣ GÚ AN.NA *šu-uq-lam*$_5$ [2] *ku-nu-ki ša Ma-ma-hi-ir* [3] *ša kà-ru-um Wa-ah-šu-ša-na* [4] *a-na ni-pí-il*$_5$*-tí Ma-ma-hi-ir* [5] *iṣ-bu-tù A-šùr-ták-lá-ku* [6] *li-mu-um ki-ma kà-ri-im* [7] AN.NA *a-na* ^{d}IM-*ṣú-lu-li* [8] *ip-qí-id um-ma A-šùr-ták-lá-ku-/ma* [9] *a-na* ^{d}IM-ANDUL-*ma* [10] 17 GÚ 47 *ma-na* [11] URUDU *ú ṣí-ba-tí-šu* [12] *Ma-ma-hi-ir i-da-na-ku-ma* [13] *ù* AN.NA *tù-šar-ni-a-[tí]* [14] *ku-nu-ki ú-lá i-pá-ṭù-ru* [15] *um-ma* ^{d}IM-ANDUL-*ma* AN.NA [16] *a-na ša-pár-tim a-na-dí-ma* [17] URUDU *ú ṣí-ba-sú a-lá-qé* [18] *um-ma A-šur-ták-lá-ku-ma* [19] *ú šu-ma . a-na ša-pár-tim* [20] *ta-na-dí . ku-nu-ki ú-lá i-pá-/ṭù-ru* [21] *um-ma* ^{d}IM-ANDUL-*ma* [22] *ú-lá i-pá-ṭù-ru*, '1 talent of tin, a *šuqlum*-load, sealed by Man-māhir, which the *kārum* of Wahšušana had seized for the balance payment due from Man-māhir, Aššur-taklāku, the *limmum*-functionary representing the *kārum*, entrusted the tin to Adad-ṣulūlī. Aššur-taklāku said to Adad-ṣulūlī: "(When) Man-māhir gives you 17 talents 47 minas of copper and (including) the interest on it, you will release the tin to us. They should not remove the seal imprints." Adad-ṣulūlī answered: "I will deposit the tin as a pledge (with somebody else who will give me the copper, so that M. has to pay that man the copper to get his tin back - J.G.D.) and (so) receive the copper and its interest." Aššur-taklāku said: "Even if you deposit it as a pledge, they should not remove the seal imprints." And Adad-ṣulūlī answered: "They will not remove them."'

324 Kt a/k 512.3-5: [3] *um-ma Lá-dí-a-ma . a-na* [4] ^{d}IM-ANDUL-*ma . ší-im* TÚG.HI-*tí-a* [5] URUDU SIG$_5$. *dí-nam*, 'Ladia said to Adad-ṣulūlī: "Give me fine copper as proceeds of my textiles (...)."'

325 Kt a/k 265.8-13: [8] ... 33 TÚG *ku-t[a-ni] ù* 1 TÚG SIG$_5$ [9] *i-Ni-na-ša-a [ta]l-qé a-na* [10] ⸢33⸣ TÚG *ša* 30 *ma-na*.TA *tal-qé-ú* [11] 10 GÍ[N.T]A KB *al-qé a-na* 1 TÚG SIG$_5$ [12] *ša* 1 [G]Ú 30 *ma-na* URUDU *tal-qé-ú* [13] 1/2 *ma-na* KB *al-qé*, (letter by Hanāya to Adad-ṣulūlī:) 'You received in Ninašša 33 *kutānum*-textiles and 1 textile of good quality. For the 33 textiles that you took (= agreed to pay me) at 30 minas (of copper) a piece, I took 10 shekels of silver a piece; for the 1 textile of good quality that you took at 90 minas of copper, I took 30 shekels of silver.'

have been able to make a profit. Textiles are also sold for copper in kt a/k 401 (see above); in another letter by Adāya, it is reported that one *kutānum*-textile was sold for 11 minas of fine copper (kt a/k 514, see above). It is impossible at the moment to determine what merchandise Adad-ṣulūlī and Aššur-damiq had been selling for which they had to pay 7 talents 30 2/3 minas of refined copper to Šumma-libb-Aššur.

Small amounts of copper were important to Adad-ṣulūlī as well. In AKT 1, 43, Ilī-Anum, one of his transporters, had borrowed 3 1/2 minas of fine copper to Sukkallum son of Iddi(n)-Suen. Adad-ṣulūlī takes action to get the copper and the interest on it back from Sukkallum in kt a/k 264.[326]

A small file is concerned with the attempt by the sons of the by then deceased Šū-Kūbim to recover an unspecified number of gold statuettes (*ilum*). After intervention by the *kārum* of Wahšušana, Adad-ṣulūlī lends 130 minas of *šikkum* copper to them and receives 2 slaves, Kēnum-waqar and Hutka, as a pledge. With the copper they satisfy an Anatolian creditor of their father, who then releases the gold *ilū* which he had received as a pledge. Subsequently, the sons of Šū-Kūbim, Ennum-Aššur, Itūr-Adad, and Aššur-ṭāb, 'pay' the gold objects for their fathers debt to the god Adad (kt a/k 447).[327] Ultimately, Adad-ṣulūlī does not receive his copper, but may sell the slaves given to him as a pledge (kt a/k 343).

In kt a/k 438, copper is used to buy donkeys or a slave-girl.[328] In kt a/k 431.8-9, Hadiāni received 1 mina of fine copper as hire. In kt a/k 259, 2 minas of copper are paid for fodder (barley and straw; see above, n.317).

---

326 Kt a/k 264.1-23: [1] [d]IM-*ṣú-lu-li* [2] *a-na Sú-kà-lim* DUMU *I-dí-Sú-in* [3] DUMU *A-ú-ur-na-hu-/uš* [4] *iṣ-ba-at-ni-a-tí-ma* [5] *um-ma* [d]IM-*ṣú-lu-li-ma* [6] *a-na Sú-kà-lim-ma* [7] 3 1/2 *ma-na* URUDU *ma-sí-a-am* [8] *a-na I-la-nim* [9] *kà-ṣa-ri-a ha-bu-lá-tí* [10] *ṭup-pá-kà ha-ar-ma-am* [11] *ú-kà-al* [12] *i-na ṭup-pì-kà* [13] *Li-la* DUMU *Dan-A-šur* [14] *ù Šu-Be-lim* DUMU *I-sà-li-a* [15] *na-lu i-na ṭup-pì* [16] *ha-ar-mì-im šu-ma* [17] *Li-lá ku-nu-ki-šu* [18] *ú-ta-dí um-ma šu-ut-ma* [19] *ku-nu-ku-a ša ṭup-pì-im* [20] *a-nim* URUDU . *ú ṣí-ba-sú* [21] *ta-ša-qá-lam lá ta-na-/pu-uš* [22] *um-ma Sú-kà-lu-ma* [23] *lá-áš-qú-lá-ku-um*, 'Adad-ṣulūlī seized us against Sukkallum son of Iddi(n)-Suen son of Awur-nahuš and said to Sukkallum: "You owe 3 1/2 minas of refined copper to my transporter Ilī-Anum; I possess your certified tablet. On your tablet Lila son of Dān-Aššur and Šū-Bēlim son of Isālia are mentioned ("lay"). If Lila identifies his seals on the certified tablet and says: "(They are) my seals", will you then pay the copper and its interest (mentioned) in this tablet, without making objections?" Sukkallum replied: "I will pay (it) to you."'

327 Kt a/k 447/b.1-15: [1] *kà-ru-um Wa-ah-šu-ša-na* [2] *ṣa-he-er* GAL *dí-nam* [3] *i-dí-ma* 1 *me-at* 30 *ma-na* [4] URUDU [d]IM-*ṣú-lu-li* [5] *iš-qúl-ma i-le-e* [6] *ša Šu-Ku-bi-im* É *nu-a-im* [7] *ú-šé-ṣí-ma a-na hu-bu-ul* [8] *Šu-Ku-bi-im a-na* [d]IM [9] *i-ší-qí-il$_5$* [10] 2 *ší-na wa-ar-du-ú* [11] *Ke-num-wa-qá-ar* [12] *ù Hu-ut-kà iš-tí* [13] 1 *me-at* 30 *ma-na* URUDU [14] *iš-tí* [d]IM-*ṣú-lu-li* [15] *uk-ta-lu-ú*, 'The plenary session of the *kārum* of Wahšušana passed verdict: Adad-ṣulūlī had paid 130 minas of [case adds: *šikkum*] copper and so redeemed the [case adds: gold] statuettes of Šū-Kūbim from the Anatolian's house (which) were paid (for Šū-Kūbim's debt) to Adad. Two slaves, Kēnum-waqar and Hutka, are being kept (as a pledge) by Adad-ṣulūlī for the 130 minas of copper.'

328 Kt a/k 438.6-22: [6] 31 *ma-na* URUDU SIG$_5$ [7] *ša Šu-A-nim ú Ha-dí-a-ni* [8] *ub-lá*[!]*-ku-ni um-ma a-na-ku-/ma* [9] ANŠE.HI.A [10] *ša-ma-am* ANŠE [11] *lá tù-uš-té-ri-am* [12] *am-tám ú-lá e-ma-ra-/am* [13] *ša-ma-ma iš-tí* [14] *Ì-lí-a-lim ú I-ku-nim* [15] *šé-bi$_4$-lá-nim* 13 *ma-na* [16] URUDU *ma-sí-am* [17] *ú* 10 *ma-na* URUDU *ší-ku-um* [18] *I-ku-nu-um na-áš-a-ku-/nu-tí* [19] *sú-ub-ra-am* [20] *ú-lá e-ma-ra-am* [21] *ša-ma-nim-ma* [22] *šé-bi$_4$-lá-nim*, (Letter to Adad-ṣulūlī, Anah-Aššur, Ilī-Anum, and Ikūnum by Pilah-Ištar:) 'About the 31 minas of fine copper that Šū-Anim and Hādiāni brought to you, I had told you to buy for me donkeys (with it), but you did not bring me a donkey! Buy a slave-girl or a donkey and send (what you buy)

Copper is sold for silver in kt a/k 259, where Šū-Bēlim reports that he sold Adad-ṣulūlī's copper at 72.5:1, but the donkeys which carried it were sold for copper. A punitive ratio of 30:1 based on a verdict by the *kārum* is used in kt a/k 486 (see above, n.318). Text kt a/k 252 is a contract stipulating that Adad-ṣulūlī will give 10 talents of refined copper to Iddi(n)-Suen son of Alāhum, who appears to be a retailer, in Purušhattum in exchange for 12 minas of refined silver (see above, p.102). This is in fact a sale of copper at the high rate of 50:1.

4.2.3 *Geography of the copper trade*

The geographical details in this archive are scarce, especially when compared with those from the following archives, and the texts available do not indicate a clear pattern of the major centres where copper was obtained or sold. It is clear that the focus of Adad-ṣulūlī's trading activities was not in Kanish, but west and north-west of it, in Wahšušana, Purušhattum (where he owned a house), and Turhumit.

In *Turhumit* Adad-ṣulūlī could sell a consignment of textiles for Puzur-Aššur, and bring the proceeds in copper from there to Purušhattum or Wahšušana (BIN 4, 36). Erišum in Turhumit owed him 15 talents 52 minas of poor copper as balance payments (kt a/k 300). Aššur-ṭāb had to pay 3 talents 20 minas of broken fine copper to him in the same city (AKT 1, 66). Adad-ṣulūlī and Aššur-damiq each had to pay 1 talent 15 minas of fine copper to Enna-Suen because of the verdict of the *kārum* of Turhumit, possibly resulting from the sale of merchandise in or near that city (kt a/k 497/b).

It is known that Adad-ṣulūlī owned a house in *Purušhattum*.[329] In this city, he receives from Iddi(n)-Suen payment in silver for the 'delivery' of refined copper (kt a/k 252). Adad-ṣulūlī and Aššur-malik receive a payment in copper destined for Puzur-Aššur; because the (market for) copper was bad in Purušhattum, they preferred to wait a while before transporting it to this city (TC 2, 18 = **13**).

The *kārum* of *Wahšušana* was involved a remarkable number of times in cases concerning Adad-ṣulūlī and his copper (CCT 1, 48; kt a/k 343; 447; 481 // 512; 486 // 509; 489). The only other verdict about copper was given by the *kārum* of Turhumit (kt a/k 497). Ṣill-Adad had to pay 40 minas of refined copper to Adad-ṣulūlī in Wahšušana without deductions (AKT 1, 42).[330] It is possible that the palace owing copper to Adad-ṣulūlī and Erišum in kt a/k 481 is that of Wahšušana. The *kārum* of Wahšušana gave tin as a pledge to Adad-ṣulūlī to force Man-māhir to pay him 17 talents 47 minas of copper plus interest (kt a/k 849).

---

to me with Ilī-Anum and Ikūnum. Ikūnum is bringing you (pl.) 13 minas of refined copper and 10 minas of *šikkum* copper; buy a *subrum*-donkey or a donkey and send (what you buy) to me!'

329 Kt a/k 403.1-5: [1] *i-na Pu-ru-uš-ha-tim* [2] *a-na* É [d]IM-*ṣú-lu-li* [3] *né-ru-ub-ma um-ma* [4] *Bu-za-zu-ma a-na* [d]IM-*ṣú-lu-li-ma*, 'In Purušhattum we entered Adad-ṣulūlī's house, and Buzāzu said to him.'

330 AKT 1, 42.1-4, 13-15: [1] 40 *ma-na* URUDU *ma-sí-am* [2] *i-ṣé-er Ṣí-lá*-[d]IM [3] [d]IM-*ṣú-lu-li* [4] *i-šu* (...) [13] URUDU *šál-ma-am i-na* [14] *Wa-ah-šu-ša-na* [15] *i-ša-qal*, 'Adad-ṣulūlī has 40 minas of refined copper as a claim on Ṣill-Adad (...) He will pay the full amount of copper (without deductions) in Wahšušana.'

In *Ulama*, agents of Adad-ṣulūlī pay his debt of 24 talents 10 minas of fine copper to Kudādum (VS 26, 1).[331]

Balṭua had taken 5 1/2 talents of refined copper from one of Adad-ṣulūlī's transporters in *Šalatuar* (kt a/k 489 and kt a/k 486 // 509).

Adad-ṣulūlī owns 53 minas of the original 83 minas of copper from *Ninašša*, which Kaza apparently owed to him (kt a/k 461).

Hanāya asks Adad-ṣulūlī to send him 2 minas of native copper and 10 minas of antimony in *Kanish* (kt a/k 265).[332]

4.2.4 *Conclusion*

Adad-ṣulūlī's archive documents the affairs of a merchant of medium rank. He was working for several other traders for whom he sold merchandise for copper. A number of texts illustrate how he himself acquired copper by contracting merchants as retailers. Compared with the rich material contained in the archives to be discussed in the following sections, the details on sale and geography in this archive are rather poor.

4.3 *Alāhum son of Sukuhum*

4.3.1 *Participants*

During excavations conducted in level II of the *kārum* area in 1950, a large rectangular house was discovered in squares N-O-P/20.[333] With a length of 16 metres and a width of 5.50 metres, its surface area was about 119 m^2. The ground floor consisted of four rooms, with a corridor along the north wall of the house. In the archive room (2), accessible from room 3, some 600 tablets, bullae and labels were found lying at the base of the north and south walls. These were labelled 'kt c/k'. According to the excavators, these tablets were originally arranged on shelves. In the centre of the room pitchers, jars, and jugs were found lying on the floor. On the basis of the contents of the documents from room 2, the owner of this house has been identified as *Alāhum son of Sukuhum*, who lived there for some time together with other members of his family.

The analysis presented below of Alāhum's archive and the role played by copper in his trade, is based on the study of about one-sixth of the epigraphical material, which contains a dossier relating to copper. The majority of the tablets found prior to the regular excavations by the Turkish Historical Society have been published, and they contain many attestations of the quite common name, Alāhum. In the majority of these, no clear relationship with the

---

331 VS 26, 1.1-7: [1] *a-na A-šur-na-da Šu-ma-li-/bi-A-šùr* [2] *ù A-lu-lá-a* [3] *qí-bi-ma um-ma* [d]IM-ANDUL-*ma* [4] 24 GÚ 10 *ma-na* URUDU SIG5 [5] *a-na Ku-da-dim a-hi-bi-il5-ma* [6] URUDU *i-na Ú-la-ma ša ki-ma* [7] *i-a-tí . Ku-da-dam ú-ša-bi-ú-/ma*, 'Speak to Aššur-nādā, Šumma-libb-Aššur, and Alulāya, thus says Adad-ṣulūlī: I had become indebted to Kudādum for 24 talents 10 minas of fine copper, but my representatives fully satisfied Kudādum with the copper in Ulama.'

332 Kt a/k 265.30-32: [30] ⌜2⌝? *ma-na* URUDU *ša ša-du-/i-šu* [31] *ù* 10 *ma-na lu-lá-am* [32] *a-na Kà-ni-iš šé-bi-lam*, 'Send 2 minas of native copper and 10 minas of antimony to Kanish.'

333 Özgüç 1959, pp.90-91; plan of the house on p.30, Fig. 38.

Alāhum known from the kt c/k archive can be discerned, and they are considered mere homonyms. Only two of these older texts belong without doubt to Alāhum's archive.[334] Some other texts relating to Alāhum are found among the tablets apparently discovered in Aššur-nādā's house, which in general deal with the affairs of his father Aššur-iddi(n). The house next to that of Alāhum, grid square O-P/20-21, belonged to Buzutāya according to the contents of the tablets found in it.[335]

How many years Alāhum's archive covered is unknown. Texts from the period in which Alāhum himself was living in Kanish may represent the oldest. More recent texts were written after he had, perhaps permanently, returned to Assur, leaving his sons and other representatives behind to take care of his affairs. Judging by the different year eponyms (*limmum*) occurring in the texts, the archive contained documents going back at least 24 years.[336]

Alāhum son of Sukuhum was the husband of Šīmat-Ištar and the father of Aššur-bēl-šadue, Aššur-mālik, and Šū-Aššur. The house in Kanish from which he directed his trade, was where he lived with his two eldest sons.[337] His enterprise was to a large extent a family business, in which father, mother, and sons participated. Apart from the trade directed by Alāhum, which constituted the bulk of the commercial references, each member of the enterprise conducted its private trade. In Anatolia, Alāhum cooperated with Aššur-imittī son of Atāya, Šū-Kūbim and Buzutāya, Aššur-ṭāb and Tahašaili; it is uncertain whether the last two persons were relatives.

The house in Assur was inhabited by Šīmat-Ištar together with some servants, but no mention is made of children still living there. One of Šīmat-Ištar's commercial activities was weaving wool into textiles for export, some or all of which she sent to her husband in Kanish.[338] Alāhum and Šīmat-Ištar invested silver in real estate by purchasing a house. Aššur-iddi(n) son of Šuli and Ayyum-kēn son of Atāya (brother of Aššur-imittī?) acted as

---

334 ICK 2, 102 is connected with kt c/k 202 // 282 and Pa 12 is connected with kt c/k 680. Read in Pa 12 line 1: 12! 1/2 *ma-na*, and line 6: 6! *ma-na*.

335 Özgüç 1959, p.xxii.

336 The year eponyms (*limmum*) are: *ša qāti* Adad-bāni = Aššur-iddi(n) (kt c/k 440.55); Alāhum (kt c/k 839.95); Alāhum son of Inah-ilī (kt c/k 839.21); Al-ṭāb (kt c/k 839.39); Amur-Aššur (kt c/k 839.69); Amur-Aššur son of Karria (kt c/k 839.8); *ša qāti* Amur-Aššur (kt c/k 789.21); Aššur-damiq (kt c/k 512.7); Aššur-damiq son of Abarsisim (kt c/k 839.14); Aššur-iddi(n) (kt c/k 442.49); Aššur-mālik son of Alāhum (kt c/k 258.5); Aššur-nādā son of Puzur-Ana (kt c/k 440.39); Buzutāya (kt c/k 258.27); Elāli (kt c/k 839.103); Elāli son of Ikūnum (kt c/k 839.4); Elāli *sangum* (kt c/k 440.32); Iddi(n)-abum (kt c/k 442.45); Iddi(n)-abum son of Narbitum (kt c/k 440.47); Iddi(n)-ahum (kt c/k 269.19); Iddi(n)-Aššur (kt c/k 839.86); Ikūppī-Ištar (kt c/k 384.22); Ilī-dān (kt c/k 839.56); Ilšu-rabi (kt c/k 512.17); Innāya (succeeds Buzutāya) (kt c/k 258.24); Kubia (kt c/k 277.8); Kubia son of Karria (kt c/k 440.13); Maṣi-ilī (kt c/k 467.8); Puzur-Nirah (kt c/k 839.33); Sallia (kt c/k 453.28); Sukkallia (kt c/k 266.4); Šū-Hubur (kt c/k 839.52); Šū-Suen son of Šū-Bēlim (kt c/k 258.18); *ša qāti* Šū-Suen (kt c/k 258.34); Ṭāb-Aššur (kt c/k 839.91).

337 Textual evidence for Alāhum's house in Kanish is found in AKT 1, 73, where the *kārum* ruled that silver belonging to Aššur-iddi(n) which had disappeared from Alāhum's house should be fully compensated.

338 Kt c/k 266.8-13: [8] ... *ki-ma* TÚG.HI.A [9] *lá né-pu-šu iš-té-na-ma* [10] *A-šur-*^{d}UTU-*ši na-áš-a-kum* [11] *a-ni iš-tí wa-ar-ki-ú-tí-im* [12] *iš-té-en*$_6$ *ù ší-na* TÚG [13] *ú-šé-ba-lá-kum*, 'Because we did not produce textiles, Aššur-šamšī is bringing you only one. Now, with the next caravan I shall send a few textiles to you,' and kt c/k 272.3-4: [3] *a-ba-ar-ni-am A-lá-hu-um* [4] DUMU *Sú-kà-li-a na-áš-a-kum*, 'Alāhum the son of Sukkallia is bringing one *abarnium*-textile to you' (both letters sent to Alāhum by Šīmat-Ištar).

Alāhum's representatives in Assur. Alāhum did not always stay in Kanish, but returned, perhaps temporarily, to Assur to negotiate with his investors (*ummiānū*), possibly because of some financial difficulties he had experienced.[339] He was registered at the *kārum*-office in Kanish as a share (*qātum*) holder.[340]

Aššur-imittī son of Atāya was one of Alāhum's business associates (see for his family firm § 4.6). Disputes over financial questions, in which an attorney (*rābiṣum*) became involved, arose between Alāhum and Aššur-imittī and their sons. Several times Alāhum appears to have been in Assur while Aššur-imittī was in Kanish: when Aššur-imittī owed silver to the *limmum* eponym in Assur, and Alāhum had been appointed guarantor together with Aninum son of Šū-Hubur and Ayyum-kēn son of Atāya, the city-council made Alāhum pay the amount of silver (kt c/k 269). Aššur-imittī took care of a consignment from Kanish to Purušhattum (kt c/k 257) consisting of tin and textiles belonging to Alāhum.

Although his name does not occcur frequently in the kt c/k texts, *Aššur-iddi(n) son of Šuli*, living in Assur, appears as one of Alāhum's trusted colleagues. In a commercial dispute about nails (*samruātum*), he and Ayyum-kēn had borrowed silver for Alāhum at a moneylender's (*bēt tamkārim*), and Alāhum's company/firm (*ellutum*), by sentence of the elders, had to repay the silver (kt c/k 261; 273; 288).

Aššur-bēl-šadue is one of Alāhum's sons and older than his brother Aššur-mālik. In Kanish he lived in his father's house (kt c/k 561). In cooperation with Aššur-mālik in Kanish, Tahašaili (in Wahšušana), and Aššur-ṭāb in Purušhattum, he took care of his father's interests in Anatolia and travelled to Assur. The proceeds in silver from sales were sent to him in Kanish, where he prepared it for transport to Assur, primarily to Alāhum and to Šīmat-Ištar, using his own seal to identify the silver's owner.[341]

Aššur-mālik is another son of Alāhum and younger brother of Aššur-bēl-šadue. He frequently stayed in Kanish while his father was in Assur. He travelled in Anatolia and also to Assur. Like his brother, Aššur-mālik coordinated the distribution of merchandise arriving from Assur in Kanish together with other business partners, like Itūr-ilum, Ilī-malak, Aššur-nādā, Aššur-ṭāb, and Tahašaili, and he sent the proceeds in silver to Assur. In contrast to his brother Aššur-bēl-šadue, Aššur-mālik did not use his own seal (for instance in kt c/k 50). It is possible that this Aššur-mālik is the same person as the Aššur-mālik son of Alāhum who acted as week eponym (*hamuštum*).[342]

---

339 In kt c/k 259 he writes to Šū-Kūbim and Buzutāya: [3] *a-na-kam a-wa-tí-a iš-tí um-me-a-/ni-a* [4] *ag-ta-mar* (......) [20] *a-hu-a-tù-nu a-na-kam* KB [21] 1 GÚ *i-lá i-dí-im il*$_5$*-ta-qé-ú* [22] *i-na ṭup-pè-a ha-ru-mu-tim* [23] *a-na ma-ma-an ṣí-ib-tám* [24] *lá ta-pá-ṭá-ra* (......), 'I have just settled my affairs with my investors ..... [After asking to send his silver to Assur he continues:] My dear brothers, for no reason they have taken here silver to an amount of 1 talent! You shall not remit interest to anybody (who appears as a debtor) in my certified tablets.' The urgent need for money is clear in several letters containing orders to sell merchandise for cash and send the proceeds at once.

340 Kt c/k 452.19-21: [19] ... 3 TÚG.HI.A [20] *a-na qá-tí-kà a-na* [21] É *kà-ri-im a-dí*, 'I deposited 3 textiles for your share to the *kārum*-office'; kt c/k 454.23-26: [23] ... 5 TÚG.HI.A *qá-tí-kà* [24] *a-na* É *kà-ri-im a-dí-i* [25] 10 TÚG.HI.A *a-na qá-tí-kà* [26] *e-zi-ib* (...), (letter to Alāhum by Aššur-ṭāb:) 'I deposited 5 textiles as your share; 10 textiles I left behind (to deposit later) for your share.'

341 Kt c/k 50.8, 12; c/k 378.4; c/k '571'.28.

342 In the first month of the year following that of Ṣilulu, CCT 5, 22b.

Šū-Aššur son of Alāhum is absent from the known c/k texts. Apart from the possibility that he is mentioned in kt c/k texts not studied here, his absence can be explained from RA 59, 23 (C. Michel no. 246), where it is mentioned that a Šū-Aššur (son of Alāhum) had died. In this letter, written by Iddi(n)-Aššur, Adad-bānī, Aššur-imittī, Ikuppī-Aššur, Puzur-Ištar, and Imdilum, who apparently were his creditors, to Pūšu-kēn, Zupa, Aššur-ṣulūlī, Innāya, and Pilah-Ištar, the addressees are asked to enter Šū-Aššur's house and entrust all items of value which he left behind (after his death) to his brother Aššur-mālik. The *naruqqum*-contract Šū-Aššur had with the senders of the letter was transferred to Aššur-mālik. In one case, Šū-Aššur was sent by his father to Kanish.[343] Alāhum and Aššur-iddi(n) agreed that Šū-Aššur, one of Alāhum's sons, would send from Assur 2 talents of Alāhum's tin to Aššur-nādā in Kanish, where the latter would choose one packet with 1 talent, sell it, and pay him 10 minas of silver when asked for it.

Aššur-ṭāb was active in various cities. He received silver and gold in Kanish, and in Purušhattum he received consignments of copper sent to him by Tahašaili, apparently in order to sell these. He was involved in the transport of textiles, at least from Zalpa in Syria to Kanish and further westward. In the kt c/k texts there are as yet no indications that he was a member of the family. However, an Aššur-ṭāb son of Alāhum, who was active in Purušhattum, occurs in other texts.[344] Aššur-ṭāb was called 'my lord' by Tahašaili (kt c/k 445), and therefore was more important and possibly also older than Aššur-bēl-šadue and Aššur-mālik (see below). The dossier about the sale of textiles and tin for copper by Tahašaili sheds some light on Aššur-ṭāb's activities: after bringing a consignment of Alāhum's textiles and tin from Kanish or somewhere further away to Tahašaili, presumably in Wahšušana, he returned to Kanish. Tahašaili sold the goods for copper and bought some gold with it, which he sent to Aššur-ṭāb. The latter travelled again to Tahašaili, collected his tin and 2 talents of copper, and went on to Purušhattum, where he received the copper in a number of consignments. Shortly afterwards, his untimely death[345] caused confusion in Kanish since he still owed silver to Alāhum's firm. In kt c/k 525, a letter to Itūr-ilum and Adada from Irmāya (= Ir(i)m-Aššur) and Šū-Ištar, the senders reported on the action they took to clear accounts with Tahašaili concerning the merchandise of Alāhum (son of Sukuhum): 'We took three witnesses against him and said to Tahaša(i)li: "Was it in accordance with Aššur-ṭāb's instructions that you have not settled accounts with us for the merchandise of Alāhum?", and he replied: "I have settled nothing with you." We took all his written instructions concerning any silver or gold which he has been sending to Aššur-ṭāb or copper that he has been transporting to Purušhattum, and someone is bringing (them) to you.'[346] This collecting of tablets written by Tahašaili concerning copper could well account

---

343 Documented by texts C 17; TC 1, 8; Pa 13; TC 3, 92.

344 CCT 2, 6.22; CCT 2, 8.28; KUG 49; TC 2, 13.

345 Kt c/k 561.1-6: [1] *a-na A-lá-hi-im qí-bi-ma* [2] *um-ma A-šur-be-el-ša-du-e-ma* [3] *a-na-kam A-šur-*UTU-*ši mu-ut A-šur-*DÙG [4] *iš-me-ma* . KB-*áp-kà a-na* [5] KB-*pì-šu iš-ta-kán um-ma* [6] *šu-ut-ma* KB *i-a-um-ma*, 'Speak to Alāhum, thus (says) Aššur-bēl-šadue: here, Aššur-šamšī has heard of Aššur-ṭāb's death, and he has taken possession of your silver, saying: "The silver is mine!"'

346 Kt c/k 525.10-24: [10] 3 *ší-be-e ni-iṣ-ba-sú-um-ma* [11] *um-ma né-nu-ma* [12] *a-na Ta-ha-ša-li-ma* [13] *a-na ma-lá té-er-tí* [14] *A-šur-*DÙG *a-na lu-qú-tim* [15] *ša A-lá-hi-im iš-tí-ni* [16] *ú-lá ta-az-ku* [17] *um-ma šu-ut-ma mì-ma iš-tí-ku-[nu]* [18] *ú-lá az-ku* [19] *ma-la lu* KB *lu* KÙ.GI [20] *ša a-na A-šur-*DÙG *uš-té-*

for the discovery of this dossier among the texts excavated in Kanish. Šū-Ištar son of Iliš-tikal and Ir(i)m-Aššur acted as representatives of Aššur-ṭāb by order of Aššur-mālik and Itūr-ilum. Šū-Ištar was a partner of Aššur-ṭāb; they occur as creditors on the same tablet (kt c/k 440), and Šū-Ištar collected money from Aššur-ṭāb's retailers.

Tahašaili can be regarded as one of the Anatolian merchants who cooperated with the Assyrians, but he could also be the child from the marriage between an Assyrian and a local woman. He managed the firm's business in the city he lived in, presumably Wahšušana, by selling textiles, tin, and copper belonging to members of the enterprise for Aššur-bēl-šadue and Aššur-mālik, and by sending the proceeds in silver and gold to Kanish. The use of his own seal is mentioned several times.[347] An interesting dossier covering a small period of time illustrates how he sold merchandise for copper which was transported in several consignments to Purušhattum, where Aššur-ṭāb saw to the sale of part of this copper. Tahašaili's place in the hierarchy of the enterprise was lower than that of Alāhum and Aššur-ṭāb, both of whom he calls 'my lord' (kt c/k 445.20), higher than that of the sons of Alāhum (kt c/k 100.1-3), and equal to that of Alāhum son of Mannua, whom he called 'my brother' (kt c/k 167.9). His private trade is attested by the transport of copper belonging to himself to Purušhattum (kt c/k 167 and 445). He cooperated with Alāhum son of Mannua, who acted as his agent.

It is not certain in which city he was active. The dossier about copper makes it clear that it was somewhere between Kanish and Purušhattum, in a city which according to kt c/k 263 possessed a *kārum*. This city was not situated on the route from Kanish to Purušhattum as described in TC 3, 165, because it appears from kt c/k 445 that, coming from Purušhattum, one could travel from Ulama either to Tahašaili or to Kanish.[348] Šalatuar and Wahšušana are possibilities, but the first had only a *wabartum*, whereas Wahšušana possessed a *kārum*. Until new evidence provides more information on this matter, it is assumed that Tahašaili lived in Wahšušana.

Alāhum occasionally had his tin sold by Iddi(n)-Ištar (son of Aššur-nādā son of Aššur-iddi(n) son of Šuli). This appears from BIN 4, 64, where Ilī-Anum, probably Adad-ṣulūlī's transporter, asks Iddi(n)-Ištar to give him tin belonging to either Aššur-nādā or Alāhum, which he wants to sell in Purušhattum (for copper?).

4.3.2 *Procurement and sale of copper*

Alāhum usually owned silver from retailers he had entrusted with merchandise, but in several cases the debt is expressed in copper. Data on Alāhum's debtors can be taken from

---

bi-lá-ni [21] *lu* URUDU *ša a-na Pu-ru-uš-/ha-tim* [22] *uš-té-bi-lu-ú* [23] *na-áš-pá-ar-tù-šu ni-il*$_5$*-qé-ma* [24] *na-áš-ú-ni-ku-nu-tí.*

[347] Kt c/k 47.20; 100.5, 7, 9; 254.8; 535.8; together with Aššur-bēl-šadue in kt c/k 50.8.

[348] Kt c/k 445.1-13: [1] *a-na A-šùr-*DÙG *qí-bi*$_4$*-ma* [2] *um-ma Ta-ha-ša-li* [3] 5 GÚ 10 *ma-na* URUDU [4] *ma-sí-am a-na* [5] *A-lá-hi-im* DUMU! *Ma-nu-/a* [6] *a-dí-in-ma a-na Pu-ru-uš-ha-tim* [7] *ú-šé-ri-ib* [8] *a-na-kam ki-ma sú-ku-<ur>-tù-ni* [9] *iš-tù Ú-lá-ma* [10] *a-na Kà-ni-iš*^{ki} [11] *i-tal-kam-ma u*$_4$*-ma-am* [12] *lá li-bi* DINGIR-*ma* [13] *A-lá-hu-um me-et*, 'Speak to Aššur-ṭāb, thus (says) Tahaša(i)li: I gave 5 talents and 10 minas of refined copper to Alāhum son of Mannua and he brought it to Purušhattum. Because over here there is a blockade, he went from Ulama to Kanish and now, unfortunately, Alāhum is dead!'

debt-notes (kt c/k 430; 432; 439; 537), and the *Sammelmemoranda* kt c/k 839 (published in Donbaz 1985a; line 113: *mehrum ša ṭuppī harrumūtim*, 'copy of certified tablets') and its duplicate kt 'e/k 80'; kt c/k 512 (a letter sent by Alāhum, probably from Assur, to Aššur-ṭāb and Alāhum?), and kt c/k 41, in content its duplicate; kt c/k 453 (*mimma ša awīlī anniūtim ṭuppūšunu harrumū*, 'the tablets of all these people are certified'); and kt c/k 258 (Alāhum as creditor in line 26, line 35: *mehrū ša ṭuppīa harrumūtim*, 'copies of my certified tablets'). By combining duplicate passages from these tablets more information on the conditions is obtained. In several instances, only then do the term of the debt, the date, or other relevant data become known. Among Alāhum's debtors the following persons had to pay in copper:

Lušnani	1 mina of copper[349]
Zuzua	5 minas of copper[350]
Alāhum son of Addellat	6 minas of fine copper to buy a textile[351]
Buzia	6 1/2 minas of copper[352]
Bēlia	22 minas of copper[353]
Alābum son of Šū-Dibar	1 mina of silver and 23 minas of fine copper[354]
Abia	30 minas of fine copper[355]
Sika	30 minas of fine copper[356]
Aššur-bēl-awātim son of Šū-Anim	3 minas of silver and 1 talent of fine copper[357]
the father-in-law of Iddi(n)-Kūbum	130 minas of copper[358]

349 Kt c/k 459.24-28: [24] 1 *ma-na* 5 *hi-dim* [25] KB *i-li-bi*$_4$ [26] *Lu-uš*[!]*-na-ni a-ha-ma* [27] 1 *ma-na* URUDU *ú* 2 *hi-dim* [28] KB *i-l[i-b]i*$_4$*-šu-ma*, 'Lušnani owes 1 mina (of copper and) 5 silver beads, furthermore he also owes 1 mina of copper and 2 silver beads.'

350 Kt c/k 444.12: 5 *ma-na* URUDU <KI> *Zu-zu-a*, 'Zuzua owes 5 minas of copper.'

351 Kt c/k 258.53-54: [53] 6 *ma-na* URUDU SIG$_5$ *a-na* TÚG *ša-a-mì-im a-na A-lá-hi-im* [54] DUMU *A-dé-lá-at a-dí-in*, 'I gave 6 minas of fine copper to Alāhum son of Addellat to buy a textile.'

352 Kt c/k 444.10-11: [10] 6 1/2 *ma-na* URUDU [11] KI *Bu-zi-a*, 'Buzia owes 6 1/2 minas of copper.'

353 Kt c/k 444.1-5: [1] 8 *ma-na* 10 GÍN KB [2] *a-na Be-li-a ù Du-da-a* [3] *a-ší-mì-im ša-a-mì-im* [4] *e-zi-ib* 22 *ma-na* [5] URUDU *iš-tí Be-lí-a*, 'I left 8 minas 10 shekels of silver with Bēlia and Dudāya to make purchases with; Bēlia owes 22 minas of copper.'

354 Kt c/k 839.71-73: [71] ... 1 *ma-na* KB *ú* 23 *ma-na* URUDU SIG$_5$ KI *A-la-bi*$_4$*-im* DUMU *Šu-Dí-bar* [72] *Puzur*$_4$*-A-na* DUMU *Ga-a-tí-im qá-ta-tum a-*3 *ha-am-ša-tim i-ša-qal* [73] *šu-ma lá iš-qúl ki-ma a-wa-at kà-ri-im ú-ṣa-áb*, 'Alābum son of Šū-Dibar owes 1 mina of silver and 23 minas of fine copper, Puzur-Ana son of Gātum being guarantor; he will pay within 3 weeks, if he doesn't pay, he will add (interest) according to the rule of the *kārum*.'

355 Kt c/k 839.99: ... 30 *ma-na* URUDU SIG$_5$ KI *A-bi-a*, 'Abia owes 30 minas of fine copper.'

356 Kt c/k 512.10': 30 *ma-na* URUDU *Sí-kà-a*, 'Sika owes 30 minas of copper." This passage is paralleled by kt c/k 453.7: 30 *ma-na* URUDU SIG$_5$ KI *Sí-kà*, 'Sika owes 30 minas of fine copper.'

357 Kt c/k 839.69-71: [69] ... 3 *ma-na* KB KI *A-šùr-be-el-a-wa-tim* [70] DUMU *Šu-A-nim ki-ma a-wa-at kà-ri-im ú-ṣa-áb ú a-ha-am-ma* 1 GÚ URUDU [71] SIG$_5$ (...), 'Aššur-bēl-awātim son of Šū-Anim owes 3 minas of silver, he will add (interest) according to the rule of the *kārum*; furthermore: 1 talent of fine copper.'

358 Kt c/k 459.30-31: [30] ... *me-at* 30 *ma-na* [31] URUDU *i-li-bi*$_4$ *e-mi I-dí-Ku-bi*$_4$, 'Iddi(n)-Kūbum's father-in-law owes 130 minas of copper.'

The background of these debts is largely unknown. When Alāhum son of Addellat was travelling to Assur and had received a *be'ūlātum* of 1 mina of silver (kt c/k 258.7), Alāhum apparently had also given him in Kanish 6 minas of copper to buy a textile, presumably in Assur. Alābum son of Šū-Dibar, Aššur-bēl-awātim son of Šū-Anim, and the father-in-law of Iddi(n)-Kūbum might be retailers having to pay in copper for the merchandise they had received; in all other instances the small amounts could have been given as *be'ūlātum* money.

The bulk of the documentation on the role played by copper in the trade of Alāhum is contained in a group of letters exchanged between Tahašaili and his colleagues. These letters relate how several consignments of textiles and tin belonging to Alāhum, were brought from Kanish and left behind by Aššur-ṭāb with Tahašaili, presumably in Wahšušana.[359] Some of these goods were subsequently transported to Purušhattum, others were sold by Tahašaili in Wahšušana for copper. Kt c/k 263 (to a lesser extent kt c/k 671 as well) can be considered Tahašaili's final account of the sale he concluded and of the way the copper had been disbursed. This accounting became necessary after the death of Aššur-ṭāb. Kt c/k 263 speaks of 277 textiles and 2 talents of tin left behind with Tahašaili. In kt c/k 671, he acknowledged that 231 textiles and 2 talents of tin, belonging to Alāhum, were left behind with him. In spite of the different amounts, the subsequent details of consignments of copper imply that they refer to the same case.

After Aššur-ṭāb had taken tin and 9 textiles to wrap it in, Tahašaili sold the remaining 268 textiles at 25 minas of copper each (= 6,700 m) and 2 talents of tin at 18 minas of copper per mina of tin (= 2,160 m) as we learn from kt c/k 263 (**4**). This transaction yielded 8,860 minas of copper. The *šaddu'atum*-tax paid by Tahašaili to the local *kārum* amounted to 150 minas. Applying the usual rate of one-sixtieth for this tax, the taxed amount was 9,000 minas, which is almost exactly the total of copper obtained by selling the textiles and tin. The *kārum* appears to have had little difficulty in levying and collecting the tax, probably because the buyer of Alāhum's merchandise was the local palace, which would after all pay through the *kārum*-office for the merchandise it had bought, and as soon as the merchant or his representative(s) came to collect the copper, the *kārum*-office could recover the tax due.

The transport minutiae demonstrate that the copper obtained was of fine and of *šikkum* quality. These two qualities differ in value, because in kt c/k 535 (see below under consignment 4) the value of *šikkum* is half that of fine copper. Nevertheless, there is no differentiation of copper apparent in the exchange ratios for textiles and tin.

Out of the amount of copper acquired in this way, the disbursements mentioned in kt c/k 263.12-end were made. Part of this copper was exchanged for 2 1/2 minas of *pašallum* gold (the passage about the exchange rate is unclear but presumably poor quality copper was paid) and sent to Aššur-ṭāb in Kanish.[360] Another part of the copper was transported in several consignments to Purušhattum.

Text kt c/k 263 details the sale of textiles and tin and the subsequent disbursements from the copper as follows (m. = minas; t. = talents):

---

359 Kt c/k 254 and 445: letters to Aššur-ṭāb from Tahaša(i)li; kt c/k 263: letter from Tahašaili to Aššur-bēl-šadue and Aššur-mālik; kt c/k 671: letter to Aššur-mālik and Itūr-ilum from Tahaša(il)i.

360 Several debtors of Aššur-ṭāb have to pay in gold in kt c/k 440 and 442.

Aššur-ṭāb left 277 textiles + 2 t. of tin with Tahašaili; he took 9 textiles for wrapping, remainder 268 textiles.

Sold at 25 m. of copper per textile:	6,700 m	
Tin sold at 18 m. of copper per m. of tin:	2,160 m	
Total copper available after sale:		
		8,860 m
subtracted from this:		
2 t. of refined copper (12)		120 m
5 t. of refined copper (16)		300 m
6 t. of refined + 14 t. of *šikkum* copper brought by A. and W. (18-19)		1,200 m
21 m. of copper rent of a wagon (24)		21 m
32 1/2 m. of copper for expenses (25)		32 m 30 g
17 t. 40 m. of refined copper taken by Aššur-bēl-šadue (26)		1,060 m
53 t .10 m. of copper for gold (33)		3,190 m
20 m. of copper for expenses (36)		20 m
2 t. 30 m. of copper as *šaddu'atum*-tax (37)		150 m
54 m. of copper as *dātum*-tax on 18 textiles (42)		54 m
		6,147 m 30 g
5 m. of silver (42) if 90:1		450 m
4 textiles (45)		c. 100 m
		6,697 m 30 g

The amount of copper disbursed is nearly that obtained by selling the 268 textiles, 6,700 minas. This could mean that the tin was sold later. The consignments to Purušhattum were:

Consignment 1) 2 talents of refined copper[361]

When Aššur-ṭāb returned to Tahašaili to bring the tin to Purušhattum, he took the 9 textiles for wrapping the tin (kt c/k 263.7-9), and 2 talents of refined copper to bring to Purušhattum (kt c/k 254.14-15; 263.12-14; 671.31-33). He also took 10 minas of copper from Tahašaili to cover expenses en route (kt c/k 254.11-13; 263.34-37 [20 minas]; 671.29-30).

Consignment 2) 5 talents of refined/fine copper[362]

---

361 Consignment 1:
kt c/k 254.11-15: [11] ... *i-nu-mì a-ta a-na* [12] *Pu-ru-uš-ha-tim té-ru-bu* [13] 10 *ma-na* URUDU *a-qá-tí-kà tal-/qé* [14] *ù a-ha-ma* 2 GÚ URUDU SIG$_5$ [15] *a-dí-na-kum*;
kt c/k 263.12-14, 34-37: [12] ... 2 GÚ URUDU *ma-sí-am* [13] *i-nu-mì* . AN.NA . *il*$_5$*-qé-<ú>* . *a-na A-šur-*DÙG [14] *a-dí-in* (...) [34] *i-nu-mì A-šur-*DÙG . *i-li-kà-ni-ma* [35] AN.NA *a-na Pu-ru-uš-ha-tim* [36] *ú-šé-ri-bu* 20 *ma-na* URUDU *a-qá-tí-šu* [37] *a-dí-šum*;
kt c/k 671.29-33: [29]*i-nu-mì A-šùr-*DÙG [30] *e-ru-bu* 10 *ma-na* URUDU *a-qá-tí-šu il*$_5$*-qé* [31] *a-ha-ma* 2 GÚ URUDU *ma-sí-am a-dí-šu-ma* [32] *iš-tí* AN.NA *a-na Pu-ru-uš-ha-tim* [33] *šu-ut ú-šé-ri-ib*.

362 Consignment 2:

Tahašaili gave 5 talents of refined/fine copper to Alāhum son of Nūr-Suen, who brought it, together with Aššur-ṭāb's tin, to Purušhattum (kt c/k 254.19-23; 263.16-17; 671.36-39). When Alāhum left, Tahašaili paid 10 minas for the rent of a donkey (kt c/k 254.32-34; 671.45-46). Alāhum son of Nūr-Suen sold the copper and might have given its proceeds in silver to Aššur-ṭāb (kt c/k 445.27-32).

Consignment 3) 17 talents 40 minas of refined copper[363]
Aššur-bēl-šadue took from Tahašaili 3 black donkeys (kt c/k 263.27) and brought, perhaps by wagon, 17 talents 40 minas of refined copper to Purušhattum (kt c/k 254.16-19; 263.26-27; 671.33-36).

Consignment 4) 6 talents of refined/fine copper + 14 talents of *šikkum* copper[364]

---

kt c/k 254.19-23, 32-34: [19] ... 5 GÚ URUDU *ma-/sí-am* [20] 10 GÍN-*tum a-ṣé-er* [21] 10 *ma-na-e . ta-ú-ur-ma* [22] *A-lá-hu-um* DUMU *Nu-ur*-ZU [23] *a-na Pu-ru-uš-ha-tim ú-šé-ri-/ib* (...) [32] *i-nu-mì A-lá-hu-um* [33] *i-li-ku* 10 *ma-na* URUDU [34] *ig-ri* ANŠE *áš-qúl*;
kt c/k 263.14-17: [14] ... *i-nu-mì . A-lá-hu-um . iš-tù* [15] *pá-zu-ur-tim . i-li-kà-ni* [16] 5 GÚ *ma-sí-am . iš-tí* AN.NA-*ki-/šu* [17] *A-lá-hu-um ú-kà-il*$_5$*-šum*;
kt c/k 445.26-32: [26] ... *ù a-ha-ma* [27] 5 GÚ URUDU *ma-sí-am* [28] *a-na A-lá-hi-im* [29] DUMU *Nu-ur*-ZU *a-dí-in-ma iš-tí* [30] AN.NA-*ki-kà ú-šé-ri-ib* [31] KB-*áp-šu i-dí-na-kum* [32] *lá i-dí-na-kum*;
kt c/k 671.36-39, 45-46: [36] ... 5 GÚ URUDU SIG$_5$ 10 GÍN-*tum* [37] *a-ṣé-er* 10 *ma-na-e ta-ur-ma* [38] *A-lá-hu-um* DUMU *Nu-ur*-ZU *a-Pu-ru-uš-ha-/tim* [39] *ú-šé-ri-ib* (...) [45] ... *i-nu-mì A-lá-hu-um i-li-ku* [46] 10 *ma-na* URUDU ⌜*ig-ri*⌝ ANŠE *áš-qúl*;
kt c/k 683: [1] *a-na A-lá-hi-im* DUMU *Ma-nu-a* [2] *ù A-lá-hi-im* DUMU [3] *Nu-ur*-ZU *qí-bi-ma* [4] *um-ma Ta-ha-ša-li-ma* [5] *a-na A-lá-hi-im* DUMU *Ma-nu-/a* [6] *qí-bi*$_4$*-ma* 2 GÚ 10 *ma-na* [7] AN.NA 5 GÚ URUDU SIG$_5$ [8] *i-a-um* 10 GÍN-*tum* [9] *a-ṣé-er* 10 *ma-na-e* [10] *ta-ú-ur A-lá-hu-um* [11] *na-ší a-hi a-ta* [12] *ki-ma i-a-tí* [13] *i-ša-ha-tí-šu . i-zi-iz* [14] AN.NA *ù* URUDU SIG$_5$ [15] *ma-lá . ta-le-e-a-ni* [16] *dí-na-ma* KB *ma-lá* [17] *a-na* AN.NA *ù* URUDU [18] *i-ma-qú-ta-ku-nu-tí-ni* [19] <*a-na*> *A-lá-hi-im pì-qí-id-ma* [20] *ù té-er-ta-kà* [21] *a-ṣé-ri-a* [22] *li-li-kam-ma* [23] *uz-ni pí-tí* 5 GÚ [24] URUDU SIG$_5$ *a-na A-lá-[hi-im]* [25] *áp-qí-id* IGI *Qar-dum-*[d]IM [26] IGI *A-šur-i-mì-tí*.
363 Consignment 3:
kt c/k 263.26-28: [26] ... 17 GÚ 40 *ma-na* URUDU [27] *ma-sí-am ù* 3 ANŠE *ṣa-lá-me* [28] *A-šur-be-el-*SA.TU . *il*$_5$*-qé*;
kt c/k 254.16-19: [16] 17 GÚ 40 *ma-na* URUDU [17] *ma-sí-am A-šur-be-el-/ša-du-e* [18] *a-na Pu-ru-uš-ha-tim* [19] *ú-šé-ri-ib*;
kt c/k 671.33-36: [33] ... 17 GÚ 40 *ma-/na* URUDU [34] *ma-sí-am A-šur-be-el-ša-du-im* [35] *a-na Pu-ru-uš-ha-tim* [36] *ú-šé-ri-ib*.
364 Consignment 4:
kt c/k 254.1-11, 27-31: [1] *a-na A-šùr*-DÙG *qí-bi-ma* [2] *um-ma Ta-ha-ša-li-ma* [3] 6 GÚ URUDU SIG$_5$ 10 GÍN-*tum* [4] *a-ṣé-er* 10 *ma-na-e* [5] *ta-ú-ur-ma* KB *wa-du* [6] URUDU *ša-bu-ur a-ha-ma* [7] 14 GÚ URUDU *ší-kam* 10 GÍN-*tum-/ma* [8] *ta-ú-ur ku-nu-ki-a A-lá-/hu-um* [9] *ù Wa-li-a a-na* [10] *Pu-ru-uš-ha-tim a-ṣé-ri-kà* [11] *ub-lu-nim* (...) [27] *i-nu-mì* URUDU *i-li-ku* [28] 30 *ma-na* URUDU *A-lá-hu-um* [29] *ù Wa-li-a a-qá-tí-šu-nu* [30] *il*$_5$*-qé-ú* 20 *ma-na* URUDU [31] *ig-ri e-ri-qí-im áš-/qúl*;
kt c/k 263.18-26: [18] *ur-kà-num* 6 GÚ *ma-sí-am* [19] *ù* 14 GÚ *ší-kam* 10 GÍN-*tum* [20] *a-ṣé-er* 10 *ma-na-e . ta-ur* [21] *i-na e-ri-qí-im . a-na* [22] *Pu-ru-uš-ha-tim A-lá-hu-um* [23] *ù Wa-li-a . ub-lu-šum* [24] 21 *ma-na ig-ri e-ri-qí-im* [25] 32 1/2 *ma-na . a-qá-tí-šu-nu* [26] *a-dí-in*;
kt c/k 535.1-10, 15-25: [1] *a-na A-šùr*-DÙG *qí-bi-ma* [2] *um-ma Ta-ha-ša-li-ma* [3] 6 GÚ URUDU SIG$_5$ 10 GÍN-*tum* [4] *a-ṣé-er* 10 *ma-na-e* [5] *ta-ú-ur-ma* KB *ú-du* [6] *a-ha-ma* 14 GÚ URUDU *ší-kam* [7] [10 GÍ]N-*tum ta-ú-ur* [8] [KB] *ú-du ku-nu-ki-a* [9] [*A-lá*]*-hu-um ù Ú-li-a* [10] *ub-lu-ni-kum* (...) [15] 32 1/2

Some time after Alāhum son of Nūr-Suen brought tin and copper to Purušhattum, Alāhum and Walia brought to Aššur-ṭāb in Purušhattum by wagon 6 talents of broken refined copper (in kt c/k 254 and 535 called 'fine copper'), and 14 talents of *šikkum* copper sealed by Tahašaili, (kt c/k 254.3-11; 263.18-23; 535.3-10; 671.23-29). Of the 6 talents of fine copper 1 1/2 mina were short, and to make up for this Tahašaili gave 3 minas of *šikkum* copper to Alāhum (kt c/k 535.18-23). Tahašaili paid 20 minas of copper for the rent of the wagon (kt c/k 254.30-31; 263.24 [21 minas]; 535.24-25) and gave them 32 1/2 minas for expenses en route, which they should pay back to Aššur-ṭāb (kt c/k 254.28-30 [30 minas]; 263.25; 535.15-17 (the proceeds of the sale of donkeys); 671.42-44 [Alāhum and Tākilāya!]).

The texts unfortunately remain silent on details which the writer considered well-known or too uninteresting to be mentioned. The buyer or buyers of the textiles and tin is/are not mentioned, but it could have been the palace in Wahšušana which was paying with copper. Nor is the reason for the exchange of the goods for copper given. Perhaps Aššur-ṭāb had left merchandise behind with Tahašaili in anticipation of a profitable deal, which was seized upon by Tahašaili. This is supported by a letter in which Tahašaili, after reporting that consignment 4 has left for Purušhattum, informs Aššur-ṭāb of the market situation, kt c/k 535.10-14: 10 ... *a-na* 11 TÚG.HI.A *ša t*[*é*]-*zi-ba-ni* 12 *a-na-kam ší-mu-um* 13 *lá-šu . mì-li-ik* 14 TÚG.HI.A *mì-li-ik*, 'For the textiles you left behind with me there is no market here. Decide what to do with the textiles.' How profitable this mechanism was is impossible to determine as data on the price fetched for the copper in Purušhattum are lacking.

The phrase 'a 10 shekel weight brought back (= added) on top of each 10 mina weight', followed by the expression *kaspū waddū*, '(its) silver (weight) established', used in consignments 2 and 4, refers to a weight correction performed when Tahašaili weighed out the copper (see above, pp.82-83).

Tahašaili provided the transporters with money in the form of copper to cover their expenses, and also some donkeys for transportation. If he lacked (enough) donkeys, they were hired, just as was the wagon. From whom is unclear, although the local palace might have had an interest in providing merchants with donkeys and wagons. The consignments were of different size and so were the transport requirements. The first needed 1 donkey to carry the 2 talents. The second consisted of 5 talents, but only 1 donkey was rented; half of the copper was either carried on a previously unloaded donkey, or divided over the donkeys carrying the tin. For the third consignment, of 17 talents 40 minas, about 7 donkeys were needed if no wagon was used; Aššur-bēl-šadue received 3 from Tahašaili. The fourth

---

ma-na URUDU *ší-im* 16 ANŠE.HI.A . *a-qá-tí-šu-nu* 17 *a-dí-in a-ma-kam lu-ba-li-ṭù-/ni-kum* 18 *i-na* 6 GÚ URUDU 19 SIG$_5$ 1 1/2 *ma-na* SIG$_5$ 20 *ba-tí-iq-ma* 3 *ma-na* URUDU 21 *ší-kam mu-lá-e-šu a-na* 22 *A-lá-hi-im a-dí-in* 23 *a-ma-kam li-dí-na-kum* 24 20 *ma-na* URUDU *ig-*[*ri*] 25 *e-ri-qí-im áš-qúl*; kt c/k 671.23-29, 41-45: 23 6 GÚ URUDU *ma-sí-am* 10 GÍN-*tum* 24 *a-ṣé-er* 10 *ma-na-e ta-ú-ur-ma* 25 KB *wa-du* URUDU *ša-bu-ra-am* 26 *a-ha-ma* 14 GÚ URUDU *ší-kam* 10 GÍN-*tum-ma* 27 *ta-ú-ur ku-nu-ki-a A-lá-hu-um* 28 *ù Wa-li-a a-na Pu-ru-uš-ha-tim* 29 *a-ṣé-ri-šu ub-lam* (...) 41 ... *i-nu-mì* 42 URUDU *i-li-ku* 30 *ma-na* URUDU 43 *A-lá-hu-um ú Ta-ki-lá-a* 44 *a-qá-tí-šu-nu il*$_5$-[*qé-ú*] ⌜20⌝ *ma-na* URUDU *ig-ri* 45 *e-ri-qí-im áš-qúl.*

consignment consisted of the largest amount, 20 talents of copper. In this case a wagon was used, rented for 20 minas of copper. The rent is one-sixtieth of the cargo, but this seems coincidental.

How much copper remained in Wahšušana with Tahašaili after the consignments had left is unknown. The copper once brought into Purušhattum was sold, but the letters are silent also on this point. Merchandise was often sold to the palace in that city for copper, which was paid through the *kārum*-office,[365] and the likelihood that Aššur-ṭāb followed this method is great.

Tahašaili sold various textiles and 114 minas of *šikkum* copper, all merchandise which Aššur-bēl-šadue had left behind with him, for silver according to kt c/k 47.[366]

An example of Tahašaili's private commerce is the despatch of 5 talents 10 minas of refined copper with Alāhum son of Mannua to Purušhattum (kt c/k 167.4-5; 445.3-7).[367] After selling it, this Alāhum brought the proceeds to Kanish, where he was at liberty to make purchases for Tahašaili.

The sale or purchase of donkeys for copper is attested in kt c/k 100; 256; 535.[368]

4.3.3 *Geography of the copper trade*

Small quantities of copper were exported to a city outside of Anatolia in kt c/k 258. Manuba son of Aduta was going to Assur and carried two packets of copper, one containing 20 1/3 minas of fine copper, which he was to leave behind in *Hahhum*, and another, with only 6 minas of *šikkum* copper, that was destined for a female relative of Alāhum in *Assur*.[369] This last amount was only worth a few shekels of silver, too small to represent

---

365 Kt c/k 528.8-9: [8] 6 *me-at* 80 *ma-na* URUDU É *kà-ri-im* [9] *Pu-ru-uš-ha-tim a-lá-qé*, 'I can collect 680 minas of copper in the *kārum*-office in Purušhattum.'

366 Kt c/k 47.14-16: [14] *ša* 1 GÚ 54 *ma-na* URUDU [15] *ší-ki-im* . *ša té-zi-ba-ni* [16] 2/3 *ma-na* 5 1/2 GÍN KB-*áp-šu*, (from Tahašaili to Aššur-bēl-šadue:) 'Of the 1 talent 54 minas of *šikkum* copper which you left with me the (proceeds) in silver are 45 1/2 shekels.'

367 Kt c/k 167.1-15: [1] *a-na A-lá-hi-im* [2] DUMU *Ma-nu-a qí-bi₄-ma* [3] *um-ma Ta-ha-ša-li-ma* [4] 5 GÚ 10 *ma-na* URUDU *ma-sí-/am* [5] *a-dí-na-kum um-ma a-ta-ma* [6] KB 1 GÍN *lu-šé-li-a-kum* [7] *a-ta* KB *a-na Kà-ni-iš* [8] *ta-ta-ba-al* [9] *a-hi a-ta šu-ma* [10] *ší-ma-am ša ba-lá-ṭí-a lá* [11] *ta-áš-a-ma-am* [12] *ší-ma-am ša ta-áš-⌜ú⌝-mu* [13] *šé-bi₄-lam šu-ma* KB [14] *i-ba-ší* KB *a-na* [15] *A-šur*-DÙG *dí-in*, 'Speak to Alāhum son of Mannua, thus (says) Tahaša(i)li: I gave 5 talents 10 minas of refined copper to you. You said: "I will earn for you the silver, every shekel (of it)." But you have just moved the silver to Kanish! My dear brother, if you did not buy profitable merchandise for me, send me whatever you bought. If silver is left over, give it to Aššur-ṭāb.'

368 Kt c/k 100.20-22: [20] *ù* 35 *ma-na* URUDU [21] *ší-im* 1 ANŠE [22] *i-li-bi₄-šu*, 'And (Adada) owes 35 minas of copper, proceeds of 1 donkey';
kt c/k 535.15-17: [15] 32 1/2 *ma-na* URUDU *ší-im* [16] ANŠE.HI.A . *a-qá-tí-šu-nu* [17] *a-dí-in a-ma-kam lu-ba-li-ṭù-/ni-kum*, 'I gave them 32 1/2 minas of copper, for the purchase of donkeys, for their expenses, let them credit it to you over there.' Kt c/k 256 is a letter from Tahaša(i)li to Aššur-mālik; the last two lines read: ANŠE-*kà* . *a-na* 27 *ma-na* URUDU, *ta-dí-in*, 'Your donkey was sold for 27 minas of copper.'

369 Kt c/k 258.40-49: [40] ... 1/3 *ma-na* 2 GÍN [41] KB *me-eh-ra-tim a-na Ma-nu-ba* DUMU *A-du-ta-a* [42] *a-dí-in-ma ku-ta-nam e-zi-ba-am* [43] *ba-tí-iq ú-tur₄ a-da-šu-ma ší-tí* KB-*pì-a* [44] *i-na a-lim*[ki] *šál-ma-am i-da-nam* [45] *a-ha-ma* 20 1/3 *ma-na* URUDU SIG₅ *ku-nu-ki-a* [46] *a-na Ma-nu-ba-ma a-dí-in i-Ha-hi-im* [47] *e-za-ba-am qá-bi-i a-ha-ma* 6 *ma-na* URUDU [48] *ší-kam i-na a-lim*[ki] *a-na kà-li-tí-ni* [49] *i-da-an* (...), 'I gave 22 shekels of silver as counter-value to Manuba son of Aduta, and he left me a *kutānum*-

the proceeds of the sale of some textile sent by the lady; it is more appropriate to regard it as a gift to the 'bride'. Also in kt c/k 444 an amount of copper was to be given to a female relative living in the capital.[370] In this case, the silver value was about 12 1/3 shekels, which could well be the proceeds from the sale of a textile sent by that lady. The 12 shekels of silver sent to Šīmat-Ištar, Alāhum's wife, can be seen in the same perspective.[371] Alāhum son of Addellat received 6 minas of fine copper for buying a textile in Assur. With a ratio between copper and silver of 90:1 in Assur, he could buy a textile for 4 shekels of silver there.

The pivotal city in this archive is in all likelihood *Wahšušana*, where Tahašaili sold textiles and tin for copper, which he partly converted into silver and gold.

The bulk of the copper was transported to *Purušhattum*. In all, 35 talents 50 minas of fine copper and 14 talents of *šikkum* copper were sent to this city, presumably in order to exchange it for silver there. But copper was obtained in Purušhattum as well; the writer of kt c/k 528, possibly Aššur-ṭāb, could collect 680 minas of copper in the *kārum*-office in Purušhattum.

It is likely that Alāhum and his enterprise imported some copper directly from *Turhumit* by selling merchandise there, but as the texts available centre around Tahašaili's business in Wahšušana, little is written about trade in this city.[372]

No major quantities of copper were sent to Kanish. It seems that of Alāhum's business in general the proceeds from the sale of copper and other merchandise arrived there, but that the city was not a regular destination for copper consignments.

4.3.4 *Conclusion*

Alāhum's archive contains an interesting collection of texts in which one specific transaction is described whereby tin and textiles were sold by the Anatolian Tahašaili for copper, that was partly exchanged for gold and partly transported to Purušhattum to be sold there.

---

textile. I will sell it at every price and the remainder of the silver (he owes) me, he will give me in Assur, the full price. Furthermore, I gave 20 1/3 minas of fine copper sealed by me to the same Manuba, he has been ordered to leave it behind in Hahhum. Furthermore, he has to give 6 minas of *šikkum* copper to our future sister-in-law in Assur.'

370 Kt c/k 444.15-16: [15] 18 1/2 *ma-na* URUDU SIG5 [16] *ma-ha-tí-a i-ša-qú-lu*, 'They shall weigh out 18 1/2 minas of fine copper (to) my *mahātum*.'

371 Kt c/k 100.8-12: [8] (......) 12 GÍN KB [9] *ku-nu-ki-a . a-na* [10] *Ší-ma-at-Išta*r [11] *Ha-da-a-ni ù*! [12] *Ša-lim-Be-lí na-áš-ú-ni-ku-nu-tí*, (from Tahašaili to Aššur-bēl-šadue and Aššur-mālik:) '12 shekels of silver for Šīmat-Ištar, sealed by me, Hadāni and Šalim-Bēlī are bringing to you.'

372 In kt c/k 279, Tahašaili writes to Aššur-nādā, Aššur-bēl-šadue, and Aššur-mālik, that a person named Ilī-malak had arrived from Turhumit.

4.4 *Imdilum son of Šū-Laban*

4.4.1 *Participants*

Documents belonging to the archive of the merchant Imdilum were discovered in the ruins of his house in Kanish during illicit diggings and during the excavation which B. Hrozný conducted in the *kārum* area in 1925 (Hrozný 1927). As most of the texts clandistinely dug up found their way into museums in Europe, notably in Prague, and the United States, and were published like the majority of the Hrozný texts, Imdilum's archive is largely accessible for research. The man and his archive received special attention in two separate studies, which were published almost simultaneously, one by M. Ichisar (1981), the other by M.T. Larsen (1982). Both studies facilitate the reconstruction of Imdilum's family and the identification of his major business contacts.

Imdilum, the eldest son of Šū-Laban, led the Anatolian branch of a family firm. M.T. Larsen suggests that after Šū-Laban's death, his brother Aššur-imittī (the father of Uṣur-ša-Ištar, see § 4.6.1) became head of the firm in Assur, and that this uncle was Imdilum's 'direct boss in Assur'. Imdilum had three brothers: Ennum-Bēlum, Puzur-Aššur, and Aššur-ṣulūlī. His sons were Puzur-Ištar and Amur-ilī. In his study of Imdilum, Larsen gave the following characterization of the merchant after he had returned to Assur: 'Imdiilum comes through to us in his correspondence as a tough but rather dull and uninspiring man, rich and successful and undoubtedly a highly competent manager of the family's business.' All texts concerned with copper that belong to Imdilum's archive appear to date from the period when he was staying in Anatolia; only CCT 4, 22b could have been written after Imdilum's return to Assur. He actively pursued his firm's interests by ordering the sale of wool for copper in several letters. Quantities of copper for his private use are brought to Imdilum on various occasions. In LB 1202, apparently in connection with the house he had just built (line 15: *annakam bētam ēpušma*, 'I built here a house'), he asks 7 persons to bring each 30 minas of fine copper which he needs for making bronze.[373] Aššur-imittī brings 30 minas of copper from Haburata 'for his gift'.[374] In CCT 6, 19b, 30 minas of Taritar copper out of 2 talents paid by Idnāya are brought ahead of the remaining 1 1/2 talent. In all three cases the copper is of fine quality and could be used to make bronze. This purpose is explicitly given in the first case and might have been envisaged in the other two as well.

In BIN 6, 29, a letter written by the firm's Assur branch, a list is given of amounts of copper and silver still owed to the firm by various persons; the purpose of this letter seems to be that Imdilum had to collect the outstanding debts in Kanish.

---

373 LB 1202 (Ichisar 1981, pp.241-243), lines 28-30: [28] 30 *ma-na*.TA URUDU SIG_5 [29] *a-na sí-pá-ri ku-lu-ku-nu-ma* [30] *le-qé-a-nim*, 'All of you must take 30 minas of fine copper for bronze (utensils).'

374 LB 1295 (Ichisar 1981, pp.338-339), reverse 5'-9': [5'] ... 30 *ma-na* URUDU [6'] *Ha-bu-ra-ta-i-am* [7'] *a-na ni-qí-i-kà* [8'] *A-šur-i-mì-tí* [9'] *na-á[š-a-kum]*, (to Imdilum from Puzur-Adad:) 'Aššur-imittī is bringing you 30 minas of copper from Haburata as your gift.'

Imdilum's uncle *Amur-Ištar* stayed for a long time in Turhumit,[375] where he organized and supervized the sale of merchandise, not only for Imdilum, but also for Uṣur-ša-Ištar. Due to his residence in Turhumit, he was often involved in the copper trade. The difference in age between Imdilum and this uncle appears to have been small, as Imdilum treats him as his equal and calls him 'my brother'. Amur-Ištar appears satisfied with his position, doing business for himself and others, and we find no complaints in letters sent by him. Where Imdilum's merchandise or money are involved, Amur-Ištar avoids taking risks and usually relies on instructions before taking a decision. The sale of wool may serve as example. Imdilum orders him to sell wool for copper of Tiritar quality in CCT 4, 47a. When this is done, Amur-Ištar asks Imdilum to decide what to do with the copper now available (CCT 6, 19b, see below for both texts), although the obvious thing to do would be to transport it for instance to Purušhattum and exchange it there for silver. He has initiative in BIN 6, 76, where he negotiates a price for wool with *pāširū* and other retailers, although this may have been done on Imdilum's request.

Amur-Ištar often cooperates with Uzua, one of Imdilum's agents. Together with Uzua he sells textiles and wool for copper to the palace of Tišmurna(?); the copper will be sold for silver (KTS 18, see below). In BIN 6, 76, he shows a willingness to help his nephew: although he had previously written that he would sell Imdilum's wool only after returning from a business trip, now, in view of Uzua meeting him, he intends to sell the wool before his departure and leave the exchange of that copper for copper of better quality to the 'maid' (*amtum*) and Uzua (see below).

Amur-Ištar is usually asked to supply fine copper. According to Ka 276, Amur-Ištar will receive the order to buy Taritar copper,[376] and Imdilum wants Taritar copper in exchange for his wool in CCT 4, 47a. Copper of a lesser quality is first exchanged for fine quality copper before being sent to Purušhattum in CCT 4, 47a and BIN 6, 76. Amur-Ištar not only obtained copper for Imdilum by selling wool and textiles, but took care of its transportation as well. He sent a donkey-load of fine copper to his nephew in Tuhpia (CCT 3, 1).[377] In CCT 4, 47a, he travels to Purušhattum with copper belonging to himself and to Imdilum. In CCT 6, 37a, Imdilum orders him to ship 9 talents of Taritar copper which Amur-Ištar owes him together with Imdilum's own copper to Pur(uš)hattum and a similar situation is described in kt n/k 1385.

Ennum-Bēlum is one of Imdilum's brothers. He is the sender of KTS 2, 44, in which he reports to Imdilum that he sold his consignment of textiles to the palace and is now

---

375 Illustrated by Uzua's complaint to Imdilum in TC 2, 36 (Ichisar 1981, pp.343-347): [34] *A-mur-Ištar a-pí-i[š] i-na Tur$_4$-hu-mì-it* [35] *iš-tí-bu*, 'Since Amur-Ištar has grown old in Turhumit.'

376 Ka 276 (quoted in *Anatolica* 12, p.133), 3-11: [3] ... URUDU [4] *Ha-bu-ra-ta-am* SIG$_5$ [5] *wa-at-ru-um ša-na-um* [6] *i-za-az Ti-ri-ta-a-am* [7] 2 1/2 *ma-na i-za-az* [8] *mì-lik we-ri-i-kà* [9] *mì-lik šu-ma* URUDU *Ti-ri-ta-ú-a-am* [10] *te$_9$-pá-áš a-na A-mur-Ištar* [11] *šu-up-ra-ma* (broken), 'Extra fine copper from Haburata stands at 2 (minas of copper per shekel of silver), copper from Taritar stands at 2 1/2 minas (each); take a decision about your copper. If you wish to obtain copper from Tiritar, write to Amur-Ištar.'

377 CCT 3, 1 (Ichisar 1981, pp.367-369), lines 18-20: [18] 2 GÚ 2 1/2 *ma-na* URUDU SIG$_5$ [19] *Kà-ki-i a-Tù-ùh-pí-a* [20] *ub-lá-ku-um*, 'Kaki brought you in Tuhpia 2 talents 2 1/2 minas of fine copper.'

entitled to 4,000 minas of copper.[378] Because he writes in the same letter that he sent some textiles on to Uššа, which is probably located west of Purušhattum, the palace which bought the textiles and will pay with copper is likely to be that of Purušhattum.

Puzur-Aššur is another of Imdilum's brothers. I assume that he is meant in the following texts connected with Imdilum. Merchandise belonging to Puzur-Aššur is sold for copper in KTS 18 (see below); outstanding claims in copper are mentioned in TC 1, 53.[379] He is in Wašhania in I 574, and asks Imdilum to send him tin from Kanish, which he will transport together with the copper stored in Wašhania to another city, presumably Purušhattum.[380] He sells copper for silver at Imdilum's request in KTS 2, 22.[381] In ATHE 63, Puzur-Aššur complains to Imdilum that Amur-Ištar refused to hand over copper to his representatives in Purušhattum; now he will himself go to that city and hopes to receive from Amur-Ištar and Lulu the copper which he will then sell for silver and send to his brother.[382] A quantity of 6 1/2 talents of poor copper sealed by Puzur-Aššur was stored in Kurara's house (in Ulama?; cf. I 484, where 2 talents of fine copper are stored in Kurara's house in Ulama), but by order of Aššur-ṭāb son of Alāhum it was given unopened to Amur-Ištar (KUG 49).[383]

---

378 KTS 2, 44.45-47: [45] TÚG.HI-*tí-a* . *ša i-šé-pí-a* . *ú-šé-ri-/ba-ni* [46] *a-na e-kál-lim a-dí-šu-nu-ma* [47] 4 *li-me* URUDU . *na-ad-a-ku*, 'I sold my textiles which I brought in with my transport, to the palace and am now credited for 4,000 (minas of) copper.'

379 TC 1, 53 (Ichisar 1981, pp.342-343), lines 1-7, 5'-11': [1] *a-na Im-dí-lim* [2] *qí-bi$_4$-ma um-ma* [3] *Ah-ša-lim-ma* [4] *i-na ba-⸢ab-tim⸣* [5] *ša Puzur$_4$-A-šur* [6] 8 GÚ 40 *ma-na* [7] *a-na ṣú-ha-ri-im* (broken) (...) [5'] *kúl* URUDU *i-ba-áb-tim* [6'] *ša Puzur$_4$-A-šur* [7'] 4 GÚ 20 *ma-na* [8'] *ù ší-im ša-áp-tí-kà* [9'] *i-tù-wa-ar* [10'] ANŠE.HI.A [11'] *a-ša-⸢qá-la-kum⸣*, 'Speak to Imdilum, thus (says) Ah-šalim: from Puzur-Aššur's outstanding claims (I gave?) 8 talents 40 minas to the servant. (...) I will pay you all copper from Puzur-Aššur's outstanding claims, 4 talents 20 minas, and the proceeds of your wool, as soon as the donkeys are back.'

380 I 574.15-24: [15] ... *a-na-kam* [16] AN.NA-*ku-ú i-da-ni-nu* [17] KB *ma-[d]u-um* [18] *ša Hu-du-ra-i* [19] *na-dí-i i-hi-id-ma* [20] AN.NA *áb-kà-ma* [21] *iš-tí* URUDU-*i-im* [22] *ša i-na Wa-áš-ha-ni-a* [23] *na-du-ú* [24] *lu-šé-ri-ib-šu*, 'Here, tin is getting scarce; much silver of the Huduraeans has been stored. Take care to bring me the tin, and I will bring it together with the copper stored in Wašhania into (city GN)' (by courtesy of K. Hecker).

381 KTS 2, 22.20-28: [20] *lu a-wi-lá-tí ki-ma a-wi-[lu-tí]-kà* [21] KB 10 *ma-na pá-ni-am-ma* [22] *ša té-pu-šu ab-kà-ma* [23] *ma-áš-kà-ta-kà a-na a-lim*[ki] [24] *ku-šu-ud* URUDU *ša i-a-tí ù Šu-A-šùr* [25] *ša i-na* É.GAL-*lim ur-da-ni* [26] *a-ta ù ša ki-ma A-šu[r-*] [27] *a-na* KB *dí-na-šu-ma* [28] *[ku-u]n-kà-ma šé-[bi-la-nim]*, 'Be a man and act accordingly by sending me the first silver, 10 minas, you acquire, and let your cash come to the City. You and Aššur-[ ]'s agent must sell my and Šū-Aššur's copper that came down from the palace for silver and send (it) to me under seal.'

382 ATHE 63 (Ichisar 1981, pp.289-290), lines 3-15: [3] *a-dí* URUDU . *ša A-mur-Ištar* [4] *a-na ša ki i-a-tí* [5] URUDU *i-na Pu-ru-uš-ha-tim* [6] *ta-da-nam ú-lá i-mu-/ú* [7] *ha-ra-ni* . *a-na* [8] *[P]u-ru-uš-ha-tim i-na* [9] *u$_4$-mì-im ṭup-pì-i* [10] *ta-ša-me-ú* [11] *na-áš-am A-mur-Ištar* [12] *ú Lu-lu li-ik-šu-⸢du⸣-/⸢ni⸣-ma* [13] URUDU *li-dí-nu-nim-ma* [14] KB *i-pá-ni-a* [15] *lu-šé-ṣí-a-ku-um*, 'As for Amur-Ištar's copper: he refused to hand it over to my representatives in Purušhattum. I will set out for Purušhattum; on the very day you read my letter, I will be on my way! I hope that Amur-Ištar and Lulu will meet me and give me the copper; I will personally send you the silver.'

383 KUG 49 (Ichisar, 1981, p.337), lines 1-13: [1] *a-na Im-dí*-DINGIR *qí-bi-ma* [2] *um-ma Ku-ra-ra-ma* [3] 6 1/2 GÚ URUDU *lá-mu-nam* [4] *ša A-zu* DUMU *A-mu-ur-Sú-[en]* [5] *ù I-ku-pì-a* DUMU *Ili$_5$-ta-ar* [6] *ku-nu-ki ša Puzur$_4$-A-šur* [7] *i-na bé-tí-a* [8] *i-ni-dí-ú-ni* [9] *A-šur*-DÙG DUMU *A-lá-hi-im* [10] *iš-pu-ra-ma* [11] *a-ma-lá té-er-tí-šu* [12] *ku-nu-ki a-na Amur-Ištar* [13] *áp-qí-id*, 'Speak to Imdilum, thus (says) Kurara: (as

Imdilum's children, who had their share in the family trade, are attested in only a few texts related to copper trading. *Amur-ilī*, apparently without his father's consent, sold merchandise for copper in Wahšušana (CCT 4, 22b). In I 531, Ah-šalim paid to Amur-ilī a total of 12 talents 20 minas of copper that he owed. Imdilum's daughter *Ištar-bāštī* had sent some textiles, and their proceeds in fine copper were to be shipped to Purušhattum and sold there (kt n/k 1385, see below). *Puzur-Ištar* had a claim on Suen-rē'ī son of Amur-Ištar for 55 minas of copper, which he was to pay in Turhumit (OIP 27, 56).[384]

Uzua did not belong to the core of the family, but according to I 467.18, he had the same boss as Imdilum and so belonged to the same firm. He is probably the same person as Uzua son of Libāya, who was married to an Anatolian woman named Zuška and who was the representative (*šazzuztum*) of Aššur-nādā son of Amur-Ištar (CCT 5, 17a and TC 3, 266).[385] Uzua is one of Imdilum's agents, and is chiefly involved in trade in northern regions like Turhumit, Tišmurna, and Kunanamit, often together with Amur-Ištar, but he went to Purušhattum as well. The cooperation between Amur-Ištar and Uzua was not always harmonious (TC 2, 36). Uzua is assisted in his commercial activities by his Anatolian wife Zuška, who is referred to as 'the maid' (*amtum*) by Amur-Ištar (BIN 6, 76), but is called 'wife'(*aššutum*) in CCT 5, 17a and TC 3, 266; in both texts Zuška lends small amounts of silver. The use of *amtum* instead of *aššutum* has no negative connotations, since both words are also used in AKT 3, 40, where the wife of Eradi is called *amtum* in line 6, and *aššutum* in line 11.[386] According to a text found in Boğazköy, ancient Hattuša, Idnāya collected 30 minas of fine copper, the proceeds of a commodity called *zikanšarru*, for Uzua's wife.[387] If Zuška was the character playing the leading role in such commercial affairs, this would explain the wordorder in BIN 6, 76.11-12, 'the maid and Uzua', where she takes precedence over Uzua.

Uzua's surviving correspondence with Imdilum consists, with the exception of KTS 17, of letters dealing with copper (CCT 4, 27a; KTS 18; LB 1225; LB 1275; TC 2, 36; VS 26, 71), apparently (part of) a dossier. Uzua sells wool for copper in Tišmurna, where the market situation was bad according to CCT 4, 27a and TC 2, 36. He entrusted merchandise

---

for) the 6 1/2 talents of poor copper, that Azu son of Amur-Suen and Ikuppia son of Ilī-tayyār had deposited in my house, under Puzur-Aššur's seals, Aššur-ṭāb son of Alāhum gave me written instructions and in accordance with his instructions I entrusted (the) sealed (copper) to Amur-Ištar.'

384 OIP 27, 56 (Ichisar 1981, pp.124-128), lines 33-40: [33] 55 *ma-na* URUDU *ma-sí-am* [34] *da-mu-qám ša-bu-ra-am i-ṣé-er* [35] *Sú-in*-SIPA DUMU *A-mur-Ištar* [36] *Puzur$_2$-Ištar* DUMU *Im-dí-lim i-šu* ITU.KAM *té-i-na-/tim* [37] *li-mu-um A-ku-tum iš-tù ha-muš-/tim* [38] *ša Li-ip-ta-nim a-na* [39] 2 *ha-am-ša-tim i-Tur$_4$-ùh-mì-it* [40] *i-ša-qal*, 'Puzur-Ištar son of Imdilum has 55 minas of broken fine refined copper as a claim on Suen-rē'ī son of Amur-Ištar. Month X, year eponym Akutum: from the week of Lip(i)t-Anim he will pay within 2 weeks in Turhumit.'

385 In these texts, Uzua had as witness Suen-rē'ī son of Amur-Ištar, while Aššur-nādā had as witness Zuzua son of Ilī-dān, cf. KTH 28 B, where the same Zuzua is witness in a settlement involving this Aššur-nādā. Aššur-nādā was married to Zuškana. His father Amur-Ištar does not seem to be the same person as Imdilum's uncle by that name.

386 Cf. Lewy 1956, pp.3-4.

387 KBo 36, 113.1-6: [1] 30 *ma-na* URUDU SIG$_5$ [2] *ší-im zi-kà-ša-ri-/im* [3] [*k*]*i-ma a-ša-at* [4] *Ú-zu-a Id-na-a* [5] *il$_5$-qé* IGI *Iš-ma-*[d]IM [6] IGI *Šu-Sú-in*.

(in Kunanamit?) to retailers, *tamkārū*, who paid him with copper (VS 26, 71).[388] Uzua appears to have transported the copper from Tišmurna and Kunanamit to Turhumit, where he could exchange poor copper for copper of better quality and hand it over to Imdilum's agents (VS 26, 71), who then sent it further to Purušhattum to sell it there (see below for TC 2, 36).[389]

Aššur-nādā occurs in the letters CCT 2, 23 and BIN 6, 133 as staying in Kunanamit.[390] In the first, Amur-Ištar reports to Imdilum that in accordance with a *kārum* judgment he had seized copper belonging to Aššur-nādā which his 'brother' Ennum-Aššur had brought out of a country named Sawit. This copper, 3 talents of fine and 5 talents 40 minas of Kunanamit copper, was entrusted to Ikuppia, probably in Turhumit. An amount of copper paid by the palace as Aššur-nādā's share, perhaps also in Turhumit, was claimed on behalf of Imdilum by various persons, including Amur-Ištar, and likewise entrusted to Ikuppia. Amur-Ištar reminds Imdilum that Aššur-nādā still has assets in copper amounting

---

388 VS 26, 71.48-51: 48 ... *lá-ma ṭup-pá-kà ša šé-ep* 49 *A-bu-[ša-lim áš]-me-ú* 5 GÚ URUDU *ma-sí-am* 50 *i-ṣé-[er* DA]M.GÀR *ke-nu-tim ša wa-ar-hi-ša-/ma* 51 URUDU 1 GÚ ⌜*i*⌝*-ša-qú-lu-ni-ni a-dí-ma*, 'Before I had heard your letter that arrived with Abu-šalim's transport, I had already given merchandise in consignment (for) 5 talents of refined copper to trustworthy retailers who pay me monthly 1 talent.'

389 VS 26, 71.1-3, 33-42: 1 *a-na Im-dí*-DINGIR . *qí-bi-ma um-ma* 2 *Ú-zu-a-ma iš-tù a-na Ku-na-na-mì-it* 3 *e-ru-bu-ni-ma* (...) 33 ... *ša* KB 10 *ma-na* 34 *qí-ip-tám ki-ma a-na u*$_4$*-me qí-ip-tàm* 35 DAM.GÀR *ta-qí-pu* . *ša* KB 10 *ma-na* 36 *ù i-a-tí* . *dí-na-ma i-na re-eš*$_{15}$ *qí-ip-tim* 37 *ša ta-da-na-ni* 10 GÚ URUDU *ma-sí-am* 38 *i-na Tur*$_4$*-hu-mì-it a-na ša ki-ma ku-a-tí* 39 *lá-dí-in-ma* URUDU *a-na Pu-ru-uš-ha-ti[m]* 40 *le-ru-ub-ma* KB *ma-lá i-ma-qú-t[ù-ni]* 41 *a-na a-lim*[ki] *šé-bi-il*$_5$*-ma* . *ší-ma-am a-[ma-lá]* 42 *qí-ip-tí-a i-na bé-tí-kà li-iš-*<*ú*>*-mu*, 'Speak to Imdilum, thus (says) Uzua: since I entered Kunanamit (...) Entrust me too with a *qīptum*-loan worth about 10 minas of silver as you are used to give on terms to a retail-seller, and at the very moment you give me the *qīptum*-loan I will give 10 talents of refined copper to your representatives in Turhumit, then let the copper enter Purušhattum and send all silver that will be obtained with it to the City, and let them make purchases in your house for the full value of the *qīptum*-loan entrusted to me.'

390 CCT 2, 23 (Ichisar 1981, pp.301-303), lines 4-13: 4 *a-šu-mì* URUDU *ša A-šur-na-da* 5 *ša En-um-A-šur a-hu-šu* 6 *iš-tù ma-at Sá-wi-it* 7 *ú-šé-ṣí-a-ni a-ma-lá dí-in* 8 *kà-ri-im En-um-A-šur* 9 *ni-iṣ-ba-at-ma* 3 GÚ URUDU <SIG$_5$> 10 5 GÚ 40 *ma-na* URUDU 11 *Ku-na-na-ma-i-am* [I]GI 2 *ší-na* 12 *me-er-e um-mì-a-ni* URUDU *a-na* 13 *I-ku-pì-a ni-ip-qí-id*, (to Imdilum from Amur-Ištar, Uṣur-ša-Aššur, and Uṣur-ša-Ištar:) 'As for Aššur-nādā's copper, which his brother Ennum-Aššur has taken out of the country of Sawit, in accordance with the *kārum* judgment we seized Ennum-Aššur and entrusted in the presence of 2 merchants 3 talents of fine copper and 5 talents 40 minas of copper from Kunanamit to Ikuppia.' This letter was followed by BIN 6, 133, where Imdilum, after quoting the relevant details from a preceding letter, orders his partners to release this copper again to Aššur-nādā:

BIN 6, 133 (Ichisar 1981, pp.210-211), lines 5-10, 1'-13': 5 *i-na ṭup-pí-ku-nu* . *ta-áš-pu-ra-nim* 6 *um-ma a-tù-nu-ma* 3 [GÚ URUDU] 7 ⌜SIG$_5$⌝ *ù* [5 GÚ] 40 [*ma-na* URUDU] 8 [*Ku-na*]*-na-ma-a-[am a-I-ku-pí-a]* 9 [*ni*]*-ip-qí-i[d* (...)] 10 [ ] x [ ] (edge is destroyed) 1' [*Hi-na*]*-a ù E[n-um-A-šur]* 2' [*i-na*] URUDU *qá-sú-nu [iš-ku-nu]* 3' ⌜*ù*⌝ *né-nu ki-ma ku-a-t[i qá-at-ni]* 4' *ni-iš-ku-un* URUDU *a-na* 5' *I-ku-pí-a-ma ni-ip-qí-i[d]* 6' *a-hu-ú-a a-tù-nu mì-ma* UR[UDU] 7' *ša ta-ap-qí-da-ni* . *a-n[a]* 8' *A-šùr-na-da i-lá-kà-ni a-n[a]* 9' *ma-ma-an lá tù-ša-ra ù i-na* 10' [*š*]*a-am-ší Ì-lí-áš-ra-ni* 11' ⌜*e*⌝*-ra-ba-ni a-na Ku-na-na-m[ì-i]t* 12' [*a-ṣ*]*é-er A-šùr-na-d[a]* 13' [*ṭù*]*-ur-d[a-š]u*, (to Amur-Ištar, Im-[ ], Uṣur-ša-Aššur, and Uṣur-ša-Ištar, from Imdilum:) 'You wrote me in your letter: "We have entrusted 3 talents of fine copper and 5 talents 40 minas copper from Kunanamit to Ikuppia (...) [Hināy]a and E[nnum-Aššur] claimed the copper, but we, acting for you, claimed it as well. We entrusted the copper to Ikuppia." My dear brothers, all copper you entrusted must go (back) to Aššur-nādā; do not release it to anybody else! When Ilī-ašranni arrives, you must send him to Aššur-nādā in Kunanamit.' Cf. also AKT 1, 17.

to 10 talents in Kunanamit. In reaction to these events, Imdilum (in BIN 6, 133) writes to Amur-Ištar and colleagues that all copper entrusted by them to Ikuppia is still property of Aššur-nādā and may not be released (sold) to anybody else. When a person called Ilī-ašranni arrives, perhaps the man who brought the letter to Amur-Ištar, they must send him to Aššur-nādā in Kunanamit. It seems that Imdilum preferred to receive full payment from Aššur-nādā at a later moment, rather than seize his assets in Turhumit.

Ah-šalim is another of Imdilum's agents. Alone or together with Puzur-Aššur he sells wool for copper (ICK 1, 135; TC 1, 53, see below). It is unclear whether in ICK 1, 135 and TC 2, 37 the proceeds of wool are in copper, that could subsequently be exchanged for silver or gold, or whether a future sale is envisaged, where copper, silver, and gold were alternatives. The repayment of a debt in copper by Ah-šalim to Amur-ilī (Imdilum's son) is described in I 531.[391] It is uncertain whether the same debt is referred to in RA 81, 16, where Ah-šalim paid 12 talents 54 minas of copper to Ikuppia and Iddi(n)-Adad.[392]

Amur-Ištar entrusted to *Ikuppia* (the son of Eddināya?), probably in Turhumit, 3 talents of fine and 5 talents 40 minas of Kunanamit copper belonging to Aššur-nādā (see above). Amur-Ištar had seized this from his brother Ennum-Aššur, apparently because Aššur-nādā owed money to Imdilum. An amount of copper due to Aššur-nādā was made available by the palace and claimed by Amur-Ištar and entrusted to Ikuppia as well (CCT 2, 23). In RA 81, 16, he receives an amount of copper from Ah-šalim. He collects 220 minas of copper from the *kārum*-office in KTH 11.[393] In TC 3, 104, he is staying with Adu in Ulama to sell copper for silver.[394]

Aššur-ṭāb son of Alāhum had sufficient influence over Imdilum's agents to give instructions to Kurara to hand over 6 1/2 talents of poor copper, sealed by Puzur-Aššur and stored in Kurara's house, to Amur-Ištar (KUG 49, see above).

Usānum is often in Purušhattum (ICK 1, 185; RA 81, 3.13'!). He collects copper from thc *kārum*-officc for Imdilum in CCT 1, 43, and copper is entrusted to him and Adu in the same text.

Idnāya son of Sallia is one of Imdilum's agents. He assumes the obligation to obtain in Turhumit and deliver in Kanish in all 6 1/2 talents of fine copper for Imdilum (ICK 2, 54 = **2**). Two talents of Tiritar copper were paid to Amur-Ištar in Turhumit (CCT 4, 47a.11-12 and CCT 6, 19b.18-19). It is probably this Idnāya who collected copper for Uzua's wife (KBo 36, 113).

---

391 I 531.8'-9': $^{8'}$ 12 GÚ 20 *ma-na* URUDU $^{9'}$ *ša li-*bi_4*-a ša-bu-a-tí*, (Ah-šalim to Amur-ilī:) 'You are now satisfied with the 12 talents 20 minas of copper I owed' (by courtesy of K. Hecker).

392 Published by C. Michel, RA 81, 16.3-7: ^{3} [*a-dí* URU]DU *ša Ah-ša-lim*! . *a-na* ^{4} [*I-k*]*u-pì-a ù I-dí-*^{d}IM ^{5} [*iš-q*]*ú-lu-ú* . 12 GÚ 54 *ma-na* ^{6} [URUDU (...)] . *qá-du-um* 6 ^{7} [ANŠE.HI.A] *x x* [ ].

393 KTH 11 (Ichisar 1981, pp.382-383), lines 4'-5' (to Imdilum, Pūšu-kēn, and Aššur-ṣulūlī by Aššur-mālik, Azu, and Iddi(n)-Adad; about Ikuppia): $^{4'}$ 2 *me-at* 20 *ma-na* URUDU $^{5'}$ [É] *kà-ri-im i-lá-qé*.

394 TC 3, 104.1'-6': $^{1'}$ *um-ma A-du-ú* $^{2'}$ *ù I-ku-pí-a-ma* $^{3'}$ *i-na Ú-lá-ma wa-áš-*[*ba-ni*] $^{4'}$ *a-ma-lá té-er-tí-kà* $^{5'}$ URUDU *a-na* KB *ni-da-m*[*a*] $^{6'}$ [*n*]*i-ta-bi-ma ni-ta-lá-*[*ak*], 'Thus Adu and Ikuppia: we are staying in Ulama. In accordance with your instructions we will sell the copper for silver and leave.'

4.4.2 *Procurement and sale of copper*

In this archive, copper is chiefly obtained in exchange for textiles and wool.

textiles

The exchange of textiles for copper is mentioned in LB 1295.[395] The palace of Purušhattum bought 32 *kutānum*-textiles from Ennu-Bēlum for 1,280 minas (CCT 2, 8),[396] and in KTS 2, 44, Ennam-Bēlum sold his own textiles to this palace, and became entitled to 4,000 minas of copper (see above). In KTS 18, a palace, presumably in Turhumit, bought *makuhum*-textiles belonging to Puzur-Aššur and Imdilum at 20 minas of copper each. Wool was sold to the same palace, and it seems the copper obtained for the textiles was of lesser quality because in line 12 a distinction is made between 'copper' and 'fine copper'.[397]

tin

References to the exchange of tin for copper are rare. It is sold incidentally together with textiles for copper in Wahšušana by Amur-ilī in CCT 4, 22b.[398]

wool

Imdilum's correspondence contains several letters dealing with the trade in wool.[399] In and around Turhumit and Tišmurna, Amur-Ištar and Uzua sold wool for Imdilum. This was

---

395 LB 1295.3'-5': 3' *a-šar ta-ša-pá-ra-ni* 4' URUDU *ší-im* TÚG.HI-*tí-kà* 5' *lá-dí-in*, (to Imdilum from Puzur-Adad:) 'I will sell the copper (which is) the proceeds of your textiles wherever you order me to do it.'

396 CCT 2, 8 (Ichisar 1981, pp.219-224), lines 63-65: 63 ... 32 *ku-ta-ni* 40 *ma-na*.TA 64 *i-na Pu-ru-uš-ha-tim . E-nu-Be-lúm* 65 *a-na* É.GAL-*lim i-din*.

397 KTS 18 (Ichisar 1981, pp.333-334), lines 1-21: 1 *a-na Im-dí-lim qí-bi-ma* 2 *um-ma A-mur-Ištar-ma* 3 *ù Ú-zu-a-ma ú-nu-tum* 4 *a-na* É.GAL-*lim e-ru-ub-ma* 5 17 TÚG *ma-ku-hi ša Puzur$_4$-A-šur* 6 3 TÚG *ku-a-ú-tim* É.GAL-*lúm* 7 20 *ma-na*.TA *a-na ší-mì-im* 8 *il$_5$-qé* 2* *me-at* 20 *ma-na* URUDU 9 *ma-as-am* É.GAL-*lúm a-na* 10 *ší-mì-im ša-na-um il$_5$-qé-ma* 11 URUDU *i-na* É *kà-ri-im ni-lá-qé-ma* 12 URUDU *ší-im* TÚG.HI.A *ù* URUDU SIG$_5$ *u$_4$-ma-ma* 13 É.GAL-*lúm ú-ša-bi-ni-a-tí-ma* 14 *mì-ma a-na ša na-am-e-dim* 15 *lá i-dí-ni-a-tí li-ba-kà* 16 *lá i-ma-ra-aṣ i-na* 17 TÚG.HI.A *ša ta-da-nim ta-ad-nu* 18 *lu* AN.NA *lu* URUDU *ni-da-ma* 19 KB 10 *ma-na né-pá-áš-ma* 20 *iš-tí wa-ar-ki-ú-tim A-du-ú* 21 *ni-ṭá-ra-dam*, 'Speak to Imdilum, thus (say) Amur-Ištar and Uzua: the merchandise entered the palace and the palace bought 17 *m*.-textiles of Puzur-Aššur and 3 of your's at 20 minas (of copper) each. (For a value of) 220 minas of refined copper the palace bought (wool) at a rate of 1:2, and (it was agreed that) we collect the copper in the *kārum*-office. On this very day the palace has satisfied us with the copper as payment for the textiles, and with the fine copper, but it did not give us anything for those of the bulk (of the merchandise). Don't worry! As many textiles as possible have been sold. We will sell both tin and copper and obtain silver, about 10 minas, and send Adu (with the silver) with the next caravan.'

398 CCT 4, 22b (Ichisar 1981, pp.227-229), 21-31: 21 ... *a-ša-me-ma* 22 *A-mur*-DINGIR . *i-na* 23 *Wa-ah-šu-ša-na* AN.NA-*ki* 24 *ù* TÚG.HI.A *a-na* URUDU 25 *i-ta-dí-in$_4$. a-hu-ú-a* 26 *a-tù-nu . A-mur*-DINGIR . *a-na* 27 *Wa-ah-šu-ša-na lá i-lá-ak* 28 *mì-ma lu-qú-tí-a . a-na* 29 *Pu-ru-uš-ha-tim* 30 *lu-šé-ri-ib-ma . a-[na]* 31 *i-ta-aṭ-lim [li-dí-in]*, 'I hear that Amur-ilī has sold piecemeal tin and textiles for copper in Wahšušana. My dear brothers, Amur-ilī must not go (a second time) to Wahšušana! He should bring all my merchandise into Purušhattum and sell it for cash.'

399 BIN 6, 76 (Ichisar 1981, pp.299-301), lines 1-19: 1 *um-ma A-mur-Ištar-ma a-na* 2 «*a-na*» *Im-dì*-DINGIR *qí-bi-ma* 3 *A-šùr-ta-ak-lá-ku kà-ṣa-ar* 4 *Hu-ra-ṣa-nim ṭup-pí ub-lá-kum* 5 *i-na ṭup-pí-a a-na* SÍG.HI.A 6 *ú-lá a-ṭá-hi i-na tù-a-ri-a* 7 *[š]a-pá-at-kà a-da-ma* KB 1 *ma-na* 8 *[ú]-šé-lá-a-kum . ki-ma Ú-zu-a* 9 *[a-n]i$^?$-e [i]k-ta-áš-dì-ni um-ma* 10 *[a]-na-ku-ma lá-ma a-tal-ku* SÍG.HI.A 11 *a-na* URUDU *li-tur$_4$-ma* GEME *ù* 12 *Ú-zu-a a-na* URUDU SIG$_5$ *lu-ta-e-ru* 13 *a-na* 60 GÚ SÍG.HI.A *lu iš-tí* 14 *ni-a-ú-tim lu iš-tí pá-ší-/ri* 15 *a-mì-gi-ir-ma* ŠÀ.BA 40 GÚ 16 *a-ni e-ru-ba-ma qá-tí-ma* 17 (erasure) *i-ta-am-ša-ah* 18 *ší-[tí* SÍG.HI.A] *i-na* 19 *T[í-iš-mu-ur]-na na-ad-/a-at*, 'Thus (says) Amur-Ištar, speak to

Imdilum: Aššur-taklāku, Huraṣānum's transporter, has brought you my letter. (I had written) in my letter: "I cannot occupy myself with the wool (now); when I return I will sell your wool and produce every mina of silver for you." But as Uzua just caught up with me, I thought: "Let the wool be exchanged for copper before I leave, and let the maid and Uzua exchange (this poor copper) for fine copper." I reached an agreement (about the price to be fetched) for 60 talents of wool, both with our men and with the retailers; of this, 40 talents just arrived (here) and were immediately taken away (by the retailers)! The remainder of the wool is stored in Tišmurna';

CCT 4, 27a (Ichisar 1981, pp.314-317), lines 16-17, left edge 1'-2': ^{16} ... *a-na-kam i-na* ^{17} *re-eš*$_{15}$ SÍG.HI.A *ú-mì-ša-am a-zu-áb* (...) [*li-ba-kà*] $^{1'}$ *a-na* URUDU *ša pá-ší-ri lá i-pá-ri-id* SÍG.HI.A *sà-am-tum*$_8$ $^{2'}$ *i-ba-ší-ma i-na Tí-iš-mur-na lu i-na Tur*$_4$*-hu-mì-it-ma* (two illegible lines), (to Imdilum from Uzua, after complaining that there is no copper and hence no market in Tišmurna:) 'Here, I am labouring daily for the sake of the wool (...) Don't worry about the copper (to be paid by) the retailers. Brown wool is available and I will (*exchange it*) in Tišmurna or in Turhumit';

CCT 4, 47a (Ichisar 1981, pp.191-192), lines 26-33: ^{26} [*a-n*]*a* SÍG.HI.A . *ší-ni-šu* ^{27} [ ] *Tí-iš-mu-ur-na* ^{28} [ANŠ]E.HI.A [ ] *x ú* ^{29} *i-tù-ru-nim-ma* ^{30} *i-ga-mu-ru-ší* . ŠÀ.BA 80 GÚ ^{31} *pá-ṣí-tum* 20 GÚ *sà-am-tum . a-hi* ^{32} *a-ta . a-na* URUDU . *Tí-ri-ta-ra-im* ^{33} [ ] *ma-na dí-ší-ma*, (to Amur-Ištar from Imdilum:) '(too broken for translation). They will return and sell (the wool) completely; from (the wool) 80 talents are white and 20 talents are brown. My dear brother, sell it for copper from Tiritar at a rate of [x] minas';

CCT 6, 19b (Ichisar 1981, pp.322-323), lines 13-17: ^{13} ... *a-na ma-lá* ^{14} *ti-ir-ti-kà ša-pá-at-kà* ^{15} *a-na i-ta-aṭ-lim ša-na-um* ^{16} *ni-dí-in-ma* URUDU*-a-kà i-ba-ší* ^{17} *mì-lik* URUDU-[*i*]*-kà mì-lik*, (Amur-Ištar and Puzur-[ ] to Imdilum:) 'We sold your wool for cash at a rate of 1:2 in accordance with your instructions and your copper is in store. Take a decision about your copper!';

ICK 1, 135 (Ichisar 1981, pp.234-235), lines 1-12: ^{1} *um-ma Im-dí-*DINGIR*-ma* ^{2} *a-na Puzur*$_4$*-A-šur* ^{3} *ù Ah-ša-lim qí-bi*$_4$*-ma a-na* ^{4} *Ah-ša-lim qí-bi*$_4$*-ma* URU[DU] ^{5} SÍG.HI.A *ší-im ša-áp-*[*tim*] ^{6} *šu-ma* KÙ.KI *za-ku ša áš-pu-ra-ku-*<*ni*> ^{7} *a-ta ú Puzur*$_4$*-A-šur* ^{8} *ku-ul-kà* ^{9} *šu-ma* KÙ.KI ^{10} *lá-šu* URUDU *ší-im* ^{11} SÍG.HI.A *a-na Puzur*$_4$*-A-šur* ^{12} *dí-in*, 'Thus (says) Imdilum, speak to Puzur-Aššur and Ah-šalim, to Ah-šalim: (concerning) the copper of the wool, the proceeds of the wool, if, as I wrote you, gold is available, you and Puzur-Aššur must (sell the copper for gold and) seal (the gold and send it to me). If there is no gold, you must give the copper, the proceeds of the wool, to Puzur-Aššur';

kt n/k 520.28-45: ^{28} ... *pá-ší-ri* ^{29} *ša* SÍG.HI*-at-kà a-dí-nu-šu-nu-tí-ni* ^{30} *ma-tí-ma mì-ma ú-la* ^{31} *iq-bi-ú-ni-im ù me-eh-ra-tim* ^{32} *a-na Ú-zu-a a-dí-in-ma* ^{33} *i-ta-lá-ak ù ku-a-tí* ^{34} *ṭup-pá-am za-ku-tám áš-pu-ra-ku-um* ^{35} *i-nu-mì mì-ší-ih-tim* ^{36} *i-na i-ki-il*$_5$*-tí e-né-šu-nu* / SÍG.HI.A ^{37} *ba-lúm sà-nu-qí-im i-na* ^{38} É*-tí-a i-dí-ú-ma ù* ^{39} *im-ta-áš-hu* ^{40} *a-na-ma* URUDU *uš-té-er-/sú-ú-ni* ^{41} *a-dí-ni-im* ^{42} *ik-tá-an-šu-nim* ^{43} *um-ma šu-nu-ma* 2 GÚ ^{44} SÍG.HI.A *i-na* É*-tí-kà* ^{45} *ni-dí-i*, (to Imdilum and Uṣur-ša-Ištar from Amur-Ištar(! Amur-Aššur on tablet):) 'As for the retailers to whom I gave your wool, they have never protested to me (about the price), and I (could) give the equivalent to Uzua, and he (would) depart, and I used to send you a full account! But at the time of the robbery, they, in their stupidity (lit. "in the darkness of their eyes"), deposited wool in my house without checking (its weight), and (robbers) have now snatched (it) away! As soon as the copper had been made ready (for transport) they submitted to the verdict, saying: "We had deposited 2 talents of wool in your house"' (by courtesy of C. Günbattı);

LB 1225.1-12: ^{1} *a-na Im-dí-lim qí-bi-ma* ^{2} *um-ma Ú-zu-a-ma a-šu-mì* ^{3} *pá-ší-ri mì-ma a-na A-mur-*[*Ištar*] ^{4} *e ta-al-pu-ta-am* ^{5} *ṭup-pu-um na-áš-pé-er-tù-*[*um*] ^{6} *ša A-mur-Ištar ù i-a-tí* ^{7} *ša pá-ší-ri nu-uš-ta-áš-qí-i*[*l*$_5$] ^{8} *i-li-kà-kum ṭup-pá-am* ^{9} *ša-ṣí-ir pá-ší-ru a-dì-ni* ^{10} *lá iš-qú-lu i-na* URUDU *ša pá-ší-/ri* ^{11} 5 GÚ URUDU SIG$_5$ *A-mur-Ištar* ^{12} [ ]*-um ša-bu-ma* URUDU *ša pá-*[*ší-ri*] (broken), 'Speak to Imdilum, thus (says) Uzua: write not a word to Amur-Ištar about the retailers! A letter was sent to you by Amur-Ištar and me telling: "We have just made the retailers pay." Guard the tablet well, because the retailers have not paid up till now. Amur-Ištar has been paid ... with 5 talents of fine copper from what the retailers owed, and the copper of the retailers ...';

LB 1275 (Ichisar 1981, pp.337-338), lines 2'-3': $^{2'}$ URUDU*-ša a-na* SIG$_5$ $^{3'}$ *nu-ta-a-ar-ma*, (to Imdilum from Uzua:) 'We will exchange (the wool's) copper for fine (copper)';

See above for TC 1, 53;

almost exclusively done in exchange for copper. Both Amur-Ištar and Uzua contracted retailers (*pāširum*), who paid in copper. They also employed people engaged more often by the firm, as the designation 'our men' in BIN 6, 76.14 indicates. The robbery of wool from Amur-Ištar's house which is described in kt n/k 520, probably took place in Turhumit, because Amur-Ištar lived there.

Wool was partly given to retailers, and partly sold for cash. Imdilum urges Ah-šalim to sell wool for silver or gold, but in TC 2, 37 he can only obtain copper for it. KTS 18 contains the single reference of a sale to a palace. The exchange ratio between wool and copper was 2:1, given as *šanā'um*, that is two parts of wool for one part of copper. The 220 minas of copper were paid through the *kārum*-office. It is not mentioned in which city this palace was situated. Tišmurna is known to have harboured during this period a small Assyrian commercial settlement, called *wabartum* (kt a/k 1429.22-23), and not a *kārum*, and is therefore probably not an option, unless one considers that the term *bēt kārim* was also used in these smaller cities instead of the non-existant *bēt wabartim*. Apparently the palace in Turhumit is meant in KTS 18.

Part of the copper obtained in this way was of poor quality. Uzua, assisted by his wife, brought such copper to a city like Turhumit, where it could be exchanged for fine copper (BIN 6, 76; LB 1275).

oxen

In kt n/k 1385.28', copper is used as equivalent of oxen.

silver

Imdilum asked Amur-Ištar to buy with silver (in Turhumit) a donkey-load of fine copper for paying his expenses (CCT 3, 1).[400] Uzua obtained copper in Tišmurna at 164:1 (CCT 2, 27a.32, see p.54).

The desire to obtain copper of fine quality is demonstrated in several texts. ICK 2, 54 (**2**) contains detailed information on the characteristics of the copper that Imdilum wished to receive from his servant Idnāya, apparently in exchange for merchandise given as *qīptum*-loan. In some instances, the copper received seems to have been of poor quality. This is inferred from attempts mentioned in several texts to exchange 'copper' for fine copper, *ana dammuqim ta''urum* (BIN 6, 76.10-12; CCT 4, 27a.22-25; kt n/k 1385; implied in VS 26, 71); see for this expression and for the related *ka''unum* § 3.2.2. The poor copper was exchanged chiefly in Turhumit. From there, the copper was brought to Purušhattum, where it was sold for silver. In CCT 4, 27a, Uzua reports from Tišmurna that he repeatedly went

---

TC 2, 37 (Ichisar 1981, pp.347-348), lines 17-20: [17] *ší-im . ša-áp-tí-kà* [18] URUDU *i-ba-ší* . KB [19] *ú* KÙ.GI . *a-na-kam . a-šé-e-ma* [20] *lá-šu*, (to Imdilum from Ah-šalim:) 'Your wool can be paid for with copper; I have searched here for silver or gold, but there is not any.'

[400] CCT 3, 1 (Ichisar 1981, pp.367-369), lines 10-20: [10] *[š]í-tí a-sà-r[i]-id Kà-ki-i* [11] ÌR-*ad-k[à]* ANŠE *ir-dí-a-ma* [12] *um-ma a-ta-ma i-na* KB-*im* [13] *a-ni-im* URUDU SIG5 *a-na* [14] *qá-tí-a* ANŠE *ma-li-a-lam* [15] *ù ší-tí* KB *šé-bi4-lam* [16] *ša* 1 2/3 *ma-na* 5 GÍN [17] KB 1 *ma-na* 10 GÍN.TA-*ma* [18] 2 GÚ 2 1/2 *ma-na* URUDU SIG5 [19] *Kà-ki-i a-Tù-ùh-pí-a* [20] *ub-lá-ku-um*, 'I will pack the remainder. Your servant Kaki brought a donkey, and you had said: "Use this silver to load a donkey with fine copper for my expenses, and send me the remainder of the silver." For 1 mina 45 shekels of silver, at a ratio of 70:1, Kaki brought you in Tuhpia 2 talents 2 1/2 minas of fine copper.' Lines 16-20 are partly duplicated by kt n/k 520, 13-16 (by courtesy of C. Günbattı).

to Turhumit to Amur-Ištar and Pūšu-kēn, and suggests bringing (poor) copper to Turhumit to exchange it for fine copper, because there was no (fine) copper to be obtained in Tišmurna. An instance of Assyrian economic policy is contained in a decree by a *kārum* (in Turhumit, or the central *kārum* in Kanish?) which prohibited importing poor copper to Turhumit with the aim of exchanging it there for fine copper (CCT 4, 27a).[401] Because a *kārum* decree (*awātum*) is of a permanent nature, this decision might have been taken in order to prevent a saturation of the local market with copper, or to avoid competition with Anatolian merchants on this point, perhaps laid down in the treaty with the local ruler. Nevertheless, this *kārum* decree was, like the prohibition on smuggling, dodged by the Assyrians as appears from lines 30-31, which suggests that at another time it would be possible to bring such copper to Turhumit, as happens in AKT 1, 17, where some 10 talents of copper are to be exchanged for a better quality in this city.

In general, the copper obtained was sent to other cities to be sold for silver. In TC 2, 36, Uzua reminds Imdilum that his representatives in Purušhattum still possess silver obtained in this way. Further references to the sale of copper for silver are contained in ATHE 63, CCT 4, 26b.26, KTS 18.18, KTS 2, 22, TC 3, 104, and KTH 11. In TC 2, 37.17-20 and ICK 1, 135, the possibilities are mentioned of exchanging the copper obtained by the sale of wool, for silver or gold.

If time was pressing, merchandise was not first exchanged for copper which was afterwards to be sold for silver, but transported directly to the city where profit or a rapid sale seemed most promising. This is illustrated by CCT 4, 22b, where Imdilum urgently needed silver and therefore demanded from his business partners that Amur-ilī should not again sell Imdilum's tin and textiles in Wahšušana for copper, but transport it directly to Purušhattum and to sell it for cash. This illustrates the steps which could be taken to exchange merchandise imported from Assur into silver: the tin and textiles could be exchanged for wool (1), which was exchanged for copper (2), which was exchanged for silver or gold (3). Steps 2 and 3 were caused by the desire to increase profit, and by the local availability of wool and copper and transport capacity. Whether steps 2 and 3 were taken, depended on the time which was available to the merchant for the proceeds of the merchandise to arrive in Kanish, and of course on the availability of wool and copper.

Information on the prices of copper of Haburata and Tiritar quality is given in Ka 276, presumably to Imdilum (see above).

---

401 CCT 4, 27a (Ichisar 1981, pp.314-317), lines 18-31: [18] *ki-ma* URUDU *i-na Tí-iš-mur-na* URUDU *a-na* [19] *kà-ú-nim lá-šu ú a-dí ma-lá ù* [20] *ší-ni-šu a-ṣé-er A-mur-Ištar* [21] *ù Pu-šu-ke-en*$_6$ *e-tí-iq-ma* [22] *um-ma a-na-ku-ma i-na Tí-iš-mur-na* [23] URUDU *a-na kà-ú-nim lá-šu* [24] URUDU *a-na Tur*$_4$*-hu-mì-it lu-šé-ri-/ba-ma* [25] *a-na* SIG$_5$ *lu nu-ta-er* [26] *um-ma šu-nu-ma a-wa-at kà-ri-im* [27] *lá tù-šé-ra-ba-am ù té-er-tí-ni* [28] *a-na A-mur-Ištar a-dí ha-am-ší-šu i-lik* [29] *um-ma a-na-ku-ma* URUDU *lu-šé-ri-ba-am* [30] *a-na* SIG$_5$ *li-tur*$_4$ *um-ma šu-ut-ma lá na-/ṭù* [31] *lá tù-šé-ra-ba-am*, (to Imdilum from Uzua:) 'Because copper - (because) there is no copper to be obtained in Tišmurna, I went repeatedly to Amur-Ištar and Pūšu-kēn, and said: "There is no copper to be obtained in Tišmurna; I will bring the copper into Turhumit, so that we can exchange (it) for fine (copper)." But they replied: "(There is a) *kārum* decree, you cannot bring it into (the city)." Well, we sent no less than five times a message to Amur-Ištar, in which I said: "I want to bring copper (into the city). It should (be) turn(ed) into fine (copper)." But he replied: "It is inappropriate to bring it (now into the city)."'

4.4.3 *Geography of the copper trade*

One of the cities where copper is obtained is *Tišmurna*, where wool and textiles are sold, presumably to the palace (CCT 4, 27a). Poor copper, presumably coming from other, neighbouring towns, could be exchanged for a better quality in Tišmurna, or if that was impossible, in Turhumit (CCT 4, 27a.22-25). The market situation in Tišmurna was at some point disturbed for an unspecified period. In CCT 4, 27a, Uzua writes that 'there is no copper to secure in Tišmurna' (*ka''unum*; also in TC 2, 36), and provides further details about the difficulties of selling various items of merchandise.

In *Kunanamit*, Aššur-nādā had outstanding claims worth 10 talents of copper (CCT 2, 23). Uzua, being in Kunanamit, asked Imdilum in VS 26, 71 to give him a *qīptum*-loan (merchandise on credit) worth 10 minas of silver, which he wanted to trade there for copper. As the locally produced copper was often of poor quality, he probably also brought the copper to Turhumit in order to exchange it for a better quality. It was in Turhumit that he would hand over 10 talents of refined copper to Imdilum's representatives, who would ship it southwards to sell it for silver in Purušhattum (VS 26, 71).

Copper belonging to Aššur-nādā was moved out of the country of *Sawit* (*māt Sawit*) by Ennum-Aššur and seized, probably in Turhumit, by Amur-Ištar who entrusted it to Ikuppia (CCT 2, 23 and BIN 6, 133).

Turhumit is attested as the city where fine copper was obtained and where copper of poor quality was exchanged for that of fine quality. In ICK 2, 54 (**2**), 6 1/2 talents of fine copper, partly copper of Taritar quality, had to be obtained in Turhumit and shipped to Kanish. Puzur-Ištar, son of Imdilum, ordered Suen-rē'ī son of Amur-Ištar to pay in Turhumit 55 minas of broken refined copper (OIP 27, 56). In CCT 3, 1, Amur-Ištar bought 2 talents 2 1/2 minas of fine copper in Turhumit and sent it to Imdilum who was staying in Tuhpia.

The bulk of the copper obtained was brought to *Purušhattum*, where it was sold for silver (TC 2, 36; VS 26, 71). Amur-Ištar, in Turhumit, wrote to Imdilum that he planned to exchange Imdilum's copper for that of fine quality, and then transport it together with his own copper to Purušhattum, apparently by wagon (from Turhumit?, CCT 4, 47a). Puzur-Aššur would sell the copper that he hoped to receive from Amur-Ištar and Lulu for silver in this city (ATHE 63). Imdilum ordered the transport of fine quality copper to this city in CCT 6, 37a and kt n/k 1385. In kt n/k 1385, any available poor copper must first be exchanged for fine copper.[402] In CCT 4, 22b, merchandise had been exchanged for copper in

---

[402] Kt n/k 1385.5'-30': [5'] ... *šu-ma* [6'] *[A]-mur-Ištar me-eh-ra-tim* [7'] [URU]DU SI[G$_5$] *i-da-an . li-dí-in* [8'] *iš-tí* URUDU-*i-a . lu-šé-ri-bu-ma* [9'] *šu-ma lá i-da-an . a-na* [10'] *i-ta-aṭ-lim . a-na* URUDU SIG$_5$ [11'] *dí-na-šu-nu . šu-ma a-na i-ta-aṭ-lim* [12'] *la i-ba-ší .* TÚG-*tí-a iš-tí* [13'] URUDU-*i-a . a-na Pu-ur-ha-tim* [14'] *lu-šé-ri-bu . mì-ma* [15'] *lá té-zi-ba-a* 10 *pì-ri-kà-nim* [16'] *na-ar-bu-tim* [17'] *ša Ištar-ba-áš-tí ší-im-šu-nu* [18'] URUDU SIG$_5$ *Ta-ri-ta-ra-i-am* [19'] *a-na Ú-zu-a . dí-in-ma* [20'] KI URUDU-*i-a . a-na Pu-ur-ha-tim* [21'] *lu-šé-ri-bu . ta-ša-pá-ra-am* [22'] *um-ma a-ta-ma* 1 GÚ URUDU [23'] *la-mu-nu-um i-ba-ší mì-šu-ba* [24'] KI URUDU *na-am-i-dim lá ta-dí-na-šu* [25'] URUDU *ší-im . ma-áš-ki-ni-a* [26'] *ù* URUDU *la-mu-num ša i-ba-ší-ú* [27'] *a-na* SIG$_5$ *. ta-e-ra* [28'] *ší-im* GUD.HI.A URUDU *ša Ta-ri-tár* [29'] *a-na Ú-zu-a dí-in-ma* [30'] KI URUDU-*i-a . lu-šé-ri-ib*, (beginning broken, but most likely sent by Imdilum:) 'If Amur-Ištar can give the equivalent in fine copper he should do so, and let them bring (this copper) in

Wahšušana; subsequently, this copper seems to have been sent to Purušhattum. CCT 1, 43 is situated in Purušhattum, where Ennam-Assur, Usānum, and Adu report that shares consisting of copper and due to Puzur-Aššur and Imdilum were received from the *kārum*-office (see § 5.6). Also in CCT 2, 8 the local palace bought textiles and paid with copper.

Only one reference is known for the sale of merchandise for copper in the city of *Wahšušana*. Amur-ilī, apparently on his own intitative, had sold tin and textiles belonging to Imdilum for copper (CCT 4, 22b).

Copper was sold in *Ulama* as well. In TC 3, 104, Adu and Ikuppia are staying there to sell copper for silver (see above). I 608 qualifies an amount of 4 minas of silver as 'the proceeds from the sale of copper of Ulama and textiles of Wahšušana'.

According to CCT 3, 1, and kt n/k 520, a donkey-load of fine copper was sent to Imdilum in *Tuhpia*.

Imdilum ordered his servant Idnāya to bring from Turhumit 6 1/2 talents of fine quality copper to *Kanish* (ICK 2, 54 = **2**).

In the letter KTS 2, 36 + Bursa 3776, Imdilum writes to his superiors in *Assur* about merchandise they had bought for him with silver he had sent.[403] The textiles, leather bags, donkey harness, and a small expense were paid in copper, the other expenses were paid in silver. This means that the men in Assur had not waited for the arrival of the silver, but had bought the merchandise and other necessary goods ahead of the servant's arrival. Imdilum

---

together with my copper. If he cannot give it, you should sell (the textiles) cash for fine copper. If it is impossible to sell for cash, let them bring my textiles together with my copper to Pur(š)hattum; do not leave anything behind! As for the 10 soft *pirikannum*-textiles of Ištar-bāštī, give their proceeds in fine Taritar copper to Uzua, that they may bring it together with my copper to Pur(š)hattum. Now you write me: "There is 1 talent of poor copper in store." How is it possible that you did not sell it together with the bulk of the copper? Exchange the copper proceeds of my "tents" and what poor copper is available for fine copper; give Taritar copper as payment for the oxen to Uzua and let him send it together with my copper' (by courtesy of S. Çeçen).

403 KTS 2, 36 can be joined with Bursa 3776 (published in Donbaz and Horasanlı 1976, Pl. 6), 1-26: [1] *a-na A-šur-i-mì-ti* [2] *Šu-Ištar En-na-nim ù* [3] *Bu-bu-ra-nim qí-bi-ma um-ma* [4] *Im-dì*-DINGIR-*ma* (erasure) 20 *ma-na* [5] KB *ni-is-ha-sú* DIRI *ša-du-a-sú* [6] *ša-bu Ku-tal-la-nu-um . ub-lá-ku-nu-ti* [7] *i-na* 20 *ma-na* KB *ša a-na* [8] *ší-mì-im ša-a-mì-im ú-šé-bi-la-*⌈*ku*⌉*-/nu-ti-ni* [9] 1 *me-at ku-ta-nu* [10] 16 GÚ 50 2/3 *ma-na* URUDU [11] *it-bu-lu* 20 *na-ru-qá-tum* 9 1/2 *ma-na* [12] URUDU *it-b[a-a]l* 6 *ma-na* URUDU *ú-nu-ut* [13] ANŠE.HI.A 4 *m[a-na* URUDU *a]-ša sà-e-dim* [14] (erased) [15] *ta-áš-qú-la* 4 ANŠE.HI.A *ṣa-la-mu* [16] 1 *ma-na* 15 GÍN KB *it-bu-lu* [17] 20 *ma-na* AN.NA *qá-tim* 14 GÍN.TA [18] KB-*ap-šu* 1 1/3 *ma-na* 5 2/3 GÍN [19] 7 GÍN KB *wa-ṣí-tum* [20] ŠUNIGIN [1]7 GÚ 10 *ma-na* URUDU *ù* 2 2/3 *ma-na* 7 2/3 GÍN [21] *ša* 17 GÚ 10 *ma-na* 10 GÍN URUDU [22] 1 2/3 *ma-na*.TA KB-*ap-šu* [23] 10 *ma-na* 18 1/2 GÍN [24] ŠUNIGIN [13 *m]a-na* 6 1/6 GÍN KB *lu-qú-tí* [25] *i[t-ba-a]l* 6 5/6 *ma-na* 4! 1/6 GÍN [26] *[ší-tí* K]Ù.B-*pí-a i-na li-bi-ku-nu*, 'Speak to Aššur-imittī, Šū-Ištar, Ennānum, and Buburānum, thus (says) Imdilum: Kutallānum brought you 20 minas of silver, its *nishātum*-tax was added and its *šaddu'atum*-tax fully paid. With the 20 minas of silver that I sent to you for the purpose of making purchases (you bought): 100 *kutānum*-textiles, costing 16 talents 50 2/3 minas of copper; 20 leather bags, costing 9 1/2 minas of copper; 6 minas of copper for the harnass of the donkeys; you paid 4 minas of copper for that of the *sa'udum*. 4 black donkeys, costing 1 mina 15 shekels of silver; 20 minas of hand tin, at a rate of 14:1, its value in silver is 1 mina 25 2/3 shekels; the export tax cost 7 shekels of silver; in all 17 talents 10 minas of copper and 2 minas 47 2/3 shekels (of silver). The value in silver of the 17 talents 10 minas 10 shekels of copper was, at a rate of 100:1, 10 minas 18 1/2 shekels. My merchandise cost in all 13 minas 6 1/6 shekels of silver, (so that) you owe me 6 minas 54 1/6 shekels, the remainder of my silver.'

uses an exchange ratio between copper and silver of 100:1 for the copper in Assur when calculating the value in silver of the purchase.

In TC 3, 53, Ennum-Bēlum writes from Assur to Imdilum about the difficulties he experiences in making Pilah-Ištar son of Uzua buy merchandise in the firm's office in Assur with 20 minas of silver claimed by Imdilum. Instead of 20 minas of silver, the son of Uzua brought 10 talents of copper into the office, but ultimately declined to make any purchase.[404]

4.4.4 *Conclusion*

Imdilum's archive is important because of the details it contains about the sale in Tišmurna of wool for copper and of its subsequent exchange for that of a better quality if poor copper was received. Imdilum's uncle Amur-Ištar and Uzua played a key role in exchanging merchandise for copper in the region in and around Turhumit and Tišmurna. The main supplier of copper was the palace. Imdilum supervized to a large extent the sale of his merchandise for copper and the further exchange of it for silver, helped by his agents who kept him well informed on the economic situation. The use of copper in Assur is illustrated by two texts.

4.5 *Puzur-Aššur son of Išar-kitt-Aššur*

4.5.1 *Participants*

Most of the texts dealing with the affairs of Puzur-Aššur son of Išar-kitt-Aššur originate from illicit diggings, and therefore, like part of Imdilum's archive, have been dispersed over a number of museums and private collections, but by now nearly all are published. It is doubtful whether all these tablets stem from Puzur-Aššur's house in Kanish, or formed part of the archive of some other merchant. It is known from documents describing the aftermath of his death, that his tablets and *tamalākum*-containers, in which possibly tablets were also kept, were removed from the strongroom (*maṣṣartum*) of his house and were ultimately handed over to Buzāzu. The reason for taking these tablets out of his house was largely financial, which means that the debt-notes were of greater importance than for instance the letters, so part of Puzur-Aššur's archive could have been left behind in his house. The popularity of the name Puzur-Aššur, for which at least 36 different patronymics are attested, makes it difficult to define which texts belong to the archive of Puzur-Aššur son of Išar-kitt-Aššur, among documents where generally no patronymic is given. An important part of his archive could be recovered by studying a group of letters exchanged with the same person, and then extending this circle to other persons who emerged from this first group as important business partners; a method also applied by C.

---

404 TC 3, 53 (Ichisar 1981, pp.342-343), lines 16-21: [16] ... *wa-ar-kà-/num* [17] 10 GÚ URUDU *ú-šé-ri-ba-ma* [18] *um-ma šu-ut-ma . ší-ma-am* [19] *i-na bé-tí-ku-nu . a-ša-am* [20] *mì-ma . ù-lá iš-a-am* [21] *i-tù-ur-ma . uš-té-ṣí*, ‘Later, he brought 10 talents of copper into (the house of our father) and said: “I will make purchases in your house.” But he did not buy anything, he retracted and took (the copper) out (of the house).’

Michel in her study of the archive of two merchants called Innāya (1991a). The correspondence between Puzur-Aššur and Ilī-wēdāku is the point of departure of this archive; and as Ilī-wēdāku was cooperating with Buzāzu, the latter's father Pūšu-kēn became involved.

Puzur-Aššur was a business partner (*tappā'um*) of Pūšu-kēn,[405] and had close ties with him and his family. In 9 out of 10 letters exchanged between both men, Puzur-Aššur was staying in Assur, during which time Pūšu-kēn acted as his representative in Kanish.[406] His stay in the capital was not permanent; after Pūšu-kēn's death he worked in Anatolia where he cooperated with Buzāzu,[407] who would play an important role in the aftermath of Puzur-Aššur's death.[408]

Pūšu-kēn possesses a higher status than Puzur-Aššur, and is called 'my father and lord' by him (CCT 2, 38; TC 2, 9). Five texts dating from the period Puzur-Aššur was absent from Kanish, deal with copper. In BIN 4, 31 (**7**), Pūšu-kēn and his son Suea report about various transports of copper belonging to Puzur-Aššur which they checked upon arrival in (probably) Purušhattum. They promise to exchange the copper for silver and send the proceeds to Assur. In CCT 2, 38, Puzur-Aššur wishes to receive 10 minas of silver in Assur, 6 minas representing the value of copper owed to him by a palace.[409] In TC 2, 9, Puzur-Aššur quotes from a letter by Pūšu-kēn who had written him to take out of 25 minas of silver brought by Šū-Ištar (as much silver as represents the value of) his copper and that of the *tamkārum*.[410] In RA 81, 1, Aššur-malik and Adad-ṣulūlī, staying in a city west of Kanish, send silver and gold to Pūšu-kēn. They ask him to send 7 minas of silver, proceeds or value of 7 talents of Puzur-Aššur's copper to him.[411] Aššur-malik, Ea-šar and Adad-

---

405 Compare BIN 6, 253, where Aššur-muttabbil (son of Pūšu-kēn) speaks to Enna-Suen, [6] *i-na mu-ú-ut Puzur$_4$-A-šùr* [7] [DUMU *I*]-*šar-ki-tá-A-šùr*, 'At the death of Puzur-Aššur son of Išar-kitt-Aššur' with OIP 27, 57, where Buzāzu interrogates Ilī-wēdāku, [5] *lá li-bi* DINGIR-*ma Puzur$_4$-A-šùr* [6] *tap-pá-a* É *a-bi$_4$-ni me-et*, 'Unfortunately Puzur-Aššur, the partner of our father's house, has died.'

406 Letters sent from Assur by Puzur-Aššur: BIN 4, 221; C 11; CCT 2, 38; TC 1, 6; TC 2, 7; TC 2, 8; TC 2, 9; VS 26, 52; sent to him: BIN 4, 31.

407 See the complaint of Pūšu-kēn's daughter Ahaha expressed to Puzur-Aššur in I 680.3-4: [3] *e-mu-uq . a-bi-ni* [4] *ku-lu-šu i-tág-ma-ar*, 'The whole capital of our father has been spent' (by courtesy of K. Hecker).

408 See Matouš 1969.

409 CCT 2, 38.7-25: [7] *um-ma a-ta-ma me-he-er-kà* [8] *e-ez-ba-ma* KB *ú-ša-áš-qal-ma* [9] *wa-ar-kà-at-kà ú-šé-ba-lá-kum* [10] *ù me-eh-ra-at* URUDU *ša* É.GAL-*lim* [11] *ù* 4 *ma-na* KB *ša ṭup-pé-e* [12] *ha-ru-mu-tim* 10 *ma-na* KB [13] *i-na ra-mì-ni-a* [14] *a-na-dí-ma ú-šé-ba-lá-kum* [15] *a-bi a-ta be-lí a-ta* [16] *i-hi-id-ma* KB 1 GÍN [17] *ša a-ma-kam e-zi-bu* [18] *za-ki-ma ú* 10 *ma-na-e* [19] KB *ša pá-kà ta-dí-na-ni* [20] *i-dí-ma* KB [21] *ku-nu-uk-ma a-na* [22] *a-lim*[ki] . *li-ik-šu-da-ma* [23] *ší-ma-am qá-tí* [24] *li-iš-a-ma* KB 1 GÍN [25] *li-li-a-am*, 'You said: "Leave your copy (of the debt-note) with me, and I will collect the silver and send it you after your departure. And as for the equivalent of the palace's copper, and the 4 minas of silver of certified debt-notes, I will bring together out of my own means 10 minas of silver and send it you." My dear father and lord, take care and make available every shekel of silver that I left behind there, and bring together the 10 minas of silver you promised me, seal the silver and send it me in the City, so that I can personally make purchases and earn some silver!'

410 TC 2, 9.9-11: [9] *ù i-na* 25 *ma-na* KB [10] *ša šé-ep Šu-Ištar* URUDU *ku-a-am* [11] *ù ša* DAM.GÀR *le-qé-ma*, 'And take from the 25 minas of silver which Šū-Ištar brings (the equivalent) of your copper and that of the merchant.'

411 RA 81, 1.14-18: [14] 7 *ma-na* KB [15] *ší-im* URUDU 7 GÚ *ša* [16] *Puzur$_4$-A-šur* [17] *a-ṣé-er Puzur$_4$-A-šur* [18] *šé-bi$_4$-il$_5$*.

ṣulūlī report to Pūšu-kēn in TC 2, 18 (**13**) that Ea-šar paid the final instalment of the copper he owed to Puzur-Aššur. They ask for donkeys to transport the copper to Purušhattum where they intend to sell it.

Buzāzu son of Pūšu-kēn cooperated with Puzur-Aššur. Not surprisingly, he possessed a lower status than Puzur-Aššur and therefore called him 'my father and lord'. Together with Ilī-wēdāku he occurs in a number of texts dealing with Puzur-Aššur's copper. He came into conflict with Ilī-wēdāku about copper that belonged to him and also to Puzur-Aššur (BIN 4, 35 = **6** and CCT 3, 32).[412] The word 'our copper' in BIN 4, 2.8 and BIN 4, 48.5 implies a form of cooperation between him and Puzur-Aššur. Together with Adad-ṣulūlī he sold *sû*-stones for copper for Puzur-Aššur (KTS 7a, see below).

A small dossier is concerned with a claim Buzāzu had on Puzur-Aššur, as a result of which he was entitled to Puzur-Aššur's share of 45 talents of *šikkum* copper from the *suhuppum* in Wahšušana.[413]

Suea, one of Buzāzu's brothers, assists his father during the inspection of copper consignments arriving in Purušhattum (BIN 4, 31 = **7**). Like his brother, he calls Puzur-Aššur 'my father' (VS 26, 56). In a letter to Puzur-Aššur he mentions copper stored in Hattum (VS 26, 56, see below).

Ah-šalim is higher in rank than Puzur-Aššur, for whom he collected copper in Wahšušana, and had Aššur-mālik bring it to him (CCT 2, 29 = **11**). It is unclear whether he is identical with Ah-šalim son of Šū-Anim, who originally had to pay 18 talents 45 minas of fine copper to 'the merchant' in Wahšušana; because Puzur-Aššur seems to have needed this copper urgently, Ilī-wēdāku thought it better to have this copper paid instead to Puzur-Aššur in Wašhania, the 'merchant''s agents would then receive the copper from Ilī-wēdāku. The way in which Puzur-Aššur and Ilī-wēdāku could use copper due to the 'merchant' indicates they belonged to the 'merchant''s firm or at least had close ties with it (TC 3, 10).[414]

Ilī-wēdāku is a junior business partner of Puzur-Aššur. Some letters demonstrate Puzur-Aššur's irritance with Ilī-wēdāku's behaviour (for instance in CCT 2, 39), whereupon the latter responds by expressing his loyalty, as in KTS 2, 34.[415] He travelled

---

412 CCT 3, 32.20-30: [20] *a-na-kam i-ṣé-er* URUDU-*im* [21] *ki-li-šu Bu-za-zu* [22] *i-ir-tí-bi*$_4$*-iṣ šé-bu-lam* [23] *la i-mu-a ṣa-al-tù-um* [24] *i-na ba-ri-šu-nu i-té-be-e* [25] *i-na* ^{d}UTU-*ši ṭup-pì-i* [26] *ta-ša-me-ú té-er-ta-kà* [27] *a-ṣé-er Ì-lí-we-da-ku* [28] *ù Bu-za-zu li-li-kam-ma* [29] *mì-ma Bu-za-zu iš-tí* [30] *Ì-lí-we-da-ku lá i-ṣa-al* !, (to Puzur-Aššur from Adad-ṣulūlī:) 'Buzāzu now sits here on top of all the copper and refuses to send it. A quarrel has arisen between them. Write, on the day you receive my letter, a letter to Ilī-wēdāku and Buzāzu in which you order Buzāzu not to quarrel with Ilī-wēdāku.'

413 CCT 2, 31a.21-27: [21] *ṭup-pá-am ša* 45 GÚ [22] URUDU *ší-ki-im ša i-na* [23] *Wa-ah-šu-ša-na i-na* [24] *sú-hu-pì-im lá-qá-a-am* [25] *qá-bi*$_4$*-a-tí-ni ṭup-pì-a* [26] *pí-tí-a-ma i-li-bi*$_4$ [27] *ṭup-pì-a šu-uk-na-šu-ma*, (to Puzur-Aššur from Buzāzu:) 'As for the document stating that you are entitled to collect 45 talents of *šikkum* copper from the *suhuppum* in Wahšušana: get access to my tablets and place (it) among them.' To this dossier belong also RA 60, MP 1; BIN 4, 112 + CCT 6, 17a; BIN 4, 160 (**9**).

414 TC 3, 10.10-15: [10] 18 GÚ 45 *ma-na* URUDU [11] *ma-sí-am ša-bu-ra-am* [12] *i-ṣé-er Ah-ša-lim* [13] DUMU *Šu-Anim* DAM.GÀR [14] *i-šu i-na Wa-ah-šu-ša-na* [15] *ša-qá-lam qá-bi*, 'Ah-šalim son of Šū-Anim owes 18 talents 45 minas of broken refined copper to the merchant; he has to pay in Wahšušana.'

415 KTS 2, 34.1'-7': [1'] ... *a-l[a-nu-kà]* [2'] *a-ba-am ša-ni-a-am* [3'] *i-šu-ma i-a-tám e-pá-áš-ma* [4'] *ku-wa-tám a-na-dí a-bi*$_4$ *a-ta* [5'] *ki-ma ša qá-qá-ad-kà* [6'] *wa-áš-bu a-na té-er-tí-kà* [7'] *a-na-hi-id*, 'Do I have

in Anatolia and to Assur. Ilī-wēdāku's role is that of an agent who stays at times in an important and strategically located city, in his case perhaps Wahšušana,[416] and from there sells copper belonging to his boss Puzur-Aššur by using retailers, or transports the copper to other cities to sell it there. Nearly half of the surviving correspondence between both merchants deals with copper, which is an indication of his (and the copper's) importance for Puzur-Aššur's trade. The memorandum TC 3, 177 illustrates his activities as a debt collector and seller of merchandise. The proceeds mentioned in the first three lines will have consisted of copper.[417] It seems Ilī-wēdāku's decisions were taken only after receiving pertinent directives from Puzur-Aššur, who was after all the owner of the metal. In turn, Puzur-Aššur played an crucial role in Ilī-wēdāku's private trade. He received silver and gold from Ilī-wēdāku's proceeds to have merchandise purchased for him in Assur or to fulfill his obligations towards the *kārum*-office in Kanish, and sent transporters carrying Ilī-wēdāku's merchandise to him. Although Ilī-wēdāku was thus involved in common commercial activities, in his correspondence he never mentions copper belonging to himself, only that of Puzur-Aššur. In CCT 4, 12b (**12**) and BIN 4, 43, Ilī-wēdāku reports problems he had with Aššur-bānī son of Kutāya. This man appears to have contracted himself as a retailer to sell for Puzur-Aššur 50 talents of refined copper, but when Ilī-wēdāku offers him this quantity with the aim of bringing it to Purušhattum to sell it there, he refuses to accept it, but is willing to transport 40 minas of silver to Puzur-Aššur (in Kanish).[418]

---

another father than you, that I would pursue my own affairs and neglect yours? My dear father, I heed your instructions as if you were present in person!'

416 In BIN 4, 43, Ilī-wēdāku was residing in Wahšušana when Aššur-bānī son of Kutāya met him.

417 TC 3, 177: [1] *ší-im* 30 *ni-ki ša* GUD.HI.A [2] *ší-im ni-ki ša* UDU.HI.A [3] *ší-im* SÍG.HI.A . *ú-ṭù-tum* [4] *ú ṣí-ba-sà . ša e-zi-bu-šu-ni* [5] *ṭup-pu-ú ha-ru-mu-tum* [6] *ša iš-tí-šu . i-ba-ší-ú-ni* [7] 10 *ma-na* KB KI . [d]IM-ANDUL [8] *il*$_5$*-qé* 25 GÚ URUDU SIG$_5$ [9] *ù* 13 GÚ SIG$_5$ [10] *ṣí-ib-tám . i-ṣé-er* [11] *Puzur*$_4$*-Sú-in il*$_5$*-qé* [12] 7 *ma-na* KB *ṣí-ib-tám* [13] *i-ṣé-er A-šur-ták-lá-ku* [14] DUMU *Am-rí-a il*$_5$*-qé* [15] *ṣí-bu-ut* . KB [16] *ša ší-im* TÚG.HI-*tí . i-na* [17] *Ú-lá-ma a-hu-ur* [18] 10 TÚG *iš-tí Ah-ša-lim* DUMU *Šu-A-/nim* [19] *ší-im* TÚG.HI-*tí . ša a-na-ku* [20] *e-zi-bu-šu-ni* [21] *mì-ma a-nim i-li-bi*$_4$ [22] *Ili*$_5$*-w[e-d]a-ku*, 'The proceeds of 30 hides(!; see below, n.427) of oxen; the proceeds of the hides(!) of sheep; the proceeds of wool; the grain and interest on it which I left behind with him; the certified debt-notes that are with him; 10 minas of silver he collected from Adad-ṣulūlī; 25 talents of fine copper and 13 talents of fine (copper) as interest he collected from Puzur-Suen; 7 minas of silver he collected as interest from Aššur-taklāku son of Amria; the interest on the silver which was paid for textiles is outstanding in Ulama; (the proceeds of) 10 textiles (are) with Ah-šalim son of Šū-Anim; the proceeds of the textiles that I had left behind with him; all this Ilī-wēdāku owes (me).' The memorandum CCT 6, 46b, mentions a 'he' to whom 25 minas 50 shekels of silver and 6 minas 18 shekels of gold were brought and who also received or owned a total of 35 talents 16 minas of fine copper. This text might have been written by Ilī-wēdāku because of the occurrence of a man named Hutka, who is recorded in a number of texts travelling between Puzur-Aššur and Ilī-wēdāku; 'he' then is Puzur-Aššur.

418 BIN 4, 43.1-27: [1] *a-na Puzur*$_4$*-A-šùr* [2] *qí-bi-ma um-ma Ili*$_5$*-we-da-/ku-ma* [3] *a-dí ša A-šur-ba-ni* [4] DUMU *Ku-ta-a ša ta-áš-pu-ra-ni* [5] *um-ma a-ta-ma* URUDU [6] 50 GÚ *ma-sí-am* [7] *[i-na] ṣé-ri-šu* [8] *[ù i-ṣ]é-er Šu-li* [9] *[i-šu]-ma* KB [10] [ ]*-ma* [11] [ ] *áš-ku-un* [12] *[iš-tù té-er-ta]-kà* [13] *[i-li-k]à-ni u*$_4$*-mu* [14] *[im-tí-du]-ma* URUDU [15] *[x - x]-uk*? *wa-ar-ki-/tám-ma* [16] *[a-n]a Wa-ah-šu-ša-na* [17] *i-li-kam-ma um-ma a-na-ku-/ma* [18] *a-ma-lá té-er-tí a-wi-lim* [19] URUDU 50 GÚ SIG$_5$ *a-Pu-ru-uš-ha-/tim*! [20] *šé-ri-ib um-ma šu-ut-ma* [21] *lá a-lá-ak um-ma* [22] *a-na-ku-ma iš-tù-ma* [23] *lá ta-ma-gu*$_5$*-*<*ru*> 5 *u*$_4$*-me* [24] *qá-i-ma* 40 *ma-na* KB [25] *lá-ak-nu-[kà-kum]* [26] *um-ma šu-ut-ma* [27] *lá a-kà-lá*, 'Speak to Puzur-Aššur, thus (says) Ilī-wēdāku: as for Aššur-bānī son of Kutāya, about whom you wrote me

Adad-ṣulūlī (see § 4.2) figures as an agent of Puzur-Aššur; he sells merchandise for copper, as in BIN 4, 36,[419] where he sells textiles in Turhumit, and collects copper from others. He takes care of the delivery of copper but lacks the means of transport for it in TC 2, 18 (to Purušhattum) and CCT 3, 49b. In the last text, he and Ikuppia intend to use porters, but as that is impossible, they ask Ilī-wēdāku for donkeys. In TC 3, 5, Ikuppia tries in vain to obtain copper owed by Adad-ṣulūlī to Puzur-Aššur from his representative Iddi(n)-Ištar.

A person called *Ikāya* sells tin for copper in Turhumit, and has to defend himself against complaints uttered by Puzur-Aššur that he is selling it for poor copper (ATHE 38).

Ikuppia is an agent as well, who sells silver and gold for copper, and transports copper to Wahšušana (RA 60, 35) which he possibly had acquired through the sale of tin and textiles belonging to Puzur-Aššur (TC 2, 33 = **5**).

Ikuppī-Aššur, if a different person from Ikuppia, is another agent, who advises Ilī-wēdāku on how to fetch a better price for Puzur-Aššur's copper (KTS 6 and C 13).[420]

Ea-šar is an agent who sells merchandise for copper (TC 2, 18 = **13**), and has this sent to Puzur-Aššur's representatives in BIN 4, 31 (**7**).

as follows: "He and Šuli owe me 50 talents of refined copper and silver ..." Many days [have passed since your letter reached] me and ... the copper. But afterwards he came to Wahšušana, and I said to him: "Bring 50 talents of fine copper to Purušhattum, in accordance with the gentleman's instructions!" But he replied: "I won't go." I told him: "As you do not agree (to this), wait 5 days and I will hand over to you 40 minas of silver under seal," and he answered: "I will not wait."'

419 BIN 4, 36.6-11: [6] ... TÚG.HI-*tí-kà* [7] *ù* 4 ANŠE.HI-*kà* . *i-šé-pì-a* [8] *a-na Tur$_4$-hu-mì-it* . *ú-šé-tí-iq* [9] *um-ma a-na-ku-ma* . *ma-tí* URUDU [10] 10 GÚ *lu ša I-ku-pì-a* . *lu ša* [11] TÚG.HI-*tí-kà* . *a-ba-kà-kum*, (to Puzur-Aššur from Adad-ṣulūlī:) 'I transported your textiles and 4 of your donkeys to Turhumit and thought: "When will I bring you 10 talents of copper, both for Ikuppia and for your textiles?"'

420 KTS 6.34-45: [34] *a-na-kam I-ku-pì-A-šur iq-bi$_4$-a-am* [35] *um-ma šu-ut-ma iš-tù-ma* [36] KB-*pu da-nu-ni* URUDU SIG$_5$ [37] 10 GÚ.TA *ša té-pu-šu* [38] *áb-kam-ma* 1 *ma-na*.TA *ù* <1 GÍN.TA> [39] *a-na* ITU.1.KAM-*im iš-té-en$_6$* [40] *ù ší-na lá i-ba-ta-qá* [41] *té-er-ta-kà li-li-kam-ma* [42] URUDU *pá-ni-a-ma* . *ša e-pu-šu* [43] *ma-lá maš-kà-ta-kà té-pu-šu* . *lá-bu-kà-ku-ma* [44] 1 *ma-na*.TA *ù* 1 GÍN.TA . *a-na* ITU.1.KAM-*im* [45] *iš-té-en$_6$ ù ší-na* . *dí-šu*, 'Here, Ikuppī-Aššur spoke to me as follows: "Because silver is scarce, bring fine copper, every time you obtain some 10 talents, (here); (and give it to a retailer at a rate of) 1 mina (of copper) per <shekel (of silver)>, for (a term of) one or two months. It must not yield less!" Send me a message, and I will bring to you the first copper that I obtain, as much as you formed assets, then give it (to an agent) for 1 shekel (of silver) for each mina of (copper) for a term of one or two months';
C 13.19-29: [19] ... *ki-ma a-na-kam* [20] KB-*pu da-nu-ni a-na-kam* [21] *I-ku-pì-A-šur iq-bi$_4$-a-am* [22] *um-ma šu-ut-ma* URUDU *ša* KB [23] 10 *ma-na ma-la ta-kà-šu-du* [24] *áb-kam-ma a-na* ITU.1.KAM-*im iš-té-en$_6$* [25] *ù ší-na a-šar ša lá ša-ha-tim* [26] *dí-šu-ma* 1 *ma-na*.TA [27] *ù* 1 GÍN.TA *la i-ba-ta-aq* [28] *i-na a-ni-tim* [29] *li-ba-am ta-da-nam*, 'Because silver is scarce here, Ikuppī-Aššur spoke here to me as follows: "Bring copper for a value of some 10 minas of silver, as you can obtain (it), and give it for a term of one or two months duration to (an agent) about whom you do not have to fear, and it must not yield less than 1 shekel per mina (of copper); encourage me by this!"'

4.5.2 *Procurement and sale of copper*

Copper is obtained in Puzur-Aššur's archive by selling a variety of commodities:

textiles

Details on the sale of textiles for copper are contained in ATHE 66, where white *lubūšum*-textiles are sold at 2 1/2 talents each for Kulumāya, and RA 60, 35, where *šulupkā'um*-textiles are sold at 50 minas of fine copper each; the sale is implied in BIN 4, 36 (in Turhumit, see above) and CCT 2, 29.[421]

tin

Tin is sold for fine copper in BIN 4, 35 (**6**), probably in Wahšušana. The distribution of tin over retailers is hinted at in KTS 2, 34, where the ultimate objective is to fetch a high price in silver for the tin. In RA 60, 35, tin is sold for fine copper at a rate of 11:1, the same rate as in ATHE 38, where tin is sold in Turhumit, but a small quantity is given for poor copper at a rate of 30:1.[422]

---

421 ATHE 66.21-37: [21] ... *a-na Ku-lu-ma-a* [22] *qí-bi-ma* 12 TÚG *lu-bu-šé-e* [23] *pá-ṣí-ú-tim* 2 1/2 GÚ.TA [24] LUGAL-*Sú-in i-dí-in-ma* [25] URUDU *a-na Ša-lim-A-šur* DUMU *A-šur*-UTU-*ši* [26] *e-zi-ib* ŠÀ.BA 3 1/2 *ma-na* KB [27] *Da-da-a a-hu-šu ú-šé-bi-lam* [28] *i-na Wa-ah-šu-ša-na Da-da-a* [29] IGI *A-hu-qar aṣ-ba-sú-ma* [30] *um-ma a-na-ku-ma ší-tí* URUDU *dí-nam* [31] *um-ma šu-ut-ma a-lá-an* KB [32] *a-nim mì-ma ší-im* TÚG.HI.A [33] *ša-nim i-li-bi$_4$-a lá i-ba-ší* [34] *a-ma-kam* IGI 2 *ší-na ra-ki-sú* [35] *a-šar* URUDU *mì-ma i-li-bi$_4$-šu* [36] *ib-tù-ru iš-tù u$_4$-mì-im ša* URUDU [37] *il$_5$-qé-ú* 1 1/2 GÍN.TA *lu-úṣ-ba-am*, (to Iddi(n)-abum and Kulumāya from Puzur-Aššur:) 'Speak to Kulumāya: Šarra-Suen sold 12 white *lubūšum*-textiles at 2 1/2 talents (of copper) each, and he left the copper behind with Šalim-Aššur son of Aššur-Šamšī. Of this his brother Dadāya sent me 3 1/2 minas of silver. I seized Dadāya in Wahšušana in the presence of Ahu-(wa)qar and said: "Give me the remainder of the copper!" He replied: "Apart from this silver, I do not owe any other proceeds of textiles." Bind him over there by contract before two witnesses so that, wherever a debt by him in copper turns up, he will add interest of 1 1/2 shekel per (mina per month) from the day he received the copper';

CCT 2, 29.21-23: [21] 1 GÚ 10 *ma-na* URUDU SIG$_5$ [22] *i-na ší-im ṣú-ba-tí-kà* [23] *Šu-Ištar ub-lá-kum*, 'Šū-Ištar brought you 1 talent 10 minas of fine copper from the proceeds of your textiles';

RA 60, 35.1-10: [1] *a-na Puzur$_4$-A-šur q[í]-bi$_4$-ma* [2] *um-ma I-ku-pí-a-ma* [3] *ta-áš-pu-ra-am um-ma* [4] *a-ta-ma ší-im* AN.NA-*ki-a* [5] *ù* TÚG *šu-lu-up-kà-e-a a-na* [6] *Wa-ah-šu-ša-na šé-bi$_4$-lá* [7] *a-na* AN.NA-*ki-kà* 11 *ma-na*.TA [8] *ù šu-lu-up-kà-e-kà* [9] 50 *ma-na*.TA URUDU SIG$_5$ [10] *ú-šé-ba-al*, 'Speak to Puzur-Aššur, thus (says) Ikuppia: you wrote me: "Send the proceeds of my tin and *šulupkā'um*-textiles to Wahšušana." I will send fine copper, for your tin 11 minas (of copper) per (mina of tin), and for your *šulupkā'um*-textiles 50 minas a piece.'

422 ATHE 38.1-18: [1] *a-na Puzur$_4$-A-šur qí-bi$_4$-ma* [2] *um-ma I-kà-a-ma mì-*[*šu*] [3] *ša ta-na-zu-mu-ú* [4] *ma-nu-um i-na Tù-*[*ur-h*]*u-/mì-/it* [5] *ša* [A]N.NA *a-na* URUDU *lá-mu-nim* [6] *⸢i⸣-ta-na-dí-nu i-nu-mì* [7] *Sú-e-a i-li-kà-ni* [8] AN.NA 9 *ma-na*.TA *ú* 10 *ma-na*./TA [9] *⸢i⸣-ta-na-dí-nu* [10] *a-na-ku* 11 *ma-na*.TA [11] *ú-šé-bi-lá-kum* [12] *A-šur ú* DINGIR-*lu a-hu-tim* [13] *li-ṭù-lá* . 4 *ma-na* [14] AN.NA *i-na li-bi* [15] AN.NA-*ki-kà* [16] *a-na Ha-na-nim a-na* [17] 30 *ma-na*.T[A URU]DU *lá-mu-nim* [18] *a-dí-šu-ma* U[RUDU?] *i-ta-ba-al*, 'Speak to Puzur-Aššur, thus (says) Ikāya: why are you complaining: "Who is selling all the time tin for poor copper in Turhumit?" When Suea arrived (with the tin), it was common to sell tin piecemeal at a rate of (only) 9:1 or 10:1, but I have sent you (copper obtained) at a rate of 11:1, Assur and the god of brotherhood be my witnesses! I gave (only) 4 minas of your tin to Hannānum at a rate of 30 minas of poor copper per mina of tin, and he has taken the copper(?) away';

BIN 4, 35.4-7: [4] ... AN.NA *lu i-a-am* [5] *lu ku-a-am a-na Ili$_5$-we-da-ku* [6] *e-zi-ib um-ma a-na-ku-ma a-dí* [7] *a-tù-ra-ni* . URUDU SIG$_5$. *e-pu-uš-ma*, 'I left tin, both mine and your's, behind with Ilī-wēdāku, and said: "Obtain fine copper before I return"';

wool

Information on the exchange rate of 2 minas of wool per mina *šikkum* copper (in a city other than Turhumit) is given in CCT 4, 38a.[423] The proceeds of wool listed in TC 3, 177 might have been in copper.[424]

silver and gold

Silver of a specific quality (*kaspum dannum*, literally 'strong' silver) and gold, in TC 2, 33 (**5**) a bead made of gold, are sold for fine copper;[425] and according to CCT 5, 13a, gold (beads) could be sold in Turhumit.[426]

grain and hides

Grain (*uṭṭatum*) and hides are sold in Turhumit for copper according to CCT 5, 13a. Wheat (*aršātum*) is sold for 20 minas of *šikkum* copper in TC 2, 33 (**5**). Copper of *šikkum* quality

---

KTS 2, 34.7-10: ^{7} *um-ma a-ta-ma* AN.NA-*ki* ^{8} *ra-bu-ma i-ta-dí-šu-ma* ^{9} *a-na* URUDU SIG$_5$ *ta-e-er-šu-ma* ^{10} *ma-la* 6 GÍN.TA *[a-kà-šu-du e-pu-uš]* (broken), (to Puzur-Aššur from Ilī-wēdāku:) 'You wrote: "Distribute the tin discreetly (over retailers) and exchange it (so) for fine copper, and [do everything so I can fetch a price of 1 shekel (of silver) per 6 shekels (of tin)]"';
RA 60, 35.7: *a-na* AN.NA-*ki-kà* 11 *ma-na*.TA.

423 CCT 4, 38a.8-14: ^{8} *um-ma a-ta-ma . ší-[im]* ^{9} *ša-áp-tí-im . k[i-ma]* ^{10} *i-za-zu . té-er-t[a-kà]* ^{11} *li-li-kam-m[a]* ^{12} *ú ša-áp-t[ám a-na-kam*$^{?}$*]* ^{13} 2 *ma-na*.TA . *[a-na* URUDU*]* ^{14} *ší-ki-im . i-za-az*, (to Puzur-Aššur from Ilī-wēdāku:) 'You wrote me: "Let me know at which rate wool stands." Well, as for wool, it stands [here] at 2:1 [for] *šikkum* [copper].'

424 TC 3, 177.3-4: ^{3} *ší-im* SÍG.HI.A . *ú-ṭù-tum* ^{4} *ú ṣí-ba-sà . ša e-zi-bu-šu-ni*.

425 TC 2, 33.1-21: ^{1} *a-na Puzur*$_4$*-A-šur qí-bi-/ma* ^{2} *um-ma I-ku-pí-a-ma* ^{3} 3 *hi-dim* KÙ.⸢GI⸣ *ù* ^{4} 1 5/6 *ma-na* 3 1/2 GÍN KB ^{5} *da-na-⸢am⸣ ša tù-šé-bi-lá-ni* ^{6} *a-na ša-lá-ší-⸢šu⸣ . ni-im-ha-sú-/ma* ^{7} *iš-té-et qá-tám* KB ^{8} *a-na-ku al-qé-ma* 1 GÍN *a-na* ^{9} 3 *ma-na ší-ki-im ù* ^{10} KÙ.GI 1 GÚ *ší-ki-im* ^{11} *[a-na] u*$_4$*-me a-na Ib-ni-lí* ^{12} *[a-dí-i]n ki-ma ša-qá-al* ^{13} [URUDU] *a-da-gu*$_5$*-lu* ^{14} *[a-ma-l]a u*$_4$*-me-e ir-ku-sú* ^{15} [ ] *i-ṣa-ba-sú ki-ma* ^{16} *[ší]-ki-im* URUDU *lá-mu-/nàm* ^{17} *[a]l-té-qé* ŠÀ.BA 4 GÚ ^{18} [URUDU] *ší-kam ù a-ha-ma* ^{19} 20 *ma-na ší-kam ší-im* ^{20} *ar-ša-tim Ta-as-ku-ul* ^{21} *na-áš-a-kum*, 'Speak to Puzur-Aššur, thus (says) Ikuppia: we have divided into three parts the 3 gold beads and the 1 mina 53 1/2 shekels of "strong" silver that you sent and I have taken one part of the silver, and gave for a specified period to Ibni-ilī (the remaining silver) at 1 shekel for 3 minas of *šikkum* copper and (the bead of) gold for 1 talent of *šikkum* copper. Because I am expecting the payment of the copper, ... he will seize$^{?}$ it (copper?) not later than the period for which he contracted (him). Instead of *šikkum* copper I have collected poor copper. Of this Taskul is bringing you (worth) 4 talents of *šikkum* copper and furthermore 20 minas of *šikkum* as proceeds of the wheat';
RA 60, 35.10-14: ^{10} ... *a-na* KB-*pì* ^{11} 1 GÚ 35 *ma-na ší-im* ^{12} KB *ù* KÙ.GI *a-dí-ni* ^{13} *Ib-ni-lí* URUDU ^{14} *lá iš-qú-lam*, 'For the silver - Ibni-ilī has not paid me up till now the 1 talent 35 minas of copper, proceeds of the silver and gold';
TC 3, 5.13-18: ^{13} ... *ší-im* ^{14} KB . *da-nim I-dí-Ištar* ^{15} *e-ri-iš-ma um-ma* ^{16} *šu-ut-ma* URUDU *a-qá-tí-a* ^{17} *la ma-qí-it . a-dí* . 10 *u*$_4$*-me* ^{18} *ší-im-a-ni*, (to Puzur-Aššur from Ikuppia:) 'I demanded from Iddi(n)-Ištar the proceeds of the "strong" silver, but he said: "I did not receive any copper; reprieve me for 10 days."'

426 CCT 5, 13a.22-30: ^{22} *ú šar-ra-nu-ma a-na Tù-ur-hu-mì-it* ^{23} *e-tí-iq-ma ší-im* ^{24} *ṣú-ba-tí-kà ú* KÙ.GI URUDU SIG$_5$ ^{25} *a-ma-kà-am lu ší-im ma-áš-ki* ^{26} *lu ší-im ú-ṭí-tí-im* ^{27} *a-na* KB *ú* KÙ.GI ^{28} *a-na ma-lá té-er-tí-<kà>* ^{29} *ú-ta-ar-ma* ^{30} *ú-šé-ba-lá-ku-um*, (to Puzur-Aššur from Ilī-wēdāku:) 'And I will *immediately* travel to Turhumit because there is fine copper (available) as payment for your textiles and for the gold. In accordance with your instructions I will exchange there the proceeds both of the hides and of the grain for silver or gold, and send it you.'

as proceeds of hides of oxen and sheep is mentioned in CCT 5, 28b and in all probability in TC 3, 177 as well.[427]

antimony

In a difficult context the sale of antimony, *lulûm*, for copper is mentioned in KTS 7a.[428]

sû-stone

Six stones of special hardness are sold for 440 Anatolian minas of copper, probably of *šikkum* quality.[429]

The copper which was obtained in this way, was exchanged for silver (for instance in RA 60, 36), and gold.[430]

4.5.3 *Geography of the copper trade*

Copper was obtained in *Turhumit* in exchange for tin (ATHE 38), for textiles (BIN 4, 36), and for gold (CCT 5, 13a). But in CCT 5, 13a, the proceeds (in copper) of hides and grain will be exchanged for silver or gold in Turhumit (see above).

The metal could be shipped to *Wahšušana* (from Turhumit in BIN 4, 36), but was also obtained there (ATHE 66 and CCT 2, 29), for instance 45 talents of *šikkum* copper from the *kārum*-office (CCT 2, 31a).

Ah-šalim sent copper to the town of *Aliaša*, which is perhaps located in the vicinity of Wahšušana.[431]

From cities like Wahšušana (CCT 4, 12b = **12**), copper could be brought either to Kanish or to *Purušhattum*, where it was sold for silver (BIN 4, 35 = **6**; TC 3, 3; TC 3, 15), after an initially bad market in TC 2, 18 (**13**). 10,000 minas were paid for merchandise by

---

427 CCT 5, 28b.11-15: [11] ... 8 *me-at* [12] URUDU *ší-kam . lu ší-im maš-ki* [13] *ša al-pí lu*$^{!}$ *ša*$^{!}$ *e-me-ri* [14] *mì-ma a-nim* [15] *Puzur*$_{7}$*-A-šùr e-zi-ba-am*, '800 (minas) *šikkum* copper, the proceeds of hides of oxen as well as of sheep; all this Puzur-Aššur has left behind with me';
TC 3, 177.1-2: [1] *ší-im* 30 *ni-ki ša* GUD.HI.A [2] *ší-im ni-ki ša* UDU.HI.A. The writing *ni-ki* instead of *maš-ki* is strange. The copy has a clear NI in both lines, which could be either a consequent scribal error for MAŠ, or if not an error, **niku* could be a synonym of *mašku*.

428 KTS 7a.1-6: [1] *a-na Puzur*$_{4}$*-A-šùr* [2] *qí-bi-ma um-ma* [3] ^{d}IM-ANDUL *ù Bu-za-zu-ma* [4] *a-na lu-li-i-kà a-na* [5] 1 GÚ-*tim* 20 *ma-na*.TA [6] *ih-lu-lu*, 'Speak to Puzur-Aššur, thus (says) Adad-ṣulūlī and Buzāzu: they *paid* 20 minas (of copper) per talent of your antimony.'

429 KTS 7a.19-22: [19] ... *ší-tí sú-a-tí-kà* [20] ⸢6⸣ *sú-a-tim* 8 GÚ LÁ 10 *ma-na* [21] *a-na* 4 *me-at* 40 *ma-na* [22] *a-ba-an ma-tim ta-ad-na*, 'The remainder of your *sû*-stones, 6 pieces weighing 7 talents 50 minas, has been sold for 440 Anatolian minas (of *šikkum* copper).'

430 ATHE 40.3-11: [3] ... *a-šu-mì Da-da-a* [4] DUMU *Tí-li-a a-na-kam qá-sú* [5] URUDU-*kà a*-GEME$^{?}$-*tí-šu-nu i-dí-i*[*n*]*-ma* [6] KÙ.GI *iš-am ni-iṣ-ba-sú-ma* [7] *um-ma né-nu-ma* KB *lu* KÙ.GI [8] *šu-qúl-ma lu ni-ik-nu-kà-kum-/ma* [9] *a-ta-ma na-an-ší-šu* [10] *um-ma šu-ut-ma a-na-ku-ma* [11] *a-ṣé-ri-šu a-la-ak*, (to Puzur-Aššur from Ilī-wēdāku and Buzāzu:) 'As for Dadāya son of Tilia, here he personally sold your copper to their maids$^{?}$ and acquired gold. We seized him, saying: "Pay (us) the silver or gold and we will put it for you under seal so that you can bring it (to him)," but he replied: "I am going myself to him"';
CCT 5, 13a.25-30: see above;
RA 60, 36.14-16: [14] ... URUDU [15] *nu-ša-áš-qal-ma a*-KB *ú* [16] KÙ.GI *nu-ta-ar-ma*, (the same as in ATHE 40:) 'We will collect the copper and exchange it for silver or gold.'

431 CCT 2, 29.36-37: [36] URUDU *a-na A-li-a-ša* [37] *a-bu-kam*, (Ah-šalim to Puzur-Aššur:) 'I brought the copper to Aliaša.' Add this text to *RGTC* 4, p.6 where the town is given as Al-Iaša, on the basis of the related text Ka 1113.

the palace in CCT 2, 13.[432] Consignments of copper were sent to Pūšu-kēn, probably in Purušhattum, during Puzur-Aššur's absence from Anatolia, and might represent the normal destination of part of the copper received as payment for merchandise. The copper mentioned in this text (BIN 4, 31 = **7**), consisted of refined and *šikkum*, and came partly from Ulama.

In a letter by Suea mention is made of copper stored in *Hattum* (VS 26, 56), the region north of Kanish.[433]

In texts TC 3, 10 and RA 58, Goudchaux 2, which have been discussed above in § 3.4.3, copper has to be shipped to Puzur-Aššur because there was a shortage of copper in the city he was staying in, which was either *Wašhania* or Kanish. He asked Ilī-wēdāku to have 'the man from Ninašša', probably the town's ruler, bring 20 talents to Wašhania by means of the *sikkum* transport system. As the procedure demanded by Puzur-Aššur could not be realised, Ilī-wēdāku had to send the amount of copper desired with donkeys or porters to Wašhania. The copper does not seem to have been to hand, because Ilī-wēdāku suggested that a person who was obliged to deliver 18 talents 45 minas of copper in Wahšušana to 'the merchant' (or his agents), be instructed by Puzur-Aššur to pay it to him in Wašhania (see above. P. Garelli offered a different interpretation of these texts in RA 58, pp.134-135).

From *Ulama* 2 talents of refined copper were sent to Kanish (BIN 4, 31 = **7**).

Puzur-Aššur uses porters from Kanish to bring 5 talents of refined copper towards Iddi(n)-abum and Kulumāya in CCT 2, 40a.

4.5.4 *Conclusion*

The significance of Puzur-Aššur's archive lies in the documentation it contains of the exchange of a variety of goods for copper, and of the geographical pattern. In trading, his business partner Pūšu-kēn and family and Ilī-wēdāku play an important role; there are no certain indications that Puzur-Aššur's own children were involved.

---

432 CCT 2, 13.5-11: [5] ... *i-nu-mì* [6] 10 *li-me* URUDU *ša Pu-ru-uš-/ha-tim* [7] É.GAL-*lam*$_5$ *i-dí-nu-šu-ni* [8] *i-na* URUDU *a-mì-im* [9] *qá-tí áš-ku-un ù Šu-Ištar* [10] *ṣú-ha-ri a-Pu-ru-uš-ha-tim* [11] *i-li-ik*, (to Puzur-Aššur from Ilī-wēdāku:) 'When they handed over (to him?) in the palace the 10,000 (minas) of Purušhattum copper, I laid a claim on that copper and my servant Šū-Ištar went to Purušhattum (to fetch it).'

433 VS 26, 56.38-41: [38] *ù a-šu-mì* URUDU [39] *ša i-na Ha-tim . na-ad-ú té-er-ta-kà* [40] *iš-tí . A-da-da .* DUB.SAR *li-li-kam-ma* [41] *a-na ša-du-a-tí-im . lá ú-ša-am-ru-ṣú-ni*, 'Let your message reach me with Adada, the scribe, about the copper stored in Hattum, so that they do not bother me on account of the *šaddu'atum*-tax.'

4.6 *Uṣur-ša-Ištar son of Aššur-imittī*

4.6.1 *Participants*

Aššur-imittī son of Amur-ilī directed a firm whose chief participants belonged largely to his own family.[434] He apparently gained a leading position after the death of Šū-Laban, when he continued the family firm with his brothers.[435] Most texts of the kt n/k-archive, discovered during the excavations in 1962 in the *kārum* area, seem to belong to the time when he had returned to Assur, from where he led his affairs in Anatolia. He had commercial links with his nephew Imdilum and with Pūšu-kēn. His eldest son *Uṣur-ša-Ištar* bore the responsibility for the Anatolian branch of his firm. Uṣur-ša-Ištar, while working for his father's firm, also conducted his private business like many other merchants, together with his brothers Iddi(n)-Aššur, Hunia, Ikuppia, and with his own son Amur-ilī and more distant relatives. His business brought him to the major Anatolian cities, like Turhumit and Wahšušana.[436] As usual, it is frequently impossible to determine whether a commercial transaction was performed on behalf of the family firm of Aššur-imittī, or for Uṣur-ša-Ištar's private business or for that of his partners. Uṣur-ša-Ištar was a *hamuštum* or week eponym in Kanish.[437]

Amur-Ištar is probably the same person as Aššur-imittī's brother by that name and thus was an uncle of Uṣur-ša-Ištar. His role in Imdilum's copper trade, which he plays largely from Turhumit, has been discussed above in § 4.4.1. In contrast to Imdilum's archive, no mention is made in that of Uṣur-ša-Ištar of Amur-Ištar's cooperation with Uzua. He organizes the sale of Uṣur-ša-Ištar's textiles for copper as he does for Imdilum, pays taxes levied on the proceeds to the *kārum*-office,[438] and sends the copper to him. In kt n/k 1153, he mentioned that an amount of refined copper sent to his nephew was first broken into smaller pieces (see below and § 3.1.4.4). He uses his house for storing copper which he acquires for several people; a small amount for Uṣur-ša-Ištar is mentioned in AKT 2, 59,[439] and it is reported to Pūšu-kēn in CCT 5, 4a.26-29, that 11 talents of his fine copper are stored in Amur-Ištar's house. The wife of Amur-Ištar (*aššat Amur-Ištar*) is mentioned in kt n/k 1289.

---

434 With patronymic, Aššur-imittī occurs on the bulla kt n/k 1709 (by courtesy of Ö. Tunca). See for the families of Amur-ilī's decendants the genealogical table in Larsen 1982, p.219.

435 Larsen 1982, p.224.

436 Kt n/k 210.25-28: [25] ... *ù a-šu-mì* [26] *ú-nu-ut* ANŠE.HI.A *ù ṭup-pì-a* [27] *ša i-na Tur₄-hu-mì-it* [28] *e-zi-ba-ku-ni*, (Ilabrat-bānī to Uṣur-ša-Ištar:) 'And as for the donkey harness and tablets of mine that I left with you in Turhumit.' Kt n/k 432.7-8: [7] *a-na Wa-ah-šu-ša-na* [8] *a-ta-lá-ak*, (Uṣur-ša-Ištar to Ikuppia:) 'I just departed for Wahšušana.'

437 AKT 1, 46 (= kt a/k 895), month VII, year Iddi(n)-ahum; TC 3, 249.21, month VII; Amsterdam, 10, month IX, year Buzutāya; kt n/k 220.4 and 14 (without month or year); kt 91/k 155.5, month XII, year Kabadia. It is probably the same person who occupied this office in the three different years because only one patronymic is known for Uṣur-ša-Ištar.

438 For instance in kt n/k 118, where he paid 40 minas of copper as *šaddu'atum*-tax to the *kārum*-office after selling Uṣur-ša-Ištar's merchandise; the amount taxed would have been 40 talents of copper.

439 In AKT 2, 59, 5 discs called *kakkartum* are mentioned, see p.59.

Amur-Ištar's son *Suen-rē'ī* is active as an agent for Uṣur-ša-Ištar selling textiles. Suen-rē'ī calls his cousin 'my father and lord' (kt n/k 482.20; 499.3), indicating Uṣur-ša-Ištar's superior rank and the difference in age between both men. In Tišmurna he sells textiles to the palace (see below). He works closely with Gallābum and an Anatolian (or the son from a mixed marriage) called Taskul. Taskul travelled with porters carrying copper for Uṣur-ša-Ištar to Ikuppia in Wahšušana (kt n/k 122; 228; 482). In kt n/k 98, Taskul is sending a total of 36 minas of *šikkum* copper for a person called Buluhala. The amount of copper due by Suen-rē'ī in kt n/k 1825 can be seen as a credit granted. In this text, he owes 5 talents of refined copper to the 'merchant' (*tamkārum*, perhaps Aššur-imittī) and has to pay within 5 years.[440] Ikuppia tries in vain to obtain 10 talents of copper from Suen-rē'ī. After receiving permission from him to take it from his representatives, Ikuppia is told by them they that do not have any copper, but that it is stored in Marithum (AKT 2, 38; kt n/k 1438).

Gallābum receives consignments of tin and textiles belonging to Uṣur-ša-Ištar and together with Suen-rē'ī exchanges these for copper (kt n/k 122; 208). He leads a caravan of 5 donkeys loaded with copper to Purušhattum (kt n/k 227). After sealing it, he despatches Uṣur-ša-Ištar's servant Huzala to Ikuppia with a consignment of copper as the proceeds of textiles (kt n/k 162).[441] On Uṣur-ša-Ištar's advice Gallābum checked a quantity of copper which Taskul kept in store (kt n/k 208). Deficits in the amounts of good and poor quality copper were established. Taskul confirmed the deficit of good quality, but denied that of the poor quality copper.[442]

Of Uṣur-ša-Ištar's brothers, *Ikuppia* seems to have been the youngest, because in the address of letters he sent to Hunia (kt n/k 234; 456; 1484), Hunia is named first. Ikuppia calls Uṣur-ša-Ištar his 'brother' (kt n/k 1334) and 'father and lord' (kt n/k 86; 379; 805), which is the usual polite address to an older brother. In turn Uṣur-ša-Ištar calls him 'my brother' (kt n/k 276; 1456). In kt n/k 379, Ikuppia is speaking about the 'firm of our father' (*bēt abīni*). In the dramatic letter kt n/k 1454, Aššur-imittī implores his sons Uṣur-ša-Ištar and Iddi(n)-Aššur to come finally to Assur and to take Ikuppia along.

Ikuppia appears in the letters as an inventive and sometimes headstrong trader. He plays an important role in the firm's copper business, which leads him frequently to the cities of Turhumit, Purušhattum, and Wahšušana. He transports copper, mainly from Turhumit, to Uṣur-ša-Ištar in Wahšušana (kt n/k 432; 1456; in 84 this was impossible due to the political situation), to Purušhattum (kt n/k 227; 1456), to Kanish (kt n/k 276), and to

---

440 Kt n/k 1825.4-11: [4] 5 GÚ URUDU *ma-sí-am* [5] *i-ṣé-er* ZU-*re-i* [6] DUMU *A-mur-Ištar* DAM.GÀR [7] *i-šu* (...) [10] *a-na* MU.5.ŠÈ [11] *i-ša-qal.*

441 Kt n/k 162.1-8: [1] *a-na I-ku-pì-a* [2] *qí-bi-ma um-ma* [3] *Ga-lá-bu-um-ma* [4] 1 GÚ <URUDU> *ma-sí-am* [5] *ku-nu-ki-a Hu-za-lá* [6] *na-áš-a-kum* [7] *ší-im* TÚG.HI.A [8] *ú* 1 *ma-nu-ni-a-ni-im*, 'Speak to Ikuppia, thus (says) Gallābum: Huzala is bringing you 1 talent of refined copper under my seal, the proceeds of textiles and 1 *manuniānum*-textile' (by courtesy of S. Bayram).

442 Kt n/k 208.27-35: [27] *a-šu-mì Ta-as-ku-ul* [28] *ša áš-pu-ra-ku-ni* URUDU *lu* SIG_5 [29] *lu lá-mu-nam ú-sà-ni-iq-ma* [30] *i-na* SIG_5 [31] 17 *ma-na ba-tí-iq* [32] *i-na lá-mu-nim* [33] 1 GÚ 15 *ma-na ba-tí-iq* [34] *ú-sà-li-šu-ma* SIG_5 *uk-ta-*[*in*] [37] *lá-mu-nam i-tí-ki-*[*ir*], 'As for Taskul, about whom I wrote you, I checked the copper, both fine and poor: from the fine copper 17 minas are missing, and from the poor copper 1 talent 15 minas are missing. I put him under pressure and he acknowledged (the deficit of) fine copper, but denied (that of) poor copper' (by courtesy of S. Bayram).

an unknown location (kt n/k 1371).[443] At his own initiative he takes silver destined for Uṣur-ša-Ištar out of Purušhattum and travels to Turhumit in order to buy copper with it, as Hunia reports to his brother in kt n/k 483 (see below). At times, Uṣur-ša-Ištar mistrusts his brother's loyalty. In kt n/k 1277, Ikuppia defends himself against accusations that he had used copper belonging to Uṣur-ša-Ištar and destined for Kanish and Purušhattum, for commercial deals (i.e. to sell it for silver or gold).[444] It appears from this text that Ikuppia

---

443 Kt n/k 84.7-17: [7] *a-dí-i* URUDU *er-sú-ú-ni-ma* [8] *lá ú-šé-bi*$_4$*-lu sí-kà-tù-šu-nu* [9] *ma-áš-ha-a . um-ma šu-nu-ma* [10] *a-na ma-at Wa-ah-šu-ša-na* [11] *i-ma-qú-tù* [12] *ú ša bi-lá-tim a-lá-kam* [13] *lá i-mu-ú . a-dí* 5 *u*$_4$*-me-e* [14] *a-šar sí-kà-tum* [15] *i-ta-qú-na-ni a-ša-me-ma* [16] *ù ša bi-lá-tim* [17] *a-gar-ma* URUDU *a-ba-ak*, (to Uṣur-ša-Ištar from Ikuppia:) 'As for the fact that I did not send the copper although it is ready (for transport), (that is because) their *pegs were removed.* People said: "They are invading the country of Wahšušana." Also the porters refused to go. Within 5 days, I will hear whether *the pegs become in good order* again, and then I will hire porters and ship the copper' (by courtesy of S. Bayram);
kt n/k 227.1-12: [1] *um-ma Ú-ṣur-ší-Ištar-ma* [2] *a-na I-ku-pí-a qí-bi*$_4$*-ma* [3] *i-na ṭup-pí-kà ta-áš-pu-ra-am* [4] *um-ma a-ta-ma* 10 ANŠE [5] DUMU *E-nu-Be-lim a-na Pu-ru-uš-ha-tim* [6] *uš-té-ri-ib* . 5 ANŠE [7] *Ga-lá-bu-um* ⸢*uš-té-ri-ib*⸣ [8] *iš-tù Pu-ru-uš-ha-tim* [9] *ṭup-pá-am ša ki-*[*m*]*a i-a-tí* [10] *ú-šé-bi*$_4$*-lu-nim um-ma šu-nu-ma* [11] 11 GÚ URUDU DUMU *E-nu-Be-lim* [12] [*ub*]*-lam*, 'Thus (says) Uṣur-ša-Ištar, speak to Ikuppia: you wrote me in your letter: "The son of Ennu-Bēlum has brought (only) 10 donkey(load)s (of copper) to Purušhattum; Gallābum has brought 5 donkey(load)s." My representatives sent me a letter from Purušhattum, in which they write: "The son of Ennu-Bēlum brought 11 talents of copper"' (10 donkeys could carry at least 20 talents; the son of Ennu-Bēlum might be Aššur-taklāku; by courtesy of S. Bayram);
kt n/k 276.9-22: [9] *lá-ma . a-li-kà-ni* [10] URUDU-*i . a-na Kà-ni-iš*^{ki} [11] *a-bu-uk-ma . lá a-kà-ša-dam* [12] *ù šu-ma .* ANŠE.HI-*ru* [13] *lá-šu iš-tí a-hi-im* [14] *ù eb-ri-im* [15] *e-ri-iš-ma . a-dí* [16] *Kà-ni-iš*^{ki} *. li-ší-ú* [17] *ù I-dí-A-šùr* [18] 2 *ù* 3 ANŠE *e-ri-iš* [19] *šu-ma* ANŠE.HI-*ru lá-šu* [20] *lá-ma . a-li-kà-ni* [21] *ku-lu .* URUDU *i-tí-ib-nim* [22] *lu ša-ki-in-ma*, (Uṣur-ša-Ištar to Ikuppia:) 'Bring my copper to Kanish before I come, I do not want to come across it. If there are no donkeys, ask brother and friend (for them), so that they will carry it till Kanish. And ask Iddi(n)-Aššur for 2 or 3 donkeys as well. If there are no donkeys (at all), let all copper be stored in straw before I come (so that I will not loose any time)' (by courtesy of S. Çeçen).
kt n/k 432.12-20: [12] URUDU *ma-lá i-ba-ší-ú* [13] *ša bi-lá-tim* [14] *ta-ru-ma a-ṣé-ri-a* [15] *a-na Wa-ah-šu-ša-na* [16] *áb-kam ú ší-tí* [17] URUDU *za-ki-ma tí-ib-a-ma* [18] *a-tal-kam a-na* [19] *Wa-«áš»-ah-šu-ša-na* [20] *ku-uš-da-ni*, (Uṣur-ša-Ištar to Ikuppia: 'As soon as you hear my tablet,) bring porters along and send all copper available to me in Wahšušana. And collect the remaining claims in copper and leave; meet me in Wahšušana' (by courtesy of C. Günbattı);
kt n/k 1371.18-20: [18] 8 GÚ URUDU *ša bi-lá-tim ub-lu-nim-kum* [19] 20 GÚ URUDU SIG$_5$ 10 ANŠE.HI.A [20] *A-šur-ták-lá-ku i-ra-dí-a-kum*, (to Uṣur-ša-Ištar from Ikuppia:) 'Porters brought you 8 talents of copper and Aššur-taklāku is bringing you 20 talents of fine copper (with) 10 donkeys' (by courtesy of S. Çeçen);
kt n/k 1456.8-12, 22-29: [8] *ki-ma té-ru-bu* ANŠE [9] 15 *ù* 16 URUDU *ma-li-ma* [10] *a-na ṣé-er ša ki-ma* [11] *i-a-tí a-na Pu-ru-uš-ha-tim* [12] *a-bu-uk* (...) [22] *ma-lá* URUDU *a-ṣé-er* [23] *ša ki-ma i-a-tí a-na* [24] *Pu-ru-uš-ha-tim tù-šé-bi*$_4$*-lu-ni* [25] *té-er-ta-kà lí-lí-kam* [26] *ma-lá i-ba-ší-ú* [27] *ša bi*$_4$*-lá-tim a-gu*$_5$*-ur-ma* [28] URUDU *a-na Wa-ah-šu-ša-na ša*$^?$*-*[*x*]*-ma* [29] *ú ší-tí* URUDU *za-ki*, (Uṣur-ša-Ištar to Ikuppia: 'I had written to you:) "Load, as soon as you enter the city, 15 or 16 donkeys with copper and send it to my representatives in Purušhattum." (...) Send me a message how much copper you have transported to my representatives in Purušhattum. Hire porters and ... all the copper available to Wahšušana, and collect the remaining claims in copper' (by courtesy of S. Çeçen).

444 Kt n/k 1277.4-26: [4] *mì-šu ša a-dí* 5-*ší-šu* [5] *ù* 10-*ri-šu . ta-áš-pu-ra-ni* [6] *um-ma a-ta-ma* [7] URUDU-*i . a-na Kà-ni-iš* [8] *ù Pu-ru-uš-ha-tim* [9] *a-na ší-a-ma-tim* [10] *ta-áš-ta-na-kà-an-ma* [11] *qá-at-kà* [12] *té-ta-na-ni* [13] *A-šùr ù* DINGIR-*kà* [14] *li-ṭù-lá . iš-tù* [15] *a-na Tur*$_4$*-hu-mì-it* [16] *e-ru-ba-ni . ma-tí-ma* [17] 10 *ma-na* URUDU *a-na* [18] *ší-a-ma-tim lá áš-ku-nu* [19] *a-bi*$_4$ *a-ta be-lí a-ta* [20] *a-ma-lá té-er-tí-kà* [21] *a-dí* 10 *u*$_4$*-me* [22] *ú-za-kà-ma a-na* [23] *Pu-ru-uš-ha-tim* [24] *e-ra-áb . ki-ma we-da-ku-ni* [25] *a-dí u*$_4$*-mì*

was at that time the only member of the firm to look after the affairs in Turhumit, which mainly consisted of selling merchandise and collecting and preparing for transport the copper which was paid for it.

During a revolt in the country of Wahšušana, Aššur-damiq declares to Uṣur-ša-Ištar that he will try to rescue Ikuppia's copper stored there and hand it over to Ikuppia's representatives, as Ikuppia himself had gone to Kuburnat (kt n/k 218).[445] Ikuppia also collects amounts of copper due to Uṣur-ša-Ištar or to the family firm and writes a debt-note if payment was delayed (kt n/k 175, owed by Uzua and Abu-šalim). In kt n/k 234, he reports to Hunia on the sale of his tin and textiles for copper.

For transport, Ikuppia uses donkeys which Uṣur-ša-Ištar suggests he can borrow from colleagues or from (his brother) Iddi(n)-Aššur (kt n/k 276; for transport to Purušhattum in 1456). He also hires porters (kt n/k 1371; to Wahšušana: kt n/k 84; 432; 1456). The use of wagons is not yet attested in letters concerning Ikuppia.

Outstanding claims of *Iddi(n)-abum* son of Iddi(n)-Ištar were collected by various persons, see below for the relevant texts.

4.6.2 *Procurement and sale of copper*

In Uṣur-ša-Ištar's archive, copper is chiefly obtained by selling tin and textiles.

textiles

More than once, 50 textiles are sold by one person, for which 10 talents of copper are paid. This is the case in kt n/k 87 (a memorandum written most likely by Uṣur-ša-Ištar), where Ennānum son of Amria owes 10 talents of refined copper for 50 *kutānum*-textiles.[446] In kt n/k 122, Suen-rē'ī obtains 10 talents of fine copper by selling 50 textiles in Tišmurna, and the same person brings 55 textiles to Tišmurna in kt n/k 208.[447] Such lots of textiles in kt

---

a-nim [26] *a-tá-am-sà-ar*, 'Why did you write me 5, even 10 times: "You constantly use my copper, (to be sent) to Kanish and Purušhattum, for commercial deals, and you keep changing your intentions!" Assur and your own god be (my) witnesses! Since I entered Turhumit, I never used even 10 minas of copper for commercial deals! My father and lord, in compliance with your instructions I will collect (all the copper) within 10 days, and then I will enter Purušhattum. I was delayed till the present day because I am alone' (by courtesy of S. Çeçen).

445 Kt n/k 218.3-16: [3] *a-dí lá a-li-kà-ni* [4] *I-ku-pí-a a-na Ku-bu-ur-na-at* [5] *a-na ší-ip-ru-tim* [6] *i-li-ik* URUDU-*ú-šu* [7] *i-na ma-at Wa-ah-šu-ša-na* [8] *na-dí-ma ki-ma* [9] *ma-tum sà-hi-at-ni* [10] *a-na* URUDU *ša-la-hi-im* [11] *a-ta-lá-ak a-na* [12] *Wa-ah-šu-ša-na* [13] *a-kà-ša-ad* [14] URUDU-*ú-šu a-na* [15] *ša ki-ma šu-a-tí* [16] *a-pá-qí-id-ma*, 'As for the fact that I did not come: Ikuppia has gone on an errand to Kuburnat, (while) his copper is stored in the country of Wahšušana. Because the country is in uproar, I will depart to rescue the copper. When I reach Wahšušana I will hand over his copper to his representatives' (by courtesy of S. Bayram).

446 Kt n/k 87.1-4: [1] *ší-im* 50 *ku-ta-ni* [2] 10 GÚ URUDU *ma-sí-um i-na* [3] *li-bi En-na-nim* [4] DUMU *Am-ri-a* (by courtesy of S. Bayram).

447 Kt n/k 122.5-30: [5] 6 GÚ URUDU *ma-sí-um* [6] *ša té-er-tí-kà* [7] *pá-ni-tim Ga-lá-bu-um* [8] *e-pu-uš* 10 GÚ URUDU SIG$_5$ [9] *ší-im* 50 TÚG.HI.A [10] *i-na Tí-ša*[!]*-mu-ur-na* [11] *am-hu-ur ší-im* AN.NA [12] *I-dí-A-šùr* [13] *nu-ša-áš-a-ma* [14] *lu ší-im* TÚG.HI.A [15] *ú* AN.NA *lá-mu-num* [16] *i-ba-ší a-na-kam* [17] *a-na kà-ú-nim* [18] *lá-šu* (...) [21] ... *iš-té-ni-iš* [22] 30 GÚ *nu-ma-lá-ma* [23] *nu-ša-*^{ba}*bar-ma* [24] *Ta-as-ku-ul* [25] *iš-tí ša bi*$_4$*-lá-tim* [26] *a-na Wa-ah-šu-ša-na* [27] *ni-ṭá-ra-ad* [28] *Šu-A-nu-um* TÚG.HI.A *mì-ma lá il*$_5$*-qé* [29] *ší-tí* TÚG-*tí-kà a-na-kam-ma* [30] *a-na* URUDU *nu-ta-ar*, (to Uṣur-ša-Ištar from Suen-rē'ī and Gallābum:) 'Gallābum has realized 6 talents of refined copper as the proceeds of your previous order. I

n/k 482 are bought by the palace of Tišmurna.[448] Amur-Ištar receives two consignments, 30 textiles in all, which he will sell for copper for Uṣur-ša-Ištar (kt n/k 1153).[449] In kt n/k 299, a retailer has not yet sold Uṣur-ša-Ištar's textiles for copper.[450] Ennānum sells 3 *kutānum*-textiles and less than 2 minas of tin for fine copper (kt n/k 99).[451]

tin

Tin belonging to Hunia was sold together with his textiles by Ikuppia for copper in kt n/k 234.[452] Small amounts of tin were paid for with copper in kt n/k 99, 167, and 208.[453]

---

received in Tišmurna 10 talents of fine copper as the proceeds of 50 textiles. We will make Iddi(n)-Aššur acquire (copper) by selling the tin. There is poor copper available to pay for both textiles and poor tin, but there is no (fine copper) to be obtained here (..) We will pay in full in total 30 talents (of copper) and we will break up (the copper ingots) and then we will send Taskul with porters (carrying the copper) to Wahšušana. Šū-Anim has taken no textiles. The remaining textiles we will exchange here for copper' (by courtesy of S. Bayram);
kt n/k 208.3-15, 19-26: [3] *i-na* 1 *me-at* 86 TÚG.HI.A [4] *ša ub-lu-ni-ni* ŠÀ.BA [5] 12 TÚG.HI.A *ša En-na-Sú-in* [6] *I-dí-A-šùr il*$_5$*-qé* [7] 55 TÚG ZU-*re-i a-na* [8] *Tí-iš-mu-ur-na ú-bi-il*$_5$ [9] 20 LÁ 1 TÚG.HI.A *a-na-ku a-dí-in* [10] *ri-ih-tum* 1 *me-at* TÚG.HI.A [11] *i-ba-ší-ú* 10 GÚ URUDU SIG$_5$ [12] *ší-im* TÚG.HI ZU-*re-i* [13] *iš-tù Tí-iš-mu-ur-na* [14] *ub-lam i-na* 1 GÚ [15] AN.NA *ša tù-šé-bi*$_4$*-lá-ni* (...) [19] ŠÀ.BA 36 1/2 *ma-na* AN.NA [20] *I-dí-A-šùr il*$_5$*-qé* URUDU *ší-im-šu* [21] *a-dí-ni lá iš-qú-lam ší-tí* [22] AN.NA (...) [23] *a-na-ku a-dí-in lu i-na ší-im* [24] *lu-qú-tim pá-ni-tim lu ší-im* [25] AN.NA *ù* TÚG.HI.A 10 GÚ URUDU [26] SIG$_5$ *i-ku-nu-ki-a i-ba-ší*, 'Out of 186 textiles which they brought me, Iddi(n)-Aššur took 12 of Enna-Suen, Suen-rē'ī brought 55 to Tišmurna, and I sold 19. The remaining 100 textiles are on hand. Suen-rē'ī brought from Tišmurna 10 talents of fine copper as proceeds of the textiles. From the talent of tin you sent me (...) Iddi(n)-Aššur took 36 1/2 minas, he has not yet paid me its proceeds in copper. I have sold the rest of the tin. Both from the proceeds of the previous consignment and (from) that of the tin and textiles there are 10 talents of fine copper under my seal in store' (by courtesy of S. Bayram).

[448] Kt n/k 482.6-10: [6] ... TÚG.HI.A *i-na* [7] *Tí-iš-mu-ur-na* É.GAL-*lúm* [8] *a-na lá-qá-im ú-hi-ir-ma* [9] *a-ší-a-tí a-dí u*$_4$*-mì-im a-nim* [10] *ak-ta-lá*, (to Uṣur-ša-Ištar from Suen-rē'ī:) 'The palace in Tišmurna was slow in accepting the textiles and therefore I was detained till to-day' (by courtesy of C. Günbattı).

[449] Kt n/k 1153.3-10, 14-18: [3] ... 20 TÚG [4] *i-pá-na ub-lu-nim* [5] 10 TÚG *wa-ar-kà-num* [6] *ub-lu-nim* 4 GÚ URUDU [7] *ma-sí-am* 2 1/2 *ma-na* [8] *i-na ša-bu-ri-im* [9] *im-ṭí ku-nu-ki-a* [10] *Hu-dur-lá na-áš-a-kum* (...) [14] TÚG.HI.A . *e-ṣú-tù-ma* [15] *a-hu-ru* TÚG.HI-*kà* [16] *a-ga-mar-ma* [17] URUDU *ma-lá i-za-ku-ú* [18] *a-ša-pá-ra-kum*, 'They brought me first 20 textiles, and later 10. Hudurla is bringing you under my seal 4 talents of refined copper, 2 1/2 minas proved to be too little from the breaking up of (the ingots). (...) Only a few textiles are left over. When I have sold all of them I will send you (report about) how much copper becomes available' (by courtesy of S. Çeçen).

[450] Kt n/k 299.14-17: [14] ... TÚG-*tù-šu* [15] *i-ba-ší-ú a-da-šu-nu-ma* [16] *ma-lá i-ma-qú-tù-ni* [17] URUDU *a-da-na-šum*, '(The retailer said to Ikuppia:) "His textiles are still in store, but I will sell them and give him as much copper as comes in"' (by courtesy of S. Çeçen).

[451] Kt n/k 99.3-15: [3] ... 1 GÚ URUDU SIG$_5$ [4] *ší-im* 3 [túg]*ku-ta-ni-kà* [5] 2 *ma-na* LÁ 7 GÍN AN.NA [6] *ku-nu-ki-kà ub-lu-nim* [7] 10 *ma-na*.TA . *la a-áb-tù-qá-ku* [8] *lu ga-am-lá-at* [9] 1 GÚ 20 LÁ 1 *ma-na* LÁ 10 GÍN [10] URUDU SIG$_5$ *ku-nu-ki-a* [11] *I-ma-ar-šu* [12] *ù Ta-ás-ku-ul* [13] *na-áš-ú-ni*[!](NU)-*ku-um* [14] ŠÀ.BA 1 *ma-na* URUDU SIG$_5$ [15] *a-na qá-tí-šu-nu a-dí-in*$_6$, '1 talent of fine copper, the proceeds of your 3 *kutānum*-textiles; they brought me 1 mina 53 shekels of tin under your seal. I have not reduced the price to (a ratio of) less than 10 minas (of copper) per (mina of tin), this is the favour (I do) you! (In total) 1 talent 18 minas 50 shekels of fine copper Imaršu and Taskul are bringing you under my seal. Thereof I gave 1 mina of fine copper to cover their expenses' (by courtesy of S. Bayram).

[452] Kt n/k 234.3-4: [3] 13 *ma-na*.TA [4] *a-na a-ni-ki-kà*, '13 minas (of copper) per (mina) for your tin' (by courtesy of S. Bayram).

[453] Kt n/k 167.3-4: [3] ... 5 *ma-na* [4] *ma-sí-am ší-im* AN.NA, '5 minas of refined (copper) as proceeds of tin' (by courtesy of S. Bayram). See above for kt n/k 99 and 208.

wool

In kt n/k 463, a consignment of wool weighing 21 talents 45 minas, and packed in 14 *maškanum*-'tents' is sold for copper. The wool, at a ratio of 2:1, yielded 10 talents 52 1/2 minas, and the *maškanū*, at 4 minas each, 56 minas of copper.[454] In AKT 2, 38, Ikuppia had to wait till the copper was paid for two consignments of Uṣur-ša-Ištar's wool.[455] Kt n/k 243 contains the instructions to sell wool for fine or poor copper, for silver or for gold, but apparently only copper was seriously envisaged.[456] None of these texts provides direct information as to the location(s) where the wool was sold.

silver

An interesting case is reported in kt n/k 483, where silver is used to buy copper. According to this text, Ikuppia brought silver from Purušhattum to Turhumit in order to buy copper there.[457] Silver is also brought to Turhumit in another text, but it is uncertain whether the copper mentioned in this letter was bought with it (kt n/k 1451). This use of silver is also attested in the *šitapkum*-operations of the *kārum*-organisation, for which see § 5.5. In kt n/k 1464 (**20**), textiles and silver are deposited with a *kārum*-office at various moments, for which Uṣur-ša-Ištar was entitled to in all 1,120 minas of copper (see p.175 for a discussion of this text).

The major purchasers of merchandise offered for sale by the Assyrians and their agents were the Anatolian palaces and their officials. In kt n/k 533, Uṣur-ša-Ištar was asked to store fine copper apparently paid by two Anatolian dignitaries, the 'head of the army', *rabi sikkitim* and the 'head of the (royal) tables', *rabi paššurī*.[458]

---

[454] Kt n/k 463.4-14: [4] 21 GÚ 45 *ma-na* [5] SÍG.HI.A 14 *maš-ku-nu . ší-im* [6] SÍG.HI.A . *ša-na-ú-ma* [7] 10 GÚ 52 1/2 *ma-na* URUDU [8] *ší-im . maš-ki-ni* [9] 56 *ma-na* URUDU [10] ŠUNIGIN 11 GÚ 50 LÁ 1 1/2 *ma-na* [11] *mì-ma . a-nim* [12] *a-na Hu-ra-ṣa-nim* [13] *Ú-ṣú-ur-ša-Ištar* [14] *e-zi-ib*, '21 talents 45 minas of wool and 14 *maškanū*: the proceeds of the wool, at a ratio of 2:1, makes 10 talents 52 1/2 minas of copper, the proceeds of the *maškanū* are 56 minas of copper. Total: 11 talents 48 1/2 minas, all this Uṣur-ša-Ištar left behind with Huraṣānum' (by courtesy of C. Günbattı).

[455] AKT 2, 38.17-20: [17] *a-dí ša a-na-kam as-hu-ru* [18] URUDU *ší-im* SÍG.HI.A *pá-ni-tí* [19] *ù wa-ar-ki-tim* [20] [UR]UDU *a-dí-ni lá iš-qú-lu*, 'As for the fact that I tarried here, they have not yet paid the copper as the proceeds of the first and subsequent consignments of wool.'

[456] Kt n/k 243.6-15: [6] SÍG.HI-*tám ša I-ku-pì-a* [7] *i-na* É *ub-ri-kà e-zi-bu* [8] *a-ta ú Šu-ma-li-ba-A-šùr* [9] *i-zi-za-ma lu a-na* URUDU [10] SIG$_5$-*im lu a-na la-mu-ni-im* [11] *lu a-na* KB *lu a-na* KÙ.KI [12] *a-hu-ú-a . a-tù-nu* [13] *dí-na-ší-ma a-dì* [14] *I-ku-pì-a i-lá-kà-ni* [15] URUDU *lu er-sú*, (Uṣur-ša-Ištar and Ikuppia to Aššur-taklāku:) 'As for the wool that Ikuppia left behind in your inn, take action, you and Šumma-libb-Aššur, and sell it eith:er for fine or poor copper or for silver or gold, let the copper be ready upon Ikuppia's arrival' (by courtesy of S. Bayram).

[457] Kt n/k 483.3-10: [3] *a-dí* KB *lá ša ú-šé-bi$_4$-lá-[ku-ni]* [4] *li-ba-kà la i-lá-mì-in* [5] *I-ku-pì-a* KB *iš-tù* [6] *Pu-ru-uš-ha-tim* [7] *ú-šé-ṣí-a-ma a-na* [8] *Tù-ur-hu-mì-it* [9] *e-ru-ma* KB *a-na* URUDU [10] *ú-ta-e-er-ma*, (Uṣur-ša-Ištar to Hunia:) 'Do not worry about the fact that I did not send you silver! Ikuppia took the silver out of Purušhattum and went to Turhumit, (where) he exchanged it for copper' (by courtesy of C. Günbattı).

[458] Kt n/k 533.3-10: [3] ... 3 GÚ URUDU SIG$_5$ [4] *Ga-nu-ša ra-bi$_4$ sí-ki-tim* [5] 50 *ma-na* URUDU SIG$_5$ [6] *Zu-pá ra-bi$_4$ pá-šu-ri* [7] *ku-nu-ki Li-ip-ta-nim* [8] DUMU *Lá-le-e-em na-áš-ú-ni-kum* [9] *a-dí té-er-tù-šu i-lá-kà-ku-ni* [10] *li-bi$_4$-ší*, (Ah-šalim to Uṣur-ša-Ištar:) 'They are bringing you under seal of Lip(i)t-Anim son of Lalum 3 talents of fine copper (of) Ganuša, the *rabi sikkatim*, and 50 minas of fine copper (of) Zupa, the *rabi paššurī*; it must remain in store until his message reaches you' (by courtesy of C. Günbattı).

The copper was sent to cities where it could be sold for silver (kt n/k 513) or gold (kt n/k 135; cf. 1689).[459] In kt n/k 513, Kurara, Ikuppia, and Hunia had not yet received, let alone had time to sell for silver, the copper the proceeds of which Usānum wished to bring to Assur. They write to Uṣur-ša-Ištar that, should the copper not be available at the time he leaves, they will borrow silver against interest and give him that.[460]

The qualities of copper occurring in this archive are mainly fine and refined, but poor (kt n/k 122; 208), and *šikkum* (AKT 2, 39, received by Uṣur-ša-Ištar but belonging to Puzur-ilī) occur as well. In kt n/k 122, Suen-rē'ī reports that poor copper is available as the proceeds of the sale of textiles and tin. Fine quality copper from Taritar is sold for silver in AKT 2, 27 and kt n/k 701. Ilabrat-bānī sends 10 minas destined for Tarīš-mātum to Uṣur-ša-Ištar, with the urgent request that Tarīš-mātum should sell it only in the presence of experts, lest she be duped by the smiths for 1 or 2 shekels of silver, that is about 10-20% of the value at the ratio of 60:1 (AKT 2, 26, see p.72 for this text). Aššur-mālik owes copper from Tapašatta to Aššur-imttī in kt n/k 1952. Uṣur-ša-Ištar collected copper from Haburata from the outstanding claims of Iddi(n)-abum (kt n/k 159; 231).

Two texts contain references to aspects of the processing and packing of copper. In kt n/k 1153, the breaking up of high-quality copper (ingots) is mentioned. In this letter Amur-Ištar reports to Uṣur-ša-Ištar that Hudurla is bringing nominally 4 talents of refined copper. But during the breaking up of this copper, an amount totalling 1% or 2 1/2 minas is lost (*ina šabburim imṭi*). It is unclear where and by whom the breaking up is carried out. The passage in kt n/k 276.21-22, 'let all copper be packed in straw' (see n.443), may point to the use of straw for packing copper to prevent possible sharp edges of pieces of broken copper from damaging the leather bags or from hurting the donkeys.

A small group of texts is concerned with the collecting by Uṣur-ša-Ištar and others of copper due by retailers to Iddi(n)-abum son of Iddi(n)-Ištar. Iddi(n)-abum left a memorandum on which debts with a total value of 24 talents 20 minas and an unknown amount of copper were itemized, which were to be paid in spring; to this were added two posts of in all 2 talents 9 minas written on *iṣurtū*. A complete list of all these claims is given in kt n/k 126, which does not specify the quality of the copper due.[461] A number of the

---

459 Kt n/k 1689.1-7, 14-17: [1] *a-na A-šur-i-mì-tí* [2] *qí-bi-ma um-ma* [3] *Hu-ra-ṣa-num-ma* [4] *ší-mì ta-dí-in-ma* [5] URUDU *i-ba-ší sí-kà-tum* [6] *ú-ṣa-ma i-na ma-hi-ri-im* [7] KB *ù* KÙ.GI *lá-šu* (...) [14] *sí-kà-tum a-na* [15] 10 *u4-me-e i-tù-ra-nim-ma* [16] KÙ.GI *a-ša-a-ma* [17] KI *a-li-ki ú-šé-ba-la-kum*, 'Speak to Aššur-imittī, thus (says) Huraṣānum: my goods have been sold and the copper is in store! The *army* is out and there is no silver or gold on the market. (...) The *army* will return within 10 days, and then I will buy gold and send it you with a caravan' (by courtesy of S. Çeçen).

460 Kt n/k 513.12-21: [12] *um-ma šu-ut-ma* [13] *ha-ra-ni* KB *dí-na-nim* [14] *um-ma né-nu-ma* URUDU [15] *ša a-wi-lim iš-qá-lá-al* [16] *šu-ma a-dí ba-áb ha-<ra>-ni-šu* [17] URUDU *lá e-ta-ar-ba-am* [18] KB *a-ṣí-ib-tim ni-lá-qé-ma* [19] KB *ni-da-šu-um* [20] *a-na* KB *Ku-ra-ra* [21] *ù Hu-ni-a e-pu-lu-šu*, '(Usānum) said: "I am about to set off, give me the silver." We answered: "The gentleman's copper can arrive at any moment." If the copper has not arrived before he is leaving, we will borrow silver at interest and give him the silver. As for the silver, Kurara and Hunia will guarantee for him' (by courtesy of C. Günbattı).

461 Kt n/k 126: the debts listed in the memorandum left behind by Iddi(n)-abum are: 1 talent 15 minas due from Gallābum; 1 talent 15 minas Hadāya and [PN]; [broken]; 2 talents 40 minas Dudāya; 40 minas Alma-[ ]; 2 talents [ ]-libašu; 4 talents [PN]; 2 talents [ ]-šaili; 1 talent [PN]; 2 talents [PN]; 2 1/2 talents Dagašušu; 5 talents Dukmana. Two debt-notes (*iṣurtum*, see Veenhof 1995): 1 talent 9 minas and 1 talent due by Tawi-Enlil (by courtesy of S. Bayram).

retailers are Anatolians. Several texts record which copper was collected and what was done with it. Uṣur-ša-Ištar used some of it to cover expenses he apparently incurred during collecting. The copper which was paid by the retailers was of high (fine, refined, Haburata), and poor quality.[462] One can only speculate what caused Iddi(n)-abum to leave the collecting of his debts in copper to Uṣur-ša-Ištar. An indication may be contained in two documents from Imdilum's archive, from which it emerges that Aššur-imittī and his nephew Imdilum jointly had a claim (22 1/2 minas of silver in CCT 6, 9a.10-17; 12 1/2 minas in CCT 2, 8.68-69) on this Iddi(n)-abum; it is possible that the debtor paid (in part) by giving Aššur-imittī's firm the right to collect copper from his outstanding claims.

4.6.3 *Geography of the copper trade*

Turhumit and *Tišmurna* are named as cities where copper is obtained. In Turhumit, an unspecified amount is bought with silver coming from Purušhattum according to kt n/k 483. In Tišmurna it is obtained in exchange for textiles according to kt n/k 122 and kt n/k 208. Refined copper is transported from Turhumit to Wahšušana in kt n/k 1451,[463] from whence it is taken to another city (probably Purušhattum) to be sold. Fine copper is transported from Tišmurna to Wahšušana in kt n/k 122. In this text, the local stocks of fine copper, probably in Tišmurna, were depleted, forcing Suen-rē'ī and his colleagues to obtain the metal elsewhere, in a location called *ma-at-hi-ru-ùh*, 'the country of Hiruh', but which can perhaps be interpreted as Marithum with metathesis.[464]

---

462 Kt n/k 231: [1] 3 GÚ 4 *ma-na* [2] *ma-sí-am* 13 *ma-na* [3] *lá-mu-nam* [4] *Ú-ṣur-ša-Ištar* [5] *a-na gam-ri-šu il$_5$-qé* [6] 2 1/2 GÚ *Ha-bu-ra-ta-i-am* [7] *a-ha-ma Ú-ṣur-ša-Ištar* [8] *il$_5$-qé* 1 GÚ 14 *ma-na* [9] *lá-mu-nam* [10] *A-šùr-ṣú-lu-li* [11] *il$_5$-qé i-na* URUDU [12] *ša I-dí-a-bi-im* [13] DUMU *I-dí-Ištar* [14] *il$_5$-qé-ú*, '3 talents 4 minas of refined copper, Uṣur-ša-Ištar took 13 minas of poor copper for his expenses; furthermore, Uṣur-ša-Ištar collected 2 1/2 talents of *Haburata* copper. Aššur-ṣulūlī collected 1 talent 14 minas of poor copper; they collected it from the copper (due to) Iddi(n)-abum son of Iddi(n)-Ištar.' In kt n/k 155, Uṣur-ša-Ištar collected 2 talents 20 minas of fine copper; kt n/k 159 records that Uṣur-ša-Ištar 'did not take' 13 1/2 minas of *Haburata* copper of the 2 1/2 talents of *Haburata* copper that were written in the memorandum and which he collected. Kt n/k 172 records, among other things, that 40 minas of refined copper were collected from Iddi(n)-abum's outstanding claims (all texts by courtesy of S. Bayram).

463 Kt n/k 1451.6-26: [6] *a-na-kam Kà-ta-a* DUMU *A-lá-ni-a* [7] *[k]à-ṣa-ar . Lulu . a-na Tur$_4$-hu-mì-it* [8] [x] KB *a-na ša ki-ma* [9] *ni-a-tí nu-šé-bi$_4$-il$_5$-šu-ma* [10] URUDU *ma-sí-a-am* 4 GÚ 30 *ma-na* [11] *ku-nu-ki . ša ki-ma* [12] *ni-a-tí . ik-nu-ku-šu-ma* [13] URUDU *iš-tù Tur$_4$-hu-mì-it* [14] *ú-šé-ṣí-a-ma* [15] *i-na Wa-ah-šu-ša-na* [16] 1 ITU.KAM *ù* ITU.2.KAM [17] *i-sú-hu-ur . ki-ma* [18] *is-hu-ru . Du-du . a-na* [19] *ṣé-ri-šu . ni-iš-pur-ma* ŠÀ.BA [20] 2 GÚ 30 *ma-na* URUDU *ma-sí-am* [21] *ù* ANŠE *ṣa-lá-ma-am* [22] *[D]u-du . ub-lam ni-ša-me-ma* [23] *Kà-ta-a . a-na Kà-ni-iš* [24] *i-tal-kam . a-ba-ú-ni . be-lu-ni* [25] *a-tù-nu . ší-tí* URUDU 2 GÚ URUDU [26] *ma-sí-um* ANŠE *ṣa-lá-mu-um*, (to Iddi(n)-abum, Uṣur-ša-Ištar, and Aššur-taklāku, from Taq(u)n-abum and Šumma-libb-Aššur:) 'Here, we had Katāya son of Alānia, Lulu's transporter, bring x silver to our representatives in Turhumit, and our representatives gave him refined copper, 4 1/2 talents, under seal. He took the copper out of Turhumit but was delayed for a few months in Wahšušana. We sent Dudu to him because he was delayed, and Dudu brought 2 1/2 talents from this copper and one black donkey. We hear that Katāya has gone to Kanish, our dear fathers and lords, the remainder of the copper, 2 talents of refined copper and one black donkey [...]' (by courtesy of S. Çeçen).

464 Kt n/k 122.19-21: [19] ... *i-na* [20] *ma-at Hi$^?$-ru-ùh* [21] *nu-kà-an-ma*, 'We will secure (copper) in the country of Hiruh' (by courtesy of S. Bayram).

In *Marithum*, apparently located in the same region as Turhumit (I 598.15),[465] hence a city where copper could be obtained, Ikuppia could, according to Suen-rē'ī's representatives, receive the copper that Suen-rē'ī owed.

Uzua had to pay 1 1/2 talents of refined copper in the town of *Zidamriš* (location unknown).[466]

Wahšušana is an important city for selling copper. Uṣur-ša-Ištar orders in kt n/k 228 that Suen-rē'ī and Gallābum should transport all the copper to Ikuppia in Wahšušana (see also kt n/k 122).[467] Internal conflicts in the country of Wahšušana prevented Ikuppia from sending copper to this city in kt n/k 84. Due to the same situation, his copper stored in that country had to be rescued by Aššur-damiq (kt n/k 218). The copper sent to Wahšušana was fine and refined in kt n/k 122. Many texts in this archive deal with consignments of copper ordered by Uṣur-ša-Ištar when he was staying in Wahšušana. The city must have offered attractive opportunities for selling copper at that time, which he wished to exploit.

Purušhattum was the destination of a consignment consisting of 15 donkeys (carrying capacity about 40 talents) that Ikuppia sent with the son of Ennu-Bēlum (= Aššur-taklāku?) and Gallābum from Turhumit to Uṣur-ša-Ištar's representatives; Ikuppia had to send the remainder of the copper to Wahšušana (kt n/k 227 and 1456). It is probable that the copper was to be sold in these cities.

Ikuppia had to send copper to *Kanish* in kt n/k 276 (from Turhumit?). If he did not succeed in finding donkeys for transport, he was to leave the copper packed in straw so that Uṣur-ša-Ištar could take the copper along immediately. In kt n/k 1277, Uṣur-ša-Ištar reproached Ikuppia for using his copper destined for Kanish and Purušhattum for commercial deals.

Although *Assur* is in general not part of the Kanish-based Anatolian copper trade, a passage from kt n/k 226 dealing with affairs in the capital should be mentioned here. Uṣur-ša-Ištar asked his brother Hunia, who was presumably staying in Assur, to use copper to buy barley if it were the right season; if not, he should use the copper to buy merchandise and send that to him (kt n/k 226).[468]

---

465 In I 598, these cities are named as alternative or subsequent destinations. Lines 14-17: ^{14} *ú a-na-ku . a-na* ^{15} *Ma-ri-it-hi-im* ^{16} *ú Tur*$_4$*-hu-mì-it* ^{17} *e-tí-iq-ma*, 'And I will travel to Marithum and/or Turhumit (and execute your order)' (by courtesy of K. Hecker).

466 Kt n/k 175.12-17: ^{12} *a-dí-i* URUDU *ša li-bi*$_4$ ^{13} *Ú-zu-a* 1 1/2 GÚ *ma-sí-am* ^{14} *Zu-ma-a Da-na-A-šùr* ^{15} *ú Hu-za-lá e-pu-ul* ^{16} *i-na Zi-da-am-ri-iš* ^{17} *i-ša-qal*, (to Uṣur-ša-Ištar from Ikuppia:) 'As for the copper that Uzua owes: Zuma [mistake for Uzua?] promised Dān-Aššur and Huzala that he will pay in Zidamriš 1 1/2 talent of refined copper' (by courtesy of S. Bayram).

467 Kt n/k 228.5-9, 7'-10': ^{5} *a-ma-lá té-er-tí-a* ^{6} URUDU *ma-lá i-ba-ší-ú* ^{7} *Ta-as-ku-ul a-na* ^{8} *Wa-ah-šu-ša-na a-ṣé-er* ^{9} [*I-ku-pí*]*-a lu-bi-il*$_5$ (...) $^{7'}$ *ù a-ta* URUDU *ma-lá ta-kà-*<*šu*>*-du* $^{8'}$ *a-na Wa-ah-šu-ša-na* $^{9'}$ *a-ṣé-er I-ku-pí-a* $^{10'}$ *šé-bi-il*$_5$, 'Let Taskul send all the copper available to Ikuppia in Wahšušana, in accordance with my instructions. (...) And you (= Suen-rē'ī), send all the copper that you acquire to Ikuppia in Wahšušana' (by courtesy of S. Bayram).

468 Kt n/k 226.21-25: ^{21} *šu-ma ša-tum ša-na-at* ^{22} *ša* URUDU *ú-ṭá-ta-am* ^{23} *ša-a-ma-am šu-ma* ^{24} *lá ki-a-am ší-ma-am* ^{25} *ša-a-am-ma šé-ṣí-a-am*, 'If it is the right season, buy for me barley with copper; if it is not, buy me merchandise (with the copper) and export it!' (by courtesy of S. Bayram). It is uncertain whether kt n/k 845 belongs to this archive. In this text, 5 minas of silver are sent to Hadāya in Assur to purchase copper and tin (instead of the expected textiles and tin): ^{8} *lu* URUDU *lu* AN.NA ^{9} *ma-lá i-ša-ú-mu* ^{10} *i-na a-lim*^{ki} ^{11} *i-na-dì*, 'He will store the copper or the tin, as much as he will buy, in the City.'

4.6.4 *Conclusion*

Uṣur-ša-Ištar's archive contains a number of specific cases which illustrate the way copper was obtained and sold. His uncle Amur-Ištar and the latter's son Suen-rē'ī were crucial in exchanging merchandise for copper and in transporting it. A fixed geographical pattern is discernable: the metal was sent from the northern to the southern half of the trading area. The choice of whether to sell it in Kanish, Wahšušana, or in Purušhattum is never mentioned in these texts, but must have been dictated by the local availability of silver and gold. Uṣur-ša-Ištar seems to have been well-informed on local market situations and he directed the movement of the copper. Unfortunately, the texts studied lack direct information on the actual sale of the metal.

4.7 *Economic aspects*

4.7.1 *Trade*

The trade consisted of two, supplementary, systems. On the one hand, there was local barter between merchants and the palace or other individuals with buying power. Merchandise, e.g. tin and textiles, were bartered for other goods (wool), or for metals, like silver, gold, and copper. This is direct exchange, in which the item obtained was the goal of the operation.

Then there was indirect barter where item *a* was bartered in place *x* for item *b*. Item *b* was brought to place *y* and there bartered for item *c*, which was brought to place *x*. This system only functions when the various items are available or accepted in a place. In the Old Assyrian situation, this could mean the exchange in Assur of silver for tin and textiles, which were bartered in Anatolia in town 1 for wool, which was brought to town 2 and there exchanged for copper, which was bartered in town 3 for silver or gold. These items were sent to Assur, and could be exchanged for tin coming via Iran and textiles originating from Southern Mesopotamia. In this overland trade with local barter, several markets were thus interconnected and supplementary. This pattern has been visualized by M.T. Larsen (1967, p.172, in a different form in 1987, p.50).

Overland trade is facilitated by the availability of a commodity functioning as a means of exchange or payment that is widely accepted and has a stable value. In other words, a commodity which functions as currency or money and can be used for indirect exchange. Various metals were used, for instance silver in Assur. Anatolia possesses several silver mines, notably in the Bolkardağ mining area in the Central Taurus, south of Niğde, which might have been the Silver Mountain mentioned in the inscriptions of Sargon of Akkad.[469] Documents from the archives of Anatolian moneylenders show that silver was provided as a loan to other Anatolians. Enišarum gave loans in silver and in cereals. In some of the so-called 'mixed loans', containing both silver and a cereal, it is possible that the debtor could

---

469 I.J. Gelb and B. Kienast, *Die altakkadischen Königsinschriften des Dritten Jahrtausends v. Chr.* (= FAOS 7), pp.164-167, Sargon C 2; Muhly et al. 1990, p.212.

choose in which item he would pay.[470] The debt-notes from the archive of Ašēd were mainly in silver.[471] Also the *kumrum*-priest of the god Hikiša provided loans in silver to his fellow townsmen.[472] Not only such debt-notes prove that silver was circulating among the local population, but also deeds recording the sale of a house or of a person as a 'slave' use silver to express the value. In some cases, silver may have been used simply to express the value of the debt, which was actually paid in another commodity. However, in most of these debt-notes the item was in fact silver, which could amount to a couple of minas. According to ICK 1, 30, the Assyrian merchant Imdilum has a claim of 10 shekels of extra fine *pašallum* gold on two Anatolians. If they do not pay at the fixed time, they will have to pay the amount in silver at the unfavourable exchange ratio between silver and gold of 10:1 (see for this text Ichisar 1981, pp.57-59). All debts owed by Anatolians to Assyrians known at the time were listed by P. Garelli (1963, p.384). In his opinion, the Assyrians loaned silver to the Anatolians, but mistrusted their solvency and therefore demanded high interest rates compared with the frequently attested rate of 30% charged among Assyrians. These Anatolians apparently were often connected with the Assyrians by business or family links.[473] The high interest rates, however, appear to have been usual among Anatolians. The Anatolian moneylender Perua charged an interest rate of 30% (kt d/k 22), 40% (kt d/k 35), and even of 60% (kt d/k 15) on loans expressed in silver provided to other Anatolians. The Assyrians used silver to pay expenses in Anatolia. It was often the money in which tax to the *kārum*-office was paid.[474]

Tin was used as an indirect means of exchange by the traders travelling between Assur and Anatolian towns. Caravan accounts mention the amount of 'hand tin' (*annak qātim*) bought in Assur for paying expenses en route. It is probable moreover that others besides Assyrians used tin as money, although no evidence for this is known at present.

Copper, in the shape of objects (sickles, axes, hammers),[475] lumps of metal, and scrap, often functioned as money as well. In Assur, it was used to buy barley (kt n/k 204) and textiles (CCT 1, 39b.1-5), and was even borrowed from a moneylender to pay for a house (VS 26, 8.13-14). Also on the journey between Assur and Anatolia it frequently served as money. The itineraries quoted in § 1.2.1 show that in many cases copper was used to pay all kind of expenses.

The metals mostly used as money were copper and tin, as can also be seen from the itineraries when expenses were recorded (see Nashef 1987). Silver was paid only occasionally, as *šaddu'atum*-tax to the *kārum*-office or as hire for freighters or guards, but

470 Veenhof 1978, pp.285-288.

471 Donbaz 1988a.

472 In kt 92/k 1038; 1047; 1048 (by courtesy of K.R. Veenhof).

473 Garelli 1963, p.262.

474 E.g. in Nashef 1987, p.35 (ICK 2, 293).

475 BIN 4, 198: [1] ⸢2⸣ *pá-ší* 1 *na-ku-pu-um* [2] *šu-qú-ul-ta-áš-nu* [3] 2 *ma-na a-na* [4] *Kur-ub-Ištar a-dí-in* [5] 1 1/3! *ma-na* 5 GÍN [6] *pá-ší a-na Zi-ki* [7] *a-dí-in* [8] 5 GÍN KB [9] *ṣa-ru-pá-am* [10] *a-na Puzur*$_4$*-Ištar* [11] DUMU *I-kà-a* [12] ⸢*a*⸣*-dí-in* IGI *Zi-ki* [13] 1 GÍN LÁ 7 1/2 ŠE KÙ.GI [14] *ù* 2 *ma-na pá-ší* [15] *a-na Nu-ur-Ištar* [16] *a-dí-in* IGI *A-bu-ša-/lim*, 'I gave 2 axes and 1 hammer - their weight is 2 minas - to Kurub-Ištar. I gave 1 mina 25 shekels in axes to Ziki. 5 shekels of refined silver I gave to Puzur-Ištar son of Ikāya, the witness was Ziki. 1 shekel minus 7 1/2 grains of gold and 2 minas in axes I gave to Nūr-Ištar, the witness was Abu-šalim.'

also for foodstuff (barley in AKT 2, 26, see p.72) and even for straw (TC 3, 164). The limited use Assyrian merchants made of silver as a means of payment is not surprising, since this metal and gold were preferably sent to Assur after having been acquired in Anatolia.

Copper was traded because alone or alloyed with tin (bronze), it was the basic material from which a variety of artefacts were made. Moreover, copper could function as money. This created a constant demand for copper, as well as for tin, which was imported by the Assyrians. A number of texts indicate that fine copper was at times offered for sale together with tin,[476] which shows that the copper sold was to a large extent used for the fabrication of bronze. It is uncertain whether the copper and tin in these texts were offered as a conditional sale (see above, p.52). The Assyrian merchants did not trade in bronze. J. Lewy already noted the absence of transactions in bronze and linked this with a text which was later published as KUG 25 by K. Hecker. In this text the words GAL ZABAR occur in line 11, which he rendered *rabi siparrim*, 'official in charge of bronze'. This would have been the Assyrian official who exercised government control over bronze, to prevent the (conquered) Anatolians from obtaining this metal from which they could make weapons.[477] P. Garelli dismissed Lewy's interpretation and concluded that the Assyrians did not forbid the Anatolians to make bronze, but instead encouraged it so as to sell more tin.[478] Lewy's (and Hecker's) interpretation of the passage from KUG 25 was challenged by K.R. Veenhof in his review of KUG, who read GAL not as *rabi*, but as *kāsum*, meaning 'a bronze cup'.[479] The absence of a trade in bronze can be explained by the fact that the alloying of tin to copper was performed locally by the smith, when he manufactured an object (often on demand). There was no commercial demand for ready bronze, only for its two components copper and tin. When bronze occurs in a text, it is in the form of artefacts, often belonging to the inventory of a household (see above, § 3.5.2). Bronze items were sometimes sent to Assur.

In his study on Old Assyrian trade, K.R. Veenhof proved that K. Polanyi had been wrong in assuming that it lacked a proper market mechanism. The texts make it clear that a market existed with price fluctuations, a drive for profit, and a demand.[480] Prices were, apart from supply and demand, fixed by the factors of transport, time, and information.[481] The complex market situation is expressed by the words *mahīrum* and *šīmum*, and constructions involving both words have been discussed by K.R. Veenhof.[482] In the texts *mahīrum* is usually employed to describe the market in both concrete and abstract sense and *šīmum* is often used for 'means of exchange'.

The copper sold by the Assyrian traders was largely bought by the palace and local dignitaries. The place where the exchange took place, was the palace, as in ATHE 32 (**14**;

476 BIN 4, 2; BIN 4, 15; BIN 4, 48; I 574; KTS 18.
477 Lewy 1958, pp.94-95. He had abandoned his earlier view, expressed in EL I, p.134 n.*a*, that *dammuqum* and *masium* were 'bronze'.
478 Garelli 1963, p.278.
479 Veenhof 1967, p.186. W. von Soden maintained the reading *rabi siparrim*, AHw, p.1048b.
480 Veenhof 1972, pp.348-357.
481 Veenhof 1988, pp.257-261.
482 Veenhof 1972, pp.358-397.

see also below). The exchange ratios depended on the supply and demand on both sides, which means that in this case the price-fixing process was in operation inside the palace. As most of the copper was bought by the palace and even inside the palace, the exchange ratios given for a certain city will often have referred to the rate current in the local palace. The rate in force outside of it in all likelihood depended upon that of the palace.

A large scale exchange of merchandise for copper requires a regular supply of this metal. The texts do not tell when the period was during which copper was mined and produced. It can be assumed that there was no mining activity during winter, which means that the mining of copper and smelting of the ore was confined to a specific period. This may have limited the possibility of obtaining copper in the mining regions outside this period, and the expression 'the time for copper has elapsed' in kt a/k 955 might refer to it.[483] Outside the mining areas, however, copper could still be acquired as long as the weather conditions did not prevent its transportation, because there must have been stocks of copper (which had been laid up for various reasons) and because there was always copper in circulation. Texts dealing with the sale of wool, tin, and textiles for copper in the regions where most of the copper originated, show that even during the period when it normally could be obtained, there were at times problems with its supply. Merchants occasionally had to wait some time for fine copper to become available,[484] or had to travel to other areas to find it. This may have been due to an unforseen disruption in the production or transport of copper or because the season for obtaining it was drawing to a close, or simply because the demand for copper of a certain quality was greater than the region could supply. When the movement of goods was interrupted due to human factors, such as war between neighbouring city-states, the closing of roads, or festivals or the influence of the weather, the trade in that area was severely hampered or even brought to a standstill. Traffic was blocked and merchants might have had to stay in a town for a considerable period. Buzāzu had to wait 5 or 6 months in Wahšušana because of a blockade (*sukurtum*) of the road to Purušhattum, possibly caused by an armed conflict (BIN 4, 35 = **6** and CTMMA 71). The ruler of Wašhania prevented caravans from leaving his city for an unknown reason (CCT 5, 15b). In addition, the absence or scarcity of silver or copper as a means of exchange could prevent a trader from exchanging his merchandise, so that he had to store it temporarily while waiting for the situation to become 'normal' again, or go to another city which was not affected and sell his merchandise there. When the supply of copper was delayed (*šadid*), it was cut off (*mahṣū*, 'smitten') from the market and only limited quantities could be obtained (JCS 14, 11).

Copper was usually exchanged for silver. It was transported to the city which offered most for it and thus provided the best opportunity for profit, either because copper was in demand there or because there was sufficient silver, both of which conditioned the exchange ratios. It was important to have the copper in the right place at the appropriate time. If a

---

483 Kt a/k 955.10-14: [10] ... *k[i-ma]* [11] *e-ma-ru . lá-šu-⌈ú⌉-ni* [12] *ù i-tí-i .* URUDU-*im* [13] *e-ta-at-qú* URUDU-*am* [14] *lá ú-šé-bi-lá-kum*, (to Asānum from Anina:) 'I did not sent you copper because there are no donkeys and the time (to obtain) copper has elapsed.'

484 KTK 21.20-22: [20] ... URUDU *lá-mu-num* [21] *i-ba-ší a-na* SIG_5 [22] *ú-qá-a*, 'The poor copper is on hand, but I am waiting for the fine copper.'

merchant himself did not know the political developments and the market situation, he was dependent on his colleagues in other cities for this important information. The archives of Imdilum and Uṣur-ša-Ištar demonstrate the importance to have someone as an agent, employee or partner (in this case Amur-Ištar) living in a city close to the main copper markets who could keep them well-informed. This enabled them to direct the movement of their copper and other goods, and to best exploit the trading opportunities in a particular area when necessary. But in KTS 6 and C 13 (see p.135), it was Ikuppī-Aššur who had to convince Ilī-wēdāku that he had better send Puzur-Aššur's copper to another city because in the area where he was at that moment, silver was in short supply (*kaspū dannū*, 'silver is hard (to obtain)') and hence expensive. Copper was sent to Šalatuar because silver was scarce (*waqar*) in Wahšušana in BIN 4, 6. When an item was scarce, a trader had to take care not to draw attention to this when trying to exchange his merchandise for what remained of it, and he would sell his ware discreetly, *rabbumma* (KTS 2, 34.8; kt n/k 131.17).

The absence of silver in a city where normally large amounts of copper were exchanged for it, forced the traders to store the copper if it could not be brought to another city as in the previous texts. In such a situation, a merchant might also entrust tin and fine copper to another trader in the hope that he would succeed in exchanging it.[485] The result of all this copper lying idle was that its price would drop, and this created a fine opportunity for acquiring it more cheaply in exchange for some other merchandise. Afterwards, when enough silver had become available, the copper could again be exchanged for it, with additional profit. This was Amur-Ištar's idea when he asked Pūšu-kēn to send him merchandise, presumably in Purušhattum: 'There is no silver on the market, and there are large stockpiles of copper. My dear father, take care to buy either fine tin or pure small-sized *amūtum* in case there are no textiles, and send it me.'[486]

Whereas an abundance of unsold copper caused by a scarcity of silver makes it cheaper to obtain copper, the opposite happens when silver is stored in abundance. The famous text ATHE 32, with the same addressee and sender as CCT 4, 34c, relates how Amur-Ištar anticipated the drop in the value of silver as a result of large scale purchases of copper by Syrian traders from Ebla who paid with silver. When the palace would be short of copper but possess abundant silver, he would sell his copper for it (**14**, see also Veenhof 1988, p.260. A tentative calculation of the profit is presented below, p.160).

---

485 BIN 4, 15.12-20: [12] *a-na* KB *ša iš-tí* [13] *En-na-Sú-in* [14] *lá ú-šé-bi$_4$-lá-ni* [15] *ší-mu-um . a-na* AN.NA [16] *ù* URUDU SIG$_5$ *lá-šu-ma* [17] *ù* AN.NA . *ma-sú-/ùh-ma* [18] *a-na* ITU.1.KAM-*im* [19] *ù* ITU.2.KAM *a-na um-me-a-/nim* [20] *ke-nim ú-šé-er*, (to Pūšu-kēn from Dān-Aššur:) 'As for the fact that I did not sent you silver with Enna-Suen: there is no means of exchange for tin or fine copper, and moreover the tin is dirty, so I have handed it over to a trustworthy merchant for a term of 1 or 2 months.'

486 CCT 4, 34c.8-17: [8] KB *i-na ma-hi-ri-im* [9] *lá-šu ú* URUDU [10] *da-ni-[š]a-ma* [11] *ú-me-ru-ú* [12] *a-bi a-ta i-hi-id-/ma* [13] *šu-ma* TÚG-*tù lá-šu* [14] *lu* AN.NA SIG$_5$ [15] *lu a-mu-tám ṣa-ha-ar-/tám* [16] *za-ku-tám ša-ma-ma* [17] *šé-bi$_4$-lam*. See Veenhof 1972, p.391, for line 11.

4.7.2 *The main copper markets*

Anatolia was divided into a number of political entities called 'country' (*mātum*). Each of these consisted of a city which gave its name to the country, where the palace (*ekallum*) was situated with the ruler (*rubā'um*) and his entourage, and a number of small villages (*ālānī*) or hamlets. The economy was based mainly on agriculture and trade, with probably mining as a seasonal alternative in some cities. The palace was the organization that ruled the city-state and control of financial resources made it the heart of the local economy.

The particulars of the copper trade are unevenly known for the main cities connected with it. The geographical aspect of these towns and regions has been discussed above, § 1.2.2, and the metallurgical aspect in § 3.1.5. For *Tišmurna* a palace and a small Assyrian trade settlement (*wabartum*) are attested.[487] The palace pays copper for textiles (kt n/k 482), *amūtum*-mineral (CCT 6, 12a), and wool (CCT 4, 27a). The copper traded there is of poor and fine/refined quality. The impossibility of obtaining fine copper is mentioned e.g. in CCT 4, 27a and kt n/k 122 (after 10 talents of fine copper had been received). Without copper the local market was paralysed (CCT 4, 27a). Fine copper once is obtained for silver at an exchange ratio between copper and silver of 164:1 (CCT 4, 27a.32, see p.54), but the use of silver in this city seems exceptional. Poor copper appears to have been brought to Turhumit for exchange (CCT 4, 27a). Wool is transported to Tišmurna in BIN 4, 58.9, and sold there according to CCT 4, 27a.11, 17.

Although *Taritar* is frequently attested as a designation of copper, the city itself is still absent from the texts. It was a mining centre which gave its name to a type of fine copper.

The country of *Sawit* is mentioned twice as provenance of copper, in AKT 3, 56 of 'pure' and in CCT 2, 23 of fine and Kunanamit copper.

The data on *Kunanamit* are scarce. From VS 26, 71, it appears that merchandise is regularly exchanged for copper in this city, whence it is shipped to Turhumit. The copper of this region was of poor quality, and could be exchanged for fine copper in Turhumit. An uproar in the country once made it impossible to collect copper (LB 1209).

For *Turhumit* a ruler and an important Assyrian *kārum* are attested. Due to its geographical location, this city was pivotal in the copper trade and its market was the link between neighbouring copper producing areas and merchants wishing to procure the metal. The available evidence on the trade conducted in Turhumit and the merchants working there has been analysed by C. Michel (1991b). The expression *mahīr Turhumit* denotes the exchange ratio for fine copper in this city,[488] and is proof of the importance of Turhumit as a market city. Poor copper from Kunanamit and Tišmurna was transported to this city and, like poor copper obtained in Turhumit itself (BIN 4, 64), could be exchanged there for fine

487 Kt a/k 1429.22-23, see AKT 2, p.38.

488 Kt n/k 661.7-10: [7] *ma-hi-ir Tur$_4$-hu-mì-it* [8] URUDU SIG$_5$. *i-na* [9] *Kà-ni-iš* [10] *i-ša-qal*, 'He will pay (the debt of 10 shekels of silver) (in) fine copper in Kanish according to the rate current in Turhumit' (by courtesy of S. Çeçen);
kt 91/k 181.13-17: [13] ... 2 GÍN KB [14] *a-na* DUB.SAR *áš-qúl* [15] *šu-ma i-na ba-áb-tí-šu* [16] *la al-té-qé ma-hi-ir* [17] *Tur$_4$-hu-mì-it* URUDU *i-ša-qal*, 'I paid 2 shekels of silver to the scribe; if I can not take it from his outstanding claims, he will pay me copper according to the rate current in Turhumit' (by courtesy of K.R. Veenhof).

or refined copper (*ana dammuqim ta"urum*). As poor copper was less in demand than fine copper, this means that a considerable amount of poor copper must have been refined in or near Turhumit to supply the trade with fine copper. Apart from this fact, however, nothing is known of the local copper processing. Tin (ATHE 38) and textiles (kt 89/k 430) were sold mainly for fine copper, and the use of silver for purchasing copper is attested in kt n/k 483 and 1451. Fine Taritar copper was obtained in Turhumit according to ICK 2, 54 (**2**), and broken fine copper had to be paid in this city in AKT 1, 66. The copper was transported to Wahšušana (kt 89/k 430), Purušhattum (VS 26, 71), and Kanish (AKT 1, 17). The Assyrians not only obtained copper in Turhumit, but occasionally they sold it there as well, as in CCT 5, 13a, where copper earned by the sale of hides and grain was sold for silver and gold. The *kārum* of Turhumit influenced the trade by having its own interest rate (kt 91/k 390.9) and possibly by issuing a decree regulating the import of poor copper into Turhumit (CCT 4, 27a). It is unclear whether in KTS 18 events in Turhumit are described; if so, this letter provides information on the sale of some textiles and of wool for copper to the palace, and on the availability of gold for paying for the merchandise, while silver is in short supply.

For *Wahšušana* a palace, a ruler, and an Assyrian *kārum* are attested. Several texts report war in this country, which seriously hindered trade (kt n/k 84; kt n/k 218; KTH 1; TC 2, 36; also AKT 3, 91?). Tin and textiles were sold for copper to the palace, which paid through the *kārum*-office. Tahašaili exchanged 268 textiles and 2 talents of tin for nearly 9,000 minas of copper, presumably in this city, according to kt c/k 263 (**4**; see § 4.3.2). Most of this was sent to Purušhattum, but part was exchanged for 2 1/2 minas of *pašallum* gold which was sent to Kanish. Once, when envoys of the *kārum* of Kanish came to Wahšušana they were unable to buy some textiles because the copper from the *kārum*'s storehouse which they intended to use as payment, contained only black copper from Turhumit which was not accepted (MNK 636). The sale of *amūtum* for copper to an important official, the *rabi sikkatim*, probably took place in Wahšušana (KTH 14 = **16**; Struwe; kt m/k 71). The city was an important link on the route from Turhumit to Purušhattum, and most of the copper brought to it came from Turhumit (kt n/k 122: 30 talents of (fine) copper; kt n/k 228). In kt 87/436 and 462, a consignment of poor copper was transported by wagon from Turhumit to Wahšušana and sold there for silver. Another consignment from Turhumit was sold for silver at a rate of 105:1 (I 507). When silver became scarce in Wahšušana, presumably as a result of the war, copper was brought to the next city, Šalatuar, to be sold there for silver (BIN 4, 6).

The town of *Šalatuar*, situated on the road from Wahšušana to Purušhattum, possessed a small Assyrian settlement (*wabartum*). Copper was transported to this place to be sold for silver, in one instance because there was not enough silver available in Wahšušana (BIN 4, 6). According to a reference in BIN 4, 148 (**8**), Ilī-Alum brought 12 talents 28 minas of copper to this town. In BIN 6, 204, it is reported that every 10 talents of refined copper that will be obtained, will be sent to Šalatuar and be exchanged there for silver. One donkey carrying 2 talents 37 minas of fine copper was brought to this town in kt a/k 425. In kt v/k 128, an exchange rate of copper for silver of 73:1 is given.

For *Purušhattum* a palace, a ruler, and an Assyrian *kārum* are attested. P. Garelli has discussed the data about this city (1989), which often had silver in sufficient quantity to attract traders wishing to exchange their merchandise. The silver was then sent to Kanish and from there to Assur. This explains why Purušhattum was the destination of most consignments of copper coming from Turhumit or Wahšušana. In kt c/k 263 (**4**), *šikkum* and fine copper were brought from Wahšušana to Purušhattum by wagon. According to kt n/k 1456, Ikuppia had to bring 15-16 donkey-loads of copper, about 38 talents or 2,280 minas, to this town, presumably from Turhumit. BIN 4, 31 (**7**) relates the arrival of several consignments of copper, probably in Purušhattum. In all, 19 talents 38 minas of *šikkum*, 19 talents 5 1/3 minas of refined, and 3 talents of black copper were brought. The exchange of copper for silver is also mentioned or implied in TC 2, 16; TC 2, 36; TC 3, 3; TTC 14. The palace acquired this copper in exchange for silver. The copper was then used to buy merchandise, e.g. tin and textiles. It bought 32 *kutānum*-textiles from Ennu-Bēlum for 1,280 minas according to CCT 2, 8.63-65. The same person reports in KTS 2, 44, that he sold for a value of 4,000 minas to the palace. This made Purušhattum also a city where copper could be obtained, and among traders information spread when the quality of (locally refined?) copper was 'good' (*ṭāb*).[489] It is related in CCT 2, 13.6, that Puzur-Aššur claimed 10,000 (minas) of copper paid by the palace to Šū-Ištar. A number of texts mention political events which influenced the market (KTH 1; KTS 2, 40). A scarcity of means of exchange is apparent from kt n/k 1464 (**20**); kt 89/k 430; TC 2, 18 (**13**).[490] A plague (*mūtānum*) apparently blocked the supply of tin and textiles to this city, which resulted afterwards in high prices for these items in silver (tin: 4:1; textile: 30 shekels).[491]

Ulama possessed an Assyrian *wabartum*. In this small city bordering on Purušhattum copper could be exchanged for silver (I 608; TC 3, 104). An exchange rate of 77:1 is known from kt v/k 128.

Wašhania was situated between Kanish and Ninašša, on the junction of the route from Kanish to Purušhattum, and of that from Kanish to Wahšušana/Turhumit and Hattum. A ruler and an Assyrian *wabartum* are attested. In this town merchandise was exchanged for copper, which could be converted into silver or gold. Copper was locally refined. The town's *šāpirum* or overseer owed 280 minas of copper for *kutānum*-textiles according to ICK 1, 92.11-15. Ennānum collected there 1,100 minas of copper from Abī-Aššur, who had apparently obtained it in exchange for merchandise (kt a/k 411.4-8). The ruler of Wašhania owed over 4,600 minas of copper to Assyrian merchants according to kt 89/k 228. In TC 3, 10, an amount of 18 talents 45 minas of copper was to be sent to Puzur-Aššur who was staying in this town and had noticed a scarcity of copper there. The local ruler at times put obstacles in the way of the traders (CCT 5, 15b; kt 89/k 228).

---

489 BIN 4, 64.12-13: [12] *a-ša-me-ma* URUDU *i-na* [13] *Pu-ru-uš-ha-tim ṭá-áb*, 'I hear that the (quality of) copper in Purušhattum is good.'

490 TC 2, 18.7-12: [7] ... *ki-ma* URUDU-*i-ú* [8] *i-na Pu-ru-uš-ha-tim* [9] *ma-ah-ṣú-ni i-re-qám* [10] *lá ni-gu₅-ur-ma* [11] *a-na Pu-ru-uš-ha-tim* [12] *lá nu-šé-ri-ib*, 'Because (the market for) copper is hit (disturbed) in Purušhattum, we did not rent a wagon to bring (the copper) into Purušhattum.'

491 Kt n/k 1339.10-13: [10] *iš-tù . mu-ta-ni* [11] *ša Pu-ru-uš-ha-tim* [12] 4 GÍN.TA AN.NA 1/2 *ma-na*.TA [13] TÚG.HI.A *i-ta-aṭ-lam* [14] *i-ta-dí-nu*. Published in Çeçen 1991, pp.57-58.

The heart of the Assyrian trade in Anatolia was the city of *Kanish*, where the main *kārum* was located. Several texts mention the transport of copper to Kanish. According to CCT 6, 46b, a total of 5 talents 16 minas of fine copper was brought to or received in this town. In some cases it is clear that it originated from Turhumit, like the 6 1/2 talents Imdilum was to receive in ICK 2, 54 (**2**).[492] The exchange of copper for *ammurum* silver at a rate of 60:1 is mentioned in kt 87/k 462. The quality of the copper sent to Kanish, as far as mentioned, was always fine or refined. Apart from the texts mentioned above, the evidence consists of: 1 talent + [ ] minas of refined (kt a/k 1080.2); fine copper for a value of 10 shekels of silver (kt n/k 661.8); 2 talents of refined (kt n/k 1451.20); 30 minas of fine copper from Haburata (kt 86/k 176.3); 18 talents 6 minas of fine copper (kt 87/k 462.9); and 40 minas of fine copper (TC 1, 75.2).[493]

Although Kanish and Purušhattum were alternative destinations,[494] more copper was sent to the last city than to the first. The distance between Turhumit and Kanish may have been greater than that from Turhumit to Purušhattum, which made it cheaper to transport copper to Purušhattum. However, the price for copper of 60:1 is attested for both cities. The decisive factor may have been a greater availability of silver or a greater demand for copper in Purušhattum, which caused the merchants to bring their goods to this city rather than to Kanish.

4.7.3 *How profitable was the trade in copper?*

The rationale behind the trade in copper was that profit could be increased by indirect barter. This could be done in different towns. The archive of Imdilum contains references to the sale of wool for copper, which was transported to e.g. Purušhattum (see § 4.4.2). It took considerably more time to obtain silver in the case of indirect barter than in that of direct barter.[495] At a moment when he apparently needed silver urgently, Imdilum ordered his partners not to let his son Amur-ilī exchange tin and textiles for copper again in Wahšušana. All his merchandise had to be sent to Purušhattum and exchanged there for cash payment in silver (CCT 4, 22b).

---

492 Kt n/k 276.9-11: [9] *lá-ma . a-li-kà-ni* [10] URUDU-*i . a-na Kà-ni-iš*^{ki} [11] *a-bu-uk-ma*, 'Bring my copper to Kanish before I come' (by courtesy of S. Çeçen);
kt n/k 661.1-10: [1] 10 GÍN KB *ṣa-ru-pá-[a]m* [2] *i-ṣé-er* [3] ^{d}IM-*ṣú-lu-li* [4] DUMU *En-na-A-šur* [5] DAM.GÀR-*ru-um i-šu* [6] *a-na* 2 ITU.KAM [7] *ma-hi-ir Tur*$_{4}$*-hu-mì-it* [8] URUDU SIG$_{5}$. *i-na* [9] *Kà-ni-iš* [10] *i-ša-qal*, 'Adad-ṣulūlī son of Enna-Aššur owes 10 shekels of refined silver to the merchant. Within 2 months he will pay in Kanish (the equivalent in) fine copper according to the exchange rate current in Turhumit' (by courtesy of S. Çeçen);
kt 87/k 462.9-12: [9] 18 GÚ 6 *ma-na* URUDU SIG$_{5}$ *a-na Ka-ni-iš* [10] *as-ri-dam* 1 *ma-na*.TA [11] *ù* 1 GÍN.TA *a-na a-mu-ri-im* [12] *ta-dí-in*, 'I loaded (and brought) 18 talents 6 minas of fine copper to Kanish, and it has been sold for *ammurum* silver at 60:1' (by courtesy of K. Hecker).

493 BIN 4, 31 (**7**) probably describes the arrival in Purušhattum of a consignment of copper.

494 CCT 4, 12b.15-21: [15] 10 ANŠE-*ri kà-aṣ-ru-/tim* [16] *ša bi*$_{4}$*-il*$_{5}$*-tim šé-ri-a-/ma* [17] URUDU-*a-kà . ša i-ba-ší-ú* [18] *lu a-na Kà-ni-iš* [19] *lu a-na Pu-ru-uš-ha-tim* [20] *a-li . té-er-ta-kà* [21] *i-lá-kà-ni lu-ub-lu*, 'Send me 10 harnessed pack-donkeys, and they will bring your available copper either to Kanish or to Purušhattum, wherever you order.'

495 Veenhof 1988, p.259.

Local barter is described in VS 26, 6 (**18**): a palace bought over 1 talent of tin and paid for it with 410 minas of copper, part of which was black copper and the rest was exchanged for silver.

What appears to be the bulk of the copper was brought from several mining areas to the first major city, Turhumit. Assyrian traders exchanged in this city tin, textiles, silver, and wool for copper. This was available in poor ('poor', 'black') and fine ('fine', 'refined') quality. Poor copper was less in demand as its copper content was low and the additional costs of transporting it to other cities reduced the profit. For this reason, it was often exchanged for fine copper in Turhumit, Tišmurna, or in other towns in or near mining regions. It follows that Turhumit for instance was an important centre for copper refining. The copper was prepared for transport by breaking up the ingots. It was then shipped to towns as far as Purušhattum and Kanish.

The price of copper increases according to the distance it is transported from its place of origin. The costs of transportation are however difficult to determine (see above, § 3.4). It emerged from the itineraries quoted in § 1.2.1 that the available evidence is too diverse and incomplete to estimate the costs per donkey. Debt-notes at times contain the remark that the full amount must be paid, without deduction of costs incurred during transportation, often specifying the city where this should take place.[496] The journey from Turhumit to Wahšušana and further to Purušhattum or Kanish caused enough expenses to make it unprofitable to transport poor copper further than Wahšušana. This can be deduced from three related kt 87/k texts (by courtesy of K. Hecker) in which the transport of copper, presumably from Turhumit, is described. In kt 87/k 436.30 and 462.4, poor copper was transported to Wahšušana and sold there for silver which was sent to Kanish. By contrast, the fine copper was sent to Kanish and exchanged there for silver (kt 87/k 453.21 and 462.9). The rate for poor copper was 80:1, possibly in Wahšušana (kt 87/k 436.18); the fine copper was sold in Kanish at a rate of 60:1 for *ammurum* silver (kt 87/k 462.10). Preferably fine copper would be transported, but it happened occasionally that retailers did not live up to their obligation to deliver this quality, and to the frustration of the merchants paid poor copper instead (as in kt 87/k 436.7-9 and 453.8-9). This could result from an temporary scarcity of refined or fine copper, as described in CCT 4, 27 (in Tišmurna) and TC 2, 36 (see pp.53 and 56).

Whereas a fair number of texts contain information about the exchange ratio between copper and silver, they often do not specify where this rate is valid, or which quality of copper or silver is concerned, nor do they inform us on circumstances which may have

---

496 Usually *šalmam išaqqal*, 'he will pay (or send) the full amount of copper':
kt a/k 1080.1-7: [1] 1 GÚ [x *ma-na*] [2] URUDU *ma-sí-a-am* [3] *i-ṣé-er A-šur*-GA[L] [4] DUMU *En-na-nim* [5] *E-lá . i-šu . šál-ma-am* [6] *a-na Kà-ni-iš* [7] *ú-ša-ak-ša-*[*d*]*a-ni*, 'Aššur-rabi son of Ennānum owes 1 talent [x mina] of refined copper to Ela; he will see to it that the full amount arrives in Kanish';
TC 1, 75.1-7, 17-21: [1] 40 *ma-na* URUDU [2] *da-mu-qá-am* [3] *i-ṣé-er* [4] *Ma-nu-um-ba-<lúm>-A-šùr* DUMU *Ku-zi-a* [5] *Ha-du-uk-lá-nu-um* [6] IR . *ša I-dí-Ku-bi-im* [7] *i-šu* (...) [17] *ba-lúm da-tim* [18] *ša ta-ší-a-tim* [19] *šál-ma-am . ú ke-na-am* [20] *i-na Kà-ni-iš*[ki] [21] *i-ša-qal*, "Mannum-bālum-Aššur son of Kuzia owes 40 minas of fine copper to Haduklānum, the servant of Iddi(n)-Kūbum (...) He will pay the full amount in Kanish without deducting transportation costs.' Also in AKT 1, 42.13-15; kt m/k 164.13-14. Cf. EL 278 (Rosenberg 1).

influenced the market. The data on prices are presented in Appendix 1. The differences in the rates for purchase and sale of copper of the same quality are often due to the different locations in which the metal was obtained and sold. A number of exchange rates are known for the major cities where copper was traded. All concern copper of fine quality:

Tišmurna: 164:1 (CCT 4, 27a)
Turhumit: 120:1 (equivalent, RA 60, Dessallien);
70:1 (equivalent, probably in T, CCT 3, 1.17)
Wahšušana: 105:1 (sale, I 507)
Šalatuar: 73:1 (kt v/k 128)
Ulama: 77:1 (kt v/k 128)
Purušhattum: 60:1 (kt m/k 131); 80:1 (abnormal situation, silver is scarce, most likely in P., kt v/k 128)
Kanish: 60:1 (sale for *ammurum* silver, kt 87/k 462)

This evidence is very meagre and only the references taken from kt v/k 128 refer to exchange rates valid at the same time. In spite of these limitations, a difference in price is discernible, with copper being cheapest in regions closest to the place of production and gradually becoming more expensive as the distance from these areas or the transport costs increase (see Map A). The average purchase price for fine/refined copper is c. 120:1, and its sale price c. 60:1, which means that a profit of 100% is possible from which a part must be deducted to pay for the costs of transportation etc. The exchange rate widely used to calculate the equivalent of fine copper in silver is 60:1, which seems to be the 'normal' rate in Kanish and perhaps also in Purušhattum. This rate fits into the 'ideal' exchange ratios between fine copper, tin, and silver in Anatolia of 60:10:1.

In Assur, copper was rated at 90:1 (TC 2, 14) and 100:1 (KTS 2, 36 + Bursa 3776), and it is self-evident that with additional costs for transportation it was economically unsound to export copper from Kanish to Assur.

Examples in which we can follow the same item from its purchase price in silver in Assur, via the exchange rate for copper in Anatolia, to the exchange rate of this copper in silver which ultimately was sent to Kanish, are not available. Instead, data on the various rates can be used to calculate what the profit might have been. See for data on prices in copper Appendix 1. The purchase prices in Assur have been taken from the data presented by K.R. Veenhof (1988, pp.253-254). The hypothetical yield when sold directly for silver is compared to the yield in silver when the item is first exchanged for copper. Three commodities have been selected: tin, *kutānum*-textiles, and wool (sh = shekels; m = minas).[497] For tin, the yield in silver appears to have been nearly 50% greater if it was first exchanged for copper. Also the indirect exchange of a textile or of wool for copper would increase the amount of silver to be obtained. It might even have doubled when compared to an immediate sale, but it must be stressed again that such calculations are tentative. There is

[497] Kt n/k texts by courtesy of S. Bayram; kt 87/k 464 and I 429 by courtesy of K. Hecker; kt 90/k 173 by courtesy of C. Michel.

at times a considerable difference in the rates of exchange. Prices for wool are very rare and those given here may be exceptional.

tin

- one donkeyload of 2 1/2 talents	
a) cost in Assur at 13 2/3:1 (kt n/k 140.7)	658.5 sh of silver
b) if sold for silver in Anatolia at 8:1 (kt n/k 102.18)	1,125 sh
c) if exchanged for fine copper at 13:1 (kt n/k 234.3) would yield	1,950 m of copper
d) if the copper was sold for silver at 60:1 this would yield	1,950 sh of silver

kutānum-texile

- 50 *k.*-s	
a) cost in Assur at 5 sh	250 sh of silver
b) if sold in Anatolia for 16 1/4 sh of silver (kt n/k 127.8)	812.5 sh
c) if exchanged for fine copper at 20 m (kt n/k 99)	1,000 m of copper
d) if the copper was sold for silver at 60:1 this would yield	1,000 sh of silver
[c2) if at 40 m (CCT 2, 8.63)	2,000 m of copper]
[d2) 2,000 m would yield at 60:1	2,000 sh of silver]

wool

- 1 talent of wool	
a) purchased in Luhusatia at 380:1 (kt 87/k 464.18), cost:	9.5 sh of silver
b) if sold for silver at 225:1 (kt 90/k 173.4), yields	16 sh
[if *naribtum* wool at 126 2/3:1 (I 429.11) yields	28 sh]
c) if exchanged for copper at 2:1 it would yield	30 m of copper
d) if the copper was sold for silver at 60:1 this would yield	30 sh of silver

The sale of tin via copper is also described in ATHE 32 (**14**). Two exchange rates are mentioned: tin for *šikkum* copper at 1:15, and 'copper' for *ammurum* silver at 140:1. In kt n/k 98, the rates of 150 and 140 (difference c. 7%) are given for *šikkum* copper, so the exchange rate of 140:1 mentioned in ATHE 32.21 might well have been for this copper. The tin may have been sold to the palace, that acquired the silver from Eblaite traders. If the Assyrian trader could have sold the tin, e.g. one donkey-load of 2 1/2 talents, elsewhere for silver, he might have obtained for this amount 18 minas 45 shekels at 8:1 or even 25 minas of silver at 6:1. However, he bartered the tin for *šikkum* copper, which would have resulted in 2,250 minas. To equal the yield of a sale directly for silver, he must have sold the *šikkum* copper at a rate of 90:1 (tin:silver = 6:1) or 120:1 (tin:silver = 8:1). The Eblaites bought copper in exchange for *ammurum* silver at a rate of 140:1. The Assyrian trader now trusts that the palace will pay him a better price for his *šikkum* copper. To be rewarded for his trouble of converting the tin first into copper, he should obtain at least a rate of 90:1 or 120:1. The main question is: was there a difference in price between *ammurum* and refined (*ṣarrupum*) silver? The relative value of silver to *šikkum* copper would by its abundance have dropped from 140 to one of the hypothetical rates, and as a drop from 140 to 90 or lower seems very great, it might have been down to 120 or lower. This 'inflation' which the Assyrian merchant apparently anticipated beforehand is still at least 14.3%.

Another text dealing with profit is CCT 3, 16b. This letter to Aššur-nādā from Iddi(n)-Ištar contains a difficult passage which runs in Assyrian: ^{6}... *iš-tù* ^{7} *u*$_4$*-mì-im ša a-Tù-*<*ur*>*-hu-mì-it* ^{8} *ta-li-kà-ni-ma ni-na-me-ru* ^{9} *ší-ma-am ša i-na ša-na-im* ^{10} *ú* 1 *ma-na*.TA *wa-at-ru ú-na-sí-/hu* ^{11} *a-ma-lá u*$_4$*-me-a* URUDU ^{12} 3*-ša-ú i-tù-ur-ma* URUDU *me-at* GÚ ^{13} *bu-tù-qá-e lu a-mu-ur*. K.R. Veenhof translated lines 9-13: 'I have always collected ("torn out") a purchase price which exceeded the double (of the price I paid) and a rate of exchange of 60:1 (for copper). Could I, as the copper always returned in time threefold, have suffered a loss of 100 talents of copper?' It would mean that Iddi(n)-Ištar had realized a profit of more than 100% or even 200% on his copper trade; *šulušā'ū* was taken to mean that the merchant received three times what he invested (a profit of 200%).[498]

In the interpretation offered by C. Michel, Iddi(n)-Ištar wanted to exchange (some of his) silver for copper in Turhumit, but the rate was at that moment 60:1 (line 10). He refused to exchange his silver at such a high rate, which was double the usual one (line 9: 120:1). Michel continues her explanation by remarking that this is a case of price inflation at the Turhumit copper market for which she gives several possible causes.[499]

The problem in this text lies in the meaning of lines 9-12a. C. Michel considers lines 9-10 to reflect the point of view of the copper traders: they asked a rate of 60:1. This interpretation focuses on '1 *mana*.TA ... *unassihu*' and ignores the rest of both lines. K.R. Veenhof had Iddi(n)-Ištar as the subject of *unassihu*, who sold his merchandise very profitably. In another publication, Veenhof translated line 12a: 'copper came back (was converted) at a triple,'[500] assuming an exchange rate of 180:1, in which he was followed by the CAD (Š/III, p.263a). I suggest that *šīmam unassihu* is said by Iddi(n)-Ištar, and means 'I collected means of exchange ("money", i.e. copper).' *Ša ina šanā'im u manā'im watru* indicates the rate at which these means of exchange were obtained and may be rendered '(a rate) exceeding that of 2:1,' like other rates expressed by means of double 'TA' (Veenhof 1972, p.450). Due to this more profitable rate of exchange, Iddi(n)-Ištar could say (with some exaggeration) that the amount of copper he had actually obtained was now threefold ('has tripled') what he was obliged by contract to deliver, and that occurred within the period of time granted to him. Line 12a contains therefore no indication of a rate of exchange. The item which Iddi(n)-Ištar had sold in Turhumit at a rate higher than 2:1 is probably wool, since the rate of 2 minas of wool per mina of copper is known from other texts (e.g. kt n/k 463.6).

498 Veenhof 1972, p.452.
499 Michel 1991b, p.259.
500 Donbaz and Veenhof 1985, p.132.

CHAPTER 5
THE PALACE AND THE *KĀRUM*

5.1 *The Assyrian position in Anatolia*

The city-state of Assur possessed during this period a virtual monopoly on the flow of tin to Anatolia, where this metal was needed for the production of bronze. This monopoly was the basis for the strong position from which Assur had negotiated concessions from the Anatolian city-states enabling it to conduct a lucrative trade with them. The mutual obligations were laid down in sworn treaties between Assur, represented by the central *kārum* in Kanish, and each city-state. As a result, Assyrian merchants were able to bring their goods to Anatolia and conduct their trade there in relative safety.

The economic goal of Assur was to obtain by this trade as much silver and gold as possible, which it needed as money in its commercial relations with the surrounding countries. This aim was threatened by possible competition from other city-states, and the city of Assur had formulated harsh regulations to prevent this as much as possible. The trade in gold was permitted among Assyrians, but it was prohibited for an Assyrian to sell gold to foreigners. In kt 79/k 101, the city-ruler of Assur states to the *kārum* of Kanish: 'Regarding gold, the rule remains unaltered: brothers (= Assyrians) can sell (it) to each other. But according to the stipulation written on the stela (recording the cities' commercial regulations) no Assyrian, whoever he is, shall sell gold to an Akkadian, an Amorite, or a Hurrian. He who does sell (gold) shall not remain alive.'[501] The prohibition on using gold as a payment for goods refers to a large extent to the market in Assur itself. The Akkadians especially were regarded as a danger to the Assyrian commercial interests in the west, and a tablet recording part of a draft treaty between the Assyrians and a local ruler stipulates that the ruler will not permit Akkadian merchants to come to his country.[502] Texts from later periods seem to indicate that the route to Kanish was still blocked for Babylonians. Traders from Mari appear not to have travelled further than Karkemish on the Euphrates, now on the border between Turkey and Syria, and a Babylonian merchant, presumably from Sippar,

---

501 This text was published by H. Sever (1990b, p.262). Kt 79/k 101.11-25: [11] *a-wa-tum ša* KÙ.GI [12] *pá-ni-a-tum-ma* [13] *a-hu-um a-na a-hi-im* [14] *a-na ší-mì-im* [15] *i-da-an* [16] *ki-ma . a-wa-at* [17] *na-ru-a-im* [18] DUMU *A-šùr šu-um-šu* [19] KÙ.GI *a-na A-ki-dí-im* [20] *A-mu-ri-im* [21] *ù Šu-bi-ri-im* [22] *ma-ma-an* [23] *la i-da-an* [24] *ša i-du-nu* [25] *ú-lá i-ba-lá-aṭ.*

502 Kt n/k 794.11-15: [11] ... *A-ki-dí-i* [12] *la tù-šé-lá-ni . šu-ma* [13] *a-na ma-tí-kà . e-té-qú-nim* [14] *lu ta-du-nu-ni-a-tí-ma* [15] *lu ni-du-ku*, 'You will not allow Akkadians to come up to your country; if they go overland to your country you will surely hand (them) over to us and we will surely kill (them).' The text was first published by E. Bilgiç in 1992 and republished in Çeçen and Hecker 1995. The damaged text of a treaty between Assur and the king of Apum, probably dating to the middle of the 18th century B.C., was found in Tell Leilan and published in Eidem1991.

went no further than Emar and Haššum (AbB 12, 51). Old Assyrian texts occasionally contain references to an Akkadian, but it is difficult to determine whether a solitary merchant is meant, who worked with the Assyrians, or whether he was a member of a larger group of Akkadian traders.[503]

The name of the country Šubaru (conventionally Subartu) is used to designate the region where a mainly Hurrian population lived in a number of city-states. Originally, this name referred to the North of Mesopotamia and the surrounding mountains. Around 2000 B.C., the areas to the east and north of the Tigris and Euphrates valley were inhabited by Hurrian-speaking peoples, who may have dominated the tin route across the mountains to Assur.[504] The Habur Triangle was a Hurrian political centre as is apparent in the title 'king of Urkiš and Nawar' borne by Atalšen (c. 2080-2040) and 'king of Urkiš' borne by Tišatal (c. 1980-1970), who also called himself 'the man from Ninua'.[505] Soon after leaving Assur, the Assyrians had to cross Hurrian territory. Ninua (= Nineveh) is known from kt n/k 931.3, 14, recording travel expenses from Ninua to Burallum, and from Assur to Ninua. The city of Urkiš, modern Tell Mozan, is absent from the Old Assyrian documents. A letter sent by the king of Mama, a city located between Maraş and Göksun, was found on the mound in Kültepe, in a layer contemporary with *kārum* level Ib. This king bore the Hurrian name Anum-hirwi, which indicates that part of Mama's population was Hurrian.[506] However, it is still unclear what the extent was of the region which the Assyrians called Subartu. The adjective 'Subaraean' is used for textiles and slaves.[507] Individuals bearing Hurrian names are mentioned in the Old Assyrian texts, where they occur as witness and/or business relation.[508] Some of these persons came from Assur, or had integrated into the Assyrian society, e.g. Ari-šarri son of Ilī-Alum who was year eponym (*limmum*) in Assur during the level Ib period, or Nubar-šarri the father of Aššur-ṭāb and Mannu-ki-Aššur.[509] Others may have been Hurrians from North Mesopotamia. A reference to Hurrian merchants as an ethnic group is only attested in kt 79/k 101, mentioned above.[510]

In spite of the fact that they are only incidentally mentioned in the Assyrian texts, we may assume that a large group of non-Assyrian merchants in Anatolia was formed by Amorites, whereby 'Amorite' refers to the population living in the area west of the

---

503 In BIN 6, 29.51, a person occurs who, judging from his name, was an Akkadian: Ir-Nanna, but the text is probably situated in Assur. In F.T. 4.29 (published in Larsen and Møller 1991, p.239), an unnamed Akkadian owes 30 minas of copper.

504 The tin was brought to Assur via the 'Low Country', *māt šapiltim*, see AKT 3, p.119.

505 Wilhelm 1989, p.11.

506 Wilhelm 1989, p.12. The letter (kt g/t 35) of Anum-hirwi to Waršama, king of Kanish, was published in Balkan 1957. See also Kammenhuber 1977, p.138.

507 CAD Š/III, p.170b; RGTC 4, p.108-109.

508 See Kammenhuber 1977, pp.142-143, for a list of these persons. Note that the text in which the name Hašim-nawar (a scribe in Hurama) occurs, has now been published as KTS 2, 29. Other Hurrian names are: *A-ra-ší-ha-tal*, kt c/k 440.33 and 442.42 (see RGTC 4, p.144); *E-wa-ri-ma-tá-al*, kt c/k 305.9; *Nu-ba-ar-ša-ri-im* father of Mannu-ki-Aššur in kt a/k 455.4; *Nu-bar*-LUGAL father of Aššur-ṭāb in kt c/k 440.41.

509 ITU.1.KAM *hu-bu-ur li-mu-um A-ri*-LUGAL DUMU *Ì-lí-a-lim*, kt n/k 22.16-18 (by courtesy of V. Donbaz). See the previous note for Nubar-šarri; Garelli 1963, p.162.

510 Kt k/k 4 is a level Ib letter written by a Hurrian trader, presumably from North Syria, see Hecker 1993b, pp.55-56. Professor Veenhof informs me that kt 91/k 539.30 mentions a scribe who can understand the Subaraean language.

Euphrates. A more sceptical view on the role played by Amorites was expressed by H. Klengel.[511] Traders from Ebla participated on the Anatolian market; a number of them bought copper from a palace and paid with *ammurum* silver in ATHE 32 (**14**; see above, p.160).[512] Assyrians had business contacts with individual Eblaites, as in kt 91/k 338.2, where an Eblaite bought wool, or in kt 90/k 187, where an Eblaite was involved in the repayment of a debt. An Assyrian reused the seal that had belonged to the Eblaite Ib-Damu.[513] Two persons are identified as the son of 'the man from Tadmor' (the later Palmyra),[514] who is perhaps the same person as the man from Tadmor who owed silver (TC 3, 82.17) and copper (F.T. 4.10). This individual appears to have joined the Assyrian merchants in Anatolia.

When the Old Assyrian merchants entered the Anatolian markets, following in the tracks of earlier generations of traders from Mesopotamia, they met well-developed economic structures. While the import of tin was an exclusively Assyrian affair, textiles were brought to Anatolia by Amorites as well. The Assyrians, as soon as they started participating in the already existing markets for wool and copper in Anatolia, had to share this trade with other merchants, most of them Anatolian. Their organization, capital and commercial skills must have made them formidable competitors in the eyes of the native traders.[515]

A number of archives belonging to Anatolians have been found during the excavations in the *kārum* of Kanish. From texts which were published, it appears that some of these Anatolians acted as money-lenders or traded in agricultural products. Other texts, like kt f/k 163 and 164 from level Ib, contain evidence of Anatolian merchants forming enterprises to do business. Little is known of the degree of cooperation between Assyrian and Anatolian merchants. Some Assyrian traders had an agent bearing an Anatolian name, who might be the child of a marriage between an Assyrian and an Anatolian, but could also be a native trader who had joined the Assyrians, or worked as their agent, if that was compatible with retaining his position as an Anatolian merchant. A number of *hamuštum* eponyms in Kanish bore non-Assyrian names (Larsen 1976, p.358). In the archive of Puzur-Aššur (see § 4.5), a man called Wawali occurs,[516] and Alāhum had an agent in Wahšušana called Tahašaili (see p.111). Occasionally, an Anatolian occurs whose connection with the Assyrians

---

511 Klengel 1989, p.267.

512 The word *ammurum* denotes silver of a specific quality, so already AHw, p.44b, which was distinguished from refined (*ṣarrupum*) silver in kt n/k 105, and from 'fine' (*dammuqum*) in kt a/k 913. In the last text, this silver is used to purchase fine copper in the *bēt ālim* in Assur, kt a/k 913.9-14: [9] *mì-ša-al* KB *da-m[u-qú-um]* [10] *[mì-iš]-lu-um . a-mu-ru-u[m]* [11] *a-na* É-*et a-lim*[ki] [12] *a-na* URUDU *ša-a-mì-im* [13] *ù-šé-ri-bu-ma* [14] URUDU SIG_5 *iṣ-bu-tù*. The writing *a-mu-ru-um* and *a-mu-ra-am* instead of *a-mu-ri-um* and *a-mu-ri-am* proves this word is not a *nisba* from Amurru, but is derived from *amārum* in the meaning 'to estimate (the purity of the silver).' *Kaspum ammurum* (only the adjective in the D-stem), would then mean 'silver considered to be of sufficient purity, though still unrefined.'

513 Seal TC 3, 14C. See for this seal Teissier 1993, p.608 (on seal 4), and 1994, p.58 (no. 529a).

514 Pilah-Adad in kt 86/k 168.5 and kt 86/k 200.19 (by courtesy of K.R. Veenhof); Puzur-Ištar in EL 303.

515 The Assyrians generally called the Anatolians *nu'ā'ū*, and used this word for the native merchants as well, see Dercksen 1992, col.793-794. See Edzard 1989 for a discussion of this word.

516 CCT 3, 49b.26; KTS 7a; KUG 41.32; LB 1255.3'; RA 59, 5.22; TC 3, 3.9; TTC 5.8; VS 26, 121.

mentioned in the text is unclear, like Amura from Kanish or Alulu from Ninašša;[517] the fact that he is identified as coming from a specific city indicates he was known among Assyrian traders.

5.2 *The palace as a buyer of merchandise*

Details on how the Anatolian palaces conducted business with merchants are almost exclusively derived from the texts written by and for Assyrians, so that many questions remain unanswered. It is assumed that agriculture formed the economic foundation of the Anatolian city-state. With the agricultural surplus, goods could be bought, and the possession of a silver or copper mine added to a country's wealth, which appears to have been distributed mainly over the ruler and his family and the dignitaries of the town, like high officials and priests. When the geographical location was favourable, trade would bring a regular flow of goods, stimulate economic activity, and so increase the town's wealth. The levying of toll on users of roads and bridges or other crossings, and of taxes on merchandise coming into the city or traded there provided the ruler without great effort with copper and silver, and also with tin and textiles. The role played by the palace in the production or refining of copper has been treated above, § 3.5.1. A variety of goods were exchanged with the palace, but for only a limited number of these does reliable documentation exist. The data on the sale of wool, silver, tin, and precious stones will not be discussed here; some can be found in the preceding chapter. Here, the sale of textiles and of *amūtum* will be used as an example.

textiles

Textiles form the object in the majority of cases where Assyrians are selling goods to the palace in exchange for copper. The textiles were either taken by pre-emption, or as a lot from the quantity offered for sale by the trader. It is unclear to which degree the merchants were able to fetch the price per textile which they had in mind, and it seems likely that at times a palace would haggle with them about it. In cities possessing a *kārum*-office, the copper was paid to this office, and the traders collected what was due to them. A complicating factor in this system was that a palace could at times run out of means of exchange, which caused a delay in payment for some time.

The quantity of textiles sold and the amount of copper due shows a wide variation. In kt a/k 391, 28 textiles were sold; in CCT 2, 8, 32 *kutānum*-textiles were sold at 40 minas a piece to the palace in Purušhattum, which makes 1,280 minas. In KTS 2, 44.45-47, textiles were sold for 4,000 minas (c. 100 textiles).

amūtum

The high value of this mineral (always purchased by the Assyrians with silver or gold) involved a financial risk which the individual merchant could hardly afford, and this caused

---

517 ICK 2, 296.21-22: [21] ... 55 *ma-na* [22] *iš-tí A-mu-ra Kà-ni-ší-im*, 'Amura from Kanish owes 55 minas (of copper)';
kt a/k 445/b.1-5, 20-22: [1] 1 GÚ 34 *ma-na* URUDU [2] *ší-kam bé-ú-lá-at* [3] *Im-li-kà-a* [4] *A-šùr-re-ṣí* [5] *ú-kà-al* (...) [20] URUDU *a-na A-lu-lu* [21] *Ni-na-ša-i-im* [22] *áš-qú-ul*, 'Aššur-rēṣī possesses 1 talent 34 minas of *šikkum* copper as working-capital from Imlikāya. (...) I paid the copper to Alulu from Ninašša.'

merchants who ventured to buy a quantity of *amūtum* in the hope of making a good profit, to form a joint enterprise (*tappā'uttum*) for this specific purpose, which raised the capital needed to buy *amūtum*. The vicissitudes of such an enterprise involving Innāya have been discussed by C. Michel (1991a, pp.173-178); four traders formed an enterprise provided by an unnamed merchant with 20 minas of silver to buy *aši'um*/KÙ.AN and 1/2 mina to cover expenses (C. Michel text 125 = ICK 1, 1).

A similar enterprise, represented by texts kt m/k 71 and the Struwe-text, was formed between three men, Šū-Ištar son of Aššur-bānī, Ennam-Aššur son of Šalim-ahum, and Mannum-ki-Aššur son of Ahu-waqar. Both texts deal with the dispute which arose between the traders over the division of the proceeds. The part of tablet kt m/k 71 where more details are given is unfortunately broken. A quantity of *amūtum* 'belonging to Mannum-ki-Aššur' had been sold to an Anatolian offical, the *rabi sikkatim* or general, and as part of the payment 3,200 minas of copper had been paid by the palace, which were stored under Mannum-ki-Aššur's seal in Šū-Ištar's house.[518] This copper and the remaining unknown amount of *amūtum* were to be sold for silver. The 3,200 minas or 53 talents 20 minas of copper paid to Mannum-ki-Aššur might be his share, and are then one-third of the total proceeds, which would yield a total of 160 talents of copper. The value in silver of 3,200 minas is more than 26 minas at a rate of 120:1.

In CCT 6, 12a, two rings made of *amūtum*-mineral were sold to the palace of Tišmurna for more than 3 talents of copper.

The buyers

In texts reporting on the sale of merchandise in the palace, the identity of the buyer or buyers remains often unmentioned. Instead, the documents state simply that the goods were bought by or in the palace. It is of some help when the title or function of a buyer is mentioned, although such data have to be used with caution, since the buyer's place of residence is often not given and it therefore remains uncertain whether the same official is involved in more than one case. The importance of the role played in the local economy by the *allahhinnum* has already been stressed by P. Garelli and K.R. Veenhof.[519] His function as manager or steward for the palace would suffice to explain why he purchased goods from the merchants, but his activities were broader and seem to have included trade. He headed an office, the *bēt allahhinnim*, which conducted trade, be it on behalf of the palace or independently, with Assyrian and presumably other merchants as well. In this office a number of officials were active. The office of the *allahhinnum* would pay copper to merchants in BIN 4, 63,[520] but it seems it had difficulties in doing so. The shortage of

---

518 Kt m/k 71.6-8: [6] 3 *li-me* 2 *me-at* URUDU *i-ší-im* [7] *a-mu-tim ša i-na* É.GAL-*lim* [8] *úṣ-a-ni-ma* (by courtesy of K. Hecker);
Struwe.12-23: [12] *a-mu-tám* [13] *ša Ma-nu-um-ki-A-šur* [14] *ša a-na* GAL *sí-kà-tim* [15] *i-ni-id-nu . lu* URUDU [16] *úṣ-a-am* . É *Šu-Ištar* [17] *i-ba-ší . lu i-ṣú-ur-tum* [18] *lu* URUDU . *i-na* É.GAL-*lim* [19] *a-hu-ur lu* URUDU *i-na* [20] É.GAL-*lim ú-ṣí-am* . 3-*šu-nu* [21] *a-na* KB . *ù-ta-ru-ma* [22] KB . *a-na Kà-ni-iš* [23] *i-lá-ak-ma* (by courtesy of K.R. Veenhof).

519 See Veenhof 1989, p.525.

520 BIN 4, 63.3-4, 12-15, 27: [3] URUDU *ša* É *a-lá-hi-nim* [4] *iš-qá-lá-al-ma* (...) [12] *i-na* URUDU [*ša* É] [13] *a-lá-hi-nim lu i-a-/am* [14] *lu ša ba-ri-ni* [15] URUDU *mì-ma lá i-dí-nu-/ni-ni* (...) [27] <URUDU> *ša* É *a-lá-hi-nim a-na* 5 *u_4-me / i-ša-qal*, (Kurub-Ištar to Azuza:) 'The copper of the

means of exchange in the *allahhinnum*'s office, perhaps in Kanish, formed the subject of a *kārum* verdict that nobody should sell goods to the office until its silver was available.[521] Little is known of the volume of the turn-over of means of exchange in the *allahhinnum*'s office to merchants, and there is thus no basis to compare it with the *rabi sikkatim* or 'general' from Wahšušana. This officer spent large amounts of copper on the purchase of *amūtum* in kt m/k 71, and in KTH 14 (**16**) he had promised 10,000 minas of copper as payment, presumably also for *amūtum*.

The goods were often bought on credit. This happened because the palace or an official did not have enough means of exchange to pay each time a merchant sold something, and preferred to let the debts mount up. The amounts indebted apparently could be recorded on a special account which the trader had. If he did not possess such an account, it could be booked on that of one of his associates, as in kt a/k 488/b, a letter to the *allahhinnum* of Ninašša from Irišum: 'And over there my partner credited (the proceeds of 2 *kutānū* and 12 shekels of *husārum*) to his *iṣurtum*; here, I took the price of my textiles and of my *husārum* out of my partner's silver. Give him his silver there!'[522] The word *iṣurtum* denotes a clay tablet written with cuneiform signs.[523] It could be written by an Assyrian for another Assyrian, as in ICK 2, 296.1-3. But in kt a/k 488/b and the following instances such a tablet was apparently kept in the *kārum*-office, and a merchant was credited on it for the copper owed to him by the palace. Copper recorded on an *iṣurtum* is distinguished from copper which is still in the palace or which had left it in lines 17-20 from the Struwe-text (see above, n.518). In kt c/k 459, 216 minas of copper were booked on the 'first tablet, the large *iṣurtum*', for which 45 1/2 shekels of silver could be collected in the *kārum*-office, perhaps in Purušhattum.[524] The amount in silver could be the equivalent of (or payment for?) the copper, which judging from the exchange rate 285:1, was probably *šikkum* copper. The *iṣurtum* was transferable to another person, as in CTMMA 1, 84, which is situated in Turhumit: 'One tablet stating that Ilī-bānī son of Iya and Aššur-mūtappil, your brother, have been satisfied with 35 talents of fine copper from the proceeds (recorded on the) *iṣurtū*, and that Amur-Aššur, your attorney, assisted them; one tablet stating that Ilī-bānī and Aššur-mūtappil, your brother, gave me 30 talents of fine copper in the form of credits recorded on *i.*'s, and that I will be responsible for (payment of) Iddi(n)-Ištar's debt of 30 talents of fine copper to Dašušu the *allahhinnum*.'[525] The summary of contents of the

---

allahhinnum's office is dangling (= still not paid) ... they have not given me any copper from what the *allahhinnum*'s office owes, be it mine or that belonging to us jointly ... <Copper> of the *allahhinnum*'s office will be paid within 5 days.'

521 Kt a/k 606/b.5-11: [5] ... *a-dí* [6] KB-*pu-šu-nu* [6] *i-ku-nu-ni* [8] *ma-ma-na* [9] *ší-ma-am* [10] *a-na* É *a-lá-hi-nim* [11] *ú-lá i-da-an*, 'Until there is certainty about (the availability of) their silver, nobody may sell goods to the *allahhinnum*'s office.'

522 Kt a/k 488/b.11-20: [11] *ù a-ma-kam* [12] *ta-pá-i a-na* [13] *i-ṣú-ur-tí-šu* [14] *ú-ṭá-hi a-na-kam* [15] *ší-im* TÚG.HI-*tí-a ù* [16] *hu-sà-ri-a* [17] *i-na* KB-*áp* [18] *tap-pá-i-a al-qé* [19] *a-ma-kam* KB-*áp-šu* [20] *dí-šum.*

523 See for *iṣurtum* Veenhof 1995.

524 Kt c/k 459.12-17: [12] ... 2 *me-at* 16 *ma-na* [13] *i-na ṭup-pì-im* [14] *pá-ni-im i-ṣú-ur-tim* [15] *ra-bi-tim* 2/3 *ma-na* 5 1/2 GÍN [16] KB É *kà-ri-im* [17] *a-lá-qé.*

525 CTMMA 1, 84.32-38: [32] 1 *ṭup-*«dub»*-pu-um ša* 30 GÚ 5 GÚ URUDU SIG$_5$ *ša ší-im i-ṣú-ra-tim* [33] *Ili*$_5$*-ba-ni* DUMU *I-a-a ù A-šur-mu-ta-pì-il*$_5$ *a-hu-kà* [34] *ša-bu-ú-ni-ma ù A-mur-A-šur ra-bi*$_4$*-iṣ-kà i-ša-ha-tí-šu-nu* [35] *i-zi-zu* 1 *ṭup-pu-um ša* 30 GÚ URUDU SIG$_5$ *i-ṣú-ra-tim* [36] *Ili*$_5$*-ba-ni ù*

first tablet demonstrates that at times a trader resorted to the judicial powers of the Assyrian system to obtain his copper from a reluctant debtor, here presumably the palace or an official in Turhumit. After appealing to the city assembly in Assur, it could instruct the *kārum* of Kanish to authorize him to hire an attorney to assist him.[526] The background to kt n/k 1343 is unknown, but it seems that the *kārum* was acting to recover the copper following an appeal made to it.[527] Uṣur-ša-Ištar, functioning as envoy, had brought the local *kārum* of an unknown town a message from the central *kārum* of Kanish (the members of the council are called his 'fathers') ordering it to collect copper from the palace. The *kārum* refused because it considered that it had nothing to do with the affair and told Uṣur-ša-Ištar to collect the copper himself, which he declined to do. It is unclear whether it was his own copper, or whether it was for a third, unmentioned, party.

The lack of means of exchange or the reluctance to pay could result in large debts of a palace to a trader. In kt a/k 583 (**15**), the palace and its officials in an unspecified town owe 24,000 minas of copper, which, even at a (hypothetical) rate of 240:1, is still 100 minas of silver. It seems that this palace had offered fields and orchards worth 20 minas and more silver to the merchant as a fulfilment of its debt, but the offer was rejected. In case of a defaulting Anatolian palace or institution, the *kārum* issued verdicts declaring a boycott, e.g. of the *allahhinnum*'s office in kt 488/b (see above), and of the *rabi simmiltim* in TMH 1, 21c (see Larsen 1976, p.327).

5.3 *The* bēt kārim *or* kārum-*office*

The commercial quarter of a city was called *kārum* or 'harbour'. The partly excavated area of Kanish which is usually referred to as the *kārum* was the walled lower city. The abundance of cuneiform texts has focused attention on the presence of Assyrian merchants there, but the simple picture of a purely Assyrian *kārum* is disturbed now excavations produce evidence that the area was populated by Assyrians, Anatolians, and 'Syrians', and

---

A-šur-mu-ta-pì-il$_5$ *a-hu-kà i-dí-nu-nim-ma* ^{37} *a-na* 30 GÚ URUDU SIG$_5$ *a-na hu-bu-ul I-dí-Ištar* ^{38} *a-na Da-šu-šu-ú a-lá-hi-nim a-za-zu-ni*.

526 See Larsen 1976, p.175, for this procedure.

527 Kt n/k 1343.3-25: ^{3} *ṭup-pá-ak-nu kà-ru-um* ^{4} *iš-ta-me-ma um-ma a-na-*[*ku-ma*] ^{5} *al-kà-nim* URUDU *mu-ùh-ra* ^{6} *um-ma kà-ru-um-ma* ^{7} *mì-nam ṭa-hu-a-ni a-lik* ^{8} *a-ta-ma mu-hu-ur-ma* [*a*]*-šar* ^{9} *li-bi-kà* URUDU *šu-ku-un*? *um-ma* ^{10} *a-na-ku-ma té-er-tí kà-ri-im* ^{11} *ta-áš-me-a* URUDU *a-ṣé-ri-ku-nu* ^{12} *ta-ur* É.GAL-*lúm* URUDU *ú-kà-al* ^{13} *al-kà-nim mu-ùh-ra* ^{14} *ma-ha-ra-am la i-mu-ú* ^{15} *ù* É.GAL-*lúm a-na* URUDU ^{16} *ma-ha-ri-im ú-ša-am-/ṭé-e* ^{17} *um-ma* É.GAL-*lúm-ma* ^{18} *mì-šu-um sà-ah-ra-tí* ^{19} *al-kam-ma* URUDU *mu-hu-ur-ma* ^{20} *a-ṣé-er a-ba-e-kà at-lá-ak* ^{21} *ki-ma* URUDU *ma-ha-ra-am lá i-mu-ú-ni* ^{22} *a-na* É.GAL-*lim um-ma* ^{23} *a-na-ku-ma té-er-tí a-ba-e-a* ^{24} *lá-ma lá i-tù-ra-am* ^{25} *be-lu-a a-tù-nu*, (to the *kārum* of Kanish from Uṣur-ša-Ištar:) 'When the *kārum* had read your letter carefully, I said: "Go and accept the copper." But the *kārum* replied: "What does it concern us? Go and accept it yourself, and place the copper wherever you please." I replied: "You have heard the order by the *kārum*, the copper has been made your responsibility. The palace is offering the copper, so go and accept it." But they refuse to accept it, and (now) the palace has warned me to accept the copper, saying: "Why are you hesitating? Come and accept the copper and then go to your fathers." Because they refused to accept the copper, I replied to the palace: "Please my lords, (I can do nothing) until my fathers' order comes back to me"' (by courtesy of S. Çeçen).

that some parts can be regarded as an industrial or artisans' quarter. The term *kārum* was used for the organization of Assyrian traders in Kanish, and this lower city or part of it was called *kārum* as well. The *kārum* of Kanish was equipped with legal powers and had its own staff (*limmū*, secretary, personnel of lower ranks).[528] Its correspondence shows that it was solely concerned with the (commercial) welfare of the Assyrian traders, and it is improbable that foreign merchants like Hurrians, Amorites or even Anatolians, formed part of this Assyrian organization, though they could have lived in the lower city of Kanish.

The *kārum*-office or *bēt kārim* occupied an important place in the trade between the local palace and Assyrian merchants, whether acting individually or as a group. In it, the *kārum*'s primary assembly (*kārum ṣaher rabi*) convened and administrators worked. If it also contained the *kārum*'s storage facilities, it must have been a large building, but no remains of such a structure have yet been found in the lower city. In texts describing the payment of copper it is said that the copper left the palace or went down from the palace before it reached the *kārum*-office, which suggests that it was not a part of the palace itself,[529] unless these verbs were used metaphorically to demonstrate the palace's importance. It may have been located on the city mound close to the palace. The great importance of the international trade makes it likely that the local ruler exercised some control over the *bēt kārim*, which even might have been a building not exclusively used by Assyrians.

Some taxes were paid in the *kārum*-office. When a consignment of tin and textiles coming from Assur or Northern Syria entered Kanish, the goods were brought up to the palace. On textiles a *nishātum*-tax of 5% and a 'tithe' (*išrātum*) were levied.[530] Textiles belonging to the *ikribū*-fund (investments granted by a temple, see Larsen 1976, p.149 with references to previous literature) were in general exempted from the tithe (but not in Medelhavsmuseet 1977/12). The *kārum*-office in Kanish also charged a *nishātum*-tax on tin of 5%. The meaning of TC 1, 23: [28] 4 *ma-na*.TA URUDU SIG$_5$ [29] *a-na* É *kà-ri-im* [30] *ša ta-ba-lim ni-dí-in*, 'We gave each 4 minas of fine copper to the *kārum*-office ...' is unclear.

If the goods were shipped from Kanish to Purušhattum, tax had to be paid there as well: e.g. in kt c/k 450: brought 154 1/2 textiles; 7 3/4 paid as 5% (*mētum hamšat*) to the *kārum*-office; 146 3/4 cleared. 2 talents of tin; road-tax as far as Purušhattum 2 1/2 minas per talent; 1/2 mina per talent paid to the *kārum*-office. Kt c/k 452: 75 textiles brought to Purušhattum; 3 3/4 paid as 5% (*mētum hamšat*) to the *kārum*-office; 71 1/4 cleared.

In Kanish the taxes on textiles were levied in the palace, but the payment of the tax on tin was performed in the *kārum*-office in kt c/k 450 and 452.[531] The *nishātum* rate of 1/60

528 See Larsen 1976, Part Three.

529 The possibility that the *kārum*-office might be part of the palace building was mentioned by E. Heinrich (1973, p.98) and K.R. Veenhof (1993, p.655).

530 See for a discussion of these forms of taxation Larsen 1967, pp.156-161.

531 Kt c/k 450: consignment of 173 textiles; 8 1/2 *nishātum*; 6[!] *išrātum*; 4 taken by owner: 154 1/2 cleared.
2 talents of tin; 2 minas *nishātum* of the *kārum*-office in Kanish.
Kt c/k 452: 125 brought to palace; 6 1/4 *nishātum*; 30 1/2 are *ikribū*; 9 1/2 *išrātum*;
the palace charged me for 2 lost textiles); 30 *ikribū* taken; 3 deposited on *kārum*-office share: 75 cleared.

on tin in kt c/k 450 is exceptional, since this rate is normally reserved for the *šaddu'atum*-tax levied on any means of exchange, for instance on silver leaving Anatolia. In Purušhattum the 5% tax called *mētum hamšat* was paid in the *kārum*-office. Although in general the tax paid was in the same currency as the amount taxed, exceptions occurred. Tax which amounted to fragments of textiles could be paid in silver.[532] In AKT 3, 113.12, the *šaddu'atum*-tax on 30 minas of silver is paid with 30 minas of refined copper, presumably to save as much silver as possible.

The 'tithe' was not a real taxation, but the amount which the palace was allowed to buy at a discount price on the basis of a right of pre-emption; the textiles were said to be 'taken for sale'.[533] In AKT 3, 61, however, the rate is 6% instead of 10%, presumably because the palace would take less if it was short of money. Payment for the 'tithe' (in the *kārum*-office) is mentioned in kt c/k 454, where the proceeds were not paid as a whole, but divided according to a distributive code (see below, n.540 and n.561).

Taxes were also levied on copper. When copper had been obtained, a *šaddu'atum*-tax was paid, e.g. in ATHE 37 or kt c/k 263 (see § 4.3).[534] After a quantity of copper had been exchanged for silver, this tax was levied on the silver. In kt a/k 1056, it amounted to 1/60,[535] but not in VS 26, 6, where 15 shekels of silver were paid on 130 1/2 shekels of silver obtained after exchanging 168 minas of copper.[536] *Nishātum*- and *šaddu'atum*-tax were also paid on amounts of copper sent to Assur (AnOr 6, 22.12-14).

5.4 *The transfer of copper from the palace to the* kārum-*office*

How and where the palace paid for the goods it had bought is not always clear. A number of texts describe in a few key-words (*warādum*, 'to come down' or *waṣā'um*, 'to leave') the transfer of copper as payment for textiles from the palace to the *kārum*-office, as in VS 26, 54: 'The proceeds of the 53 *kutānum*-textiles for which copper left the palace and is now in the *kārum*-office.'[537] This means the copper left the palace's warehouse and was brought to the *kārum*-office, perhaps near the palace.

---

2 talents of tin; 6 minas paid as *nishātum* in Kanish to the *kārum*-office.

532 Veenhof 1972, p.95.

533 Veenhof 1972, pp.369-370.

534 ATHE 37.12-14: [12] ... 8 GÚ 40 LÁ 2/3 *ma-na* [13] *e-zi-ib ša-du-e-tim* [14] *ša a-ta ta-áš-a-ma-ni*, '8 talents 39 1/3 minas (of copper) which you bought for me, apart from (the amount paid as) *šaddu'atum*-tax.'

535 Kt a/k 1056.26-27: [26] ... ŠÀ.BA 3 1/3 GÍN KB [27] *a-na kà-ri-im ša-du-a-tám a-dí-/in*, '(2 1/2 talents of copper sold for 3 1/3 minas of silver), of this I gave 3 1/3 shekels of silver to the *kārum* as *šaddu'atum*-tax.'

536 VS 26, 6.11-18: [11] ... *ší-tí* [12] 1 *me-at* 68 *ma-na* [13] 1 2/3 *ma-na* 6 1/2 GÍN.TA [14] KB-*ap-šu* ⸢2⸣ *ma-na* 10 1/2 GÍN [15] ŠÀ 15 GÍN *ša-du-a-tám* [16] *ni-dí ší-tí* 1 5/6 *ma-na* 6 1/2 GÍN [17] KB *ku-nu-ki-ni Dan-A-šùr* [18] *na-áš-a-ku-um*, 'The remainder (of your copper) is 168 minas. At a rate of 106.5:1 its equivalent in silver is 2 minas 10 1/2 shekels. Thereof we paid 15 shekels as *šaddu'atum*. The remainder, 1 mina 56 1/2 shekels of silver, is Dān-Aššur bringing you under our seals.'

537 VS 26, 54.12-15: [12] ... *ší-im* 53 *ku-ta-ni* [13] *ša* URUDU . *iš-tù* É.GAL-*lim* [14] *ú-ṣa-ni-ma* . URUDU *i-na* [15] É *kà-ri-im i-ba-ší-ú*.

When the copper had arrived in the *kārum*-office, the trader was able to claim what was his due. If the textiles had belonged to different persons, each collected his part of the copper. The palace might pay in instalments. In BIN 4, 156, the proceeds of 70 textiles bought by the palace ultimately came into the hands of a single merchant, but their payment was collected by three persons. The merchant himself collected the copper paid for 9 1/2 textiles (20 minas per textile) in the *kārum*-office, while two other Assyrians, who after collecting the copper apparently would exchange it for silver, had to pay him silver for 50 1/2 and for 10 textiles.[538]

According to the letter kt n/k 506 (courtesy of C. Günbattı), out of 113 textiles, 35 were sold for 1,024 minas of copper to a palace, perhaps in Ulama. Half of this amount of copper became available, half was to be given later. The 512 minas of copper which had become available, presumably in the *kārum*-office, were not paid to the trader in copper, but in 48 1/3 shekels of *pašallum* gold. We do not know who ordered the exchange of the copper for gold, or whether it was usual to pay in this way. 40 other textiles were sold for 20 1/2 minas of silver, but no mention is made of the buyer. Part of this silver was used to purchase gold, making a total of 1 mina of *pašallum* gold which was sent to Assur. A further 12 textiles were sold for 20 minas of copper a piece; when this copper became available, it would be exchanged for silver. The buyer of the 40 and 12 textiles was not the palace. The purchase of *pašallum* gold with copper obtained by selling textiles to a palace which occurs in this text, is also attested in kt c/k 263.29-33, where 2 1/2 minas of this type of gold were bought in Wahšušana for the huge amount of 53 talents 10 minas of copper, that is at a rate of 1276:1, which is double the rate of 635.5:1 used in kt n/k 506. This makes it likely that a poor quality of copper was used in the exchange in Wahšušana.

In CCT 2, 24, a palace buys 20 textiles of normal quality at 40 Anatolian minas of copper each (see above, p.88). The proceeds are divided in two unequal parts: for two-thirds of the textiles the trader is credited in the *kārum*-office for 487 minas, which he can collect, and payment for the remaining one-third will be in silver, at 9 shekels of silver per textile, which can be collected within 10 days, presumably in the *kārum*-office as well. It is uncertain why this office paid silver or gold instead of copper in kt c/k 459, kt n/k 506, and kt n/k 1464 (**20**). It may have been customary between the Assyrians who after all were eager to receive silver. The (only administrative?) exchange of the copper within the walls of the *kārum*-office would then have been a service to them, perhaps performed when the copper was still not available.

Payment for the 'tithe' by the palace was likewise done via the *kārum*-office. In KTS 18.3-11, 20 textiles were taken by the palace *ana šīmim* at 20 minas of copper each, and this copper, together with 440 minas of refined copper taken as a 'tithe', presumably on wool, could be collected in the *kārum*-office. In TC 3, 211, it is said that 'I deposited' 16 textiles in the *kārum*-office. The amount is almost 10% of the textiles which had been brought to the palace, and hence represents the 'tithe'. The 'depositing' means that the trader was entitled

---

538 BIN 4, 156: [1] *i-na* 70 TÚG.HI./A [2] *ša i-na* É./GAL-*lim* [3] *il*$_5$-*qé-ú-ni* [4] *ša* 9! 1/2 TÚG.HI.A [5] 20 *ma-na*.TA URUDU [6] É *kà-ri-im* [7] *a-lá-qé* [8] *ša* 50 1/2 TÚG.HI.A [9] KB *Ku-ra-ra* [10] *i-da-nam* [11] *ša* 10 TÚG.HI.A KB [12] *Šu-li i-da-nam*.

to the proceeds for these 16 textiles, which the palace would pay to the *kārum*-office.[539] In kt c/k 454, the proceeds of the 'tithe' were divided in two unequal parts which were booked on two different 'accounts'. The last 7 lines of this text detail how a total of 5 minas 37 1/2 shekels of silver was to be collected, two-thirds from the enterprise's stock (*mišittum*) for 'tithe' and one-third from a fund called 'the thirds' (*šalšātum*). The textiles were apparently bought for about 21 shekels of silver a piece.[540]

In kt n/k 1464 (**20**), an amount of 200 minas of copper could be collected from the *suhuppum*. J. Lewy interpreted this word as 'accumulation', here large stocks of copper which were kept in the *kārum*-office (1958, p.93), in which he was followed by M.T. Larsen (1976, p.314 'the stocks'), W. von Soden (AHw, p.1054b 'etwa "Vorrat"?'), and CAD S, p.350 'stock(?)'. The last to write about *suhuppum* was K.R. Veenhof, who considered it 'eine kollektive Lagerung von Kupfer, an der einzelne Händler/Firmen und auch der *kārum* als solcher Anteile haben und deren Buchführung im *kārum*-Kontor erledigt wurde' (VS 26, p.22 on text 54). Lewy was correct in implying that *suhuppum* does not denote a warehouse or special storage room in the building. We can narrow down its meaning to the local assets of the *kārum* as an organization (as in BIN 4, 112 + CCT 6, 17a.2-7), of a firm or of a merchant registered in the *kārum*-office, which had been built up on various occasions. Although only the *suhuppum* in the *kārum*-office in Wahšušana is explicitly mentioned (BIN 4, 112.4; CCT 2, 31a.23; kt v/k 169), it is likely that similar funds existed in other cities, and even in Assur. In kt c/k 269, Alāhum was given permission by the city council in Assur to recover his claim from the silver of the debtor and two guarantors from the *suhuppum* either in the City or outside of it.[541] A creditor trying to recover claims needed special permission from the *kārum* to seize part of the *suhuppum* of his debtor. The permission in writing was given in the form of a *ṭuppum ša* (*dīn*) *kārim*, as in BIN 4, 41.41-43: 'Obtain a document with a verdict of the *kārum*, stating that you can take from the *suhuppum* from whatever he possesses, (as much as necessary) to compensate for the gold of Adad.' And kt 86/k 218.4-12 (verdict by the *kārum*) read: 'Amur-Ištar son of Amur-ilī can take 50 shekels of silver from the *suhuppum* at the expense of Šū-Suen son of Puzur-šadue, either from his "house" or from whatever else he possesses.'[542] In most cases the item taken from the *suhuppum* is copper, but in some it is silver, as in the passage just quoted from kt 86/k 218. It is unclear whether the amount of silver in such texts merely served to indicate the value of the copper which was to be collected, or whether silver was actually paid.

---

539 TC 3, 211.55-57: ^{55} ... *a-na me-at* 55 TÚG.HI ^{56} *ša a-na* É.GAL-*lim e-ru-bu-ni* 16 TÚG.HI-*tí* ^{57} É *kà-ri-im na-ad-a-ku.*

540 Kt c/k 454.29-35: ^{29} ... 3 2/3 *ma-na* ^{30} 5 GÍN KB *i-na mì-ší-it* ^{31} *e-lá-tim ša iš-ra-tim* ^{32} *a-ta ú a-na-ku ni-lá-qé* ^{33} 1 5/6 *ma-na* 2 1/2 GÍN ^{34} *i-na ša-al-ša-tim* ^{35} *ni-lá-qé.*

541 Kt c/k 269.21-25: *lu i-na a-lim*^{ki} *lu* [*i-na* GA]NA$_2$-*lim* ^{22} *a-li* KB-*áp A-*[*šur-i-mì*]-*tí* ^{23} *A-ni-nim ú A-*[*ú-ke-e*]*n*$_6$ *i-bu-*[*ru-ni*] ^{24} *A-lá-hu-um i-*x-x ^{25} *i-sú-hu-pì-im i-lá-qé.*

542 BIN 4, 41.41-43: ^{41} *ṭup-pá-am ša dí-in kà-ri-im le-qé-a-/am* ^{42} *ša i-mì-ma i-šu-ú a-na* KÙ.GI *ša* ^{d}IM ^{43} *i-sú-hu-pì-im a-ta ta-lá-qé-ú;*
kt 86/k 218.4-12: ^{4} 5/6 *ma-na* KB ^{5} *A-mur-Ištar* DUMU *A-mur-*DINGIR ^{6} *i-na sú-hu-pí-im* ^{7} *i-ṣé-er* ^{8} *Šu-Sú-in* ^{9} DUMU *Puzur*$_4$-<SA>.DU ^{10} *lu i-É-tí-šu* ^{11} *lu i-mì-ma* ^{12} *i-šu-ú i-lá-«ki»-/qé* (by courtesy of K.R. Veenhof).

5.5 *The* kārum-*office and the* šitapkum-*system*

A number of texts contain references to a share (*qātum*) that a trader possessed in copper coming from the palace.[543] Other texts mention amounts of copper which someone deposited (*nadā'um*) as his share in the *kārum*-office in fulfilment of his obligation to pay *dātum*; the copper was credited after depositing of copper, silver, or textiles. While in the first type of text the word 'share' might be interpreted as that part in the proceeds to which each member of a group of traders was entitled, the structure of the second type is very similar to texts concerning the depositing of shares in connection with an activity called *šitapkum*, whereby values are usually expressed in silver.

Making deposits for *šitapkum* was part of collective commercial activities conducted by the *kārum* as an organization.[544] It is uncertain whether the copper which was to be secured (*ša ka''unim*) in the *kārum*-office resulted from the same or from another *kārum*-led collective activity (see above, § 3.2.2). The *kārum* acquired the merchandise by fixing for each registered merchant an amount expressed in copper or silver, called *dātum* or 'charge', that had to be deposited during a specific period.[545] The trader met his obligation by depositing his share of a specified commodity at various opportunities which were created for this purpose, and were called *šitapkum*. Silver and textiles were mainly deposited,[546] but occasionally also other commodities,[547] and perhaps even copper.[548]

The aim of *šitapkum* was to obtain copper or silver under favourable conditions for the merchant. To realize this the merchandise deposited was used to purchase wool according to two texts, but it remains unclear whether this was the normal procedure or happened only in those particular instances.[549] How the *kārum* traded this wool is not mentioned. The payment of copper as the proceeds of a merchants' share in such a wool transaction may be mentioned in kt a/k 230.[550] It is conceivable that deposited goods could also be sold directly to the local palace in exchange for copper.

It is not clear why merchants did not sell this merchandise themselves, but instead handed it over to the *kārum* that sold it to the palace for copper. They were eager to deposit their share, as it meant no risk, little trouble and guaranteed proceeds. On the other hand, it

---

543 Kt a/k 481.8-12: [8] *i-na wa-ṣa* URUDU [9] *ša* É-*kà-lim i-na* [10] *qá-tí-a . ù qá-tí* [11] *I-ri-šum lá-qá-am* [12] *qá-bi-a-tí*, 'You have been allowed to take (the copper) from my share and from that of Irišum as soon as the copper from the palace becomes available.'
544 See Larsen 1976, pp.338-342.
545 See for *dātum* with this meaning Veenhof 1972, pp.228 last seven lines, and 270-272.
546 CCT 2, 5.21-28: [21] ... *lu* KB [22] (erasure) *lu* TÚG.HI.A [23] *iš-ta-pu-ku* [24] *qá-tí a-na* É [25] *kà-ri-im i-ta-dí* [26] *a-na ni-kà-sí* [27] *qá-tí* [28] *ša-ak-li-il*$_{5}$, 'Deposit my share (of silver or textiles) to the *kārum* office (if) they *š*. either silver or textiles, make my share complete.'
547 ATHE 31.27-28: [27] *lu* KB *lu* TÚG.HI.A *lu mì-ma* [28] *iš-ta-pu-ku-ni qá-tí i-ta-dí-ma*.
548 CCT 2, 6.42: [x x *l*]*u*$^{?}$ URUDU *ma-lá iš-tap-ku-ni*.
549 BIN 6, 25.4-9: [4] *šu-ma* KB *a-na* SÍG.HI.A [5] *iš-ta-pu-ku qá-tí* [6] *qá-at Pu-šu-ke-en* [7] *qá-at* DINGIR-*na-da* [8] *ù qá-tí* DUMU *Lá-li-im* [9] *id-a* (to Iddi(n)-abum and Kulumāya from Puzur-Aššur);
TC 3, 15.28-31: [28] ... *šu-ma* . É [29] *kà-ri-im* KB . (erasure) [30] *ší-im . ša-áp-tim* [31] *iš-ta-pu-ku* KB *šu-qú-ul* (to Puzur-Aššur from Buzāzu).
550 Kt a/k 230.8-10: [8] *ší-im* 3 *me-at* SÍG.HI.A [9] 1 *me-at* 37 *ma-na* URUDU [10] *qá-sú al-qé*, 'I collected 137 minas of copper as proceeds of 300 (minas of) wool as his share.'

seems that those participating in this depositing were more or less obliged to do so because of the *dātum* fixed for them.

During the *nikkassū ša kārim*, 'the accounting of the *kārum*', often called simply the *nikkassū*, the balances were drawn up, and if a merchant had not deposited the full amount of the *dātum* fixed for him, he had to make a balance payment (*nipiltum*). Text kt c/k 1500 records three financial transactions conducted by or on behalf of Aššur-emūqī son of Buzutāya with two different *kārū*. They provide chronological details from which the date of each accounting can be derived. The first is an acknowledgment by the *kārum* of Purušhattum that Aššur-emūqī paid them an amount of copper as a balance payment for another merchant, which is dated to month V of the eponym year of Aššur-imittī; the second and third involve payments on behalf of Aššur-emūqī to the *kārum* of Kanish connected with the accounting, and both are dated to month XI of the eponym year of Ennamānum. The fourth records a debt that could have originated from a balance payment, dated to month V of the eponym year of Aššur-mālik.[551] In the letter kt a/k 474, Adad-ṣulūlī is told that during the accounting in the *kārum*-office it was established that he had to make a balance payment to another person. As this payment is expressed in copper, it is possible that the location of the accounting was Purušhattum. It is difficult to determine how far all these payments were connected with *šitapkum*. There are thus at least two months in which the *nikkassū ša kārim* was held, month V attested for Purušhattum(?), and month XI for Kanish. It is possible that an accounting took place in both towns during boths months, so that every half year the accounts were prepared. Alternatively, each town, including those not mentioned in the two texts, could have held its accounting once a year in a particular month. Several references in other texts demonstrate that the accounting took place periodically ('the next *n.*', CCT 2, 6.41; 'there will be a *n.* in four months', BIN 4, 28.24), and the passage 'I assumed liability for the accounting instead of our father in the year when he dicd' in BIN 6, 91, suggests that the accounting was in fact held once a year.[552]

---

551 Kt c/k 1500.1-44:
1 [1] KIŠIB *kà-ri-im . Pu-<ru>-uš-ha-tim* [2] 6 *me-at* 61 1/2 *ma-na* URUDU [3] *i-na áb-na-tí-ni . ni-pì-il*$_5$*-tám* [4] *ša ni-kà-sí . ša En-na-nim* [5] DUMU *Šu-Hu-bur . A-šùr-e-mu-qí* [6] *a-na kà-ri-im-«im» . Pu-ru-uš-ha-/tim* [7] *iš-qú-ul . iš-tù ha-muš-tim* [8] *ša A-šur-iš-tí-kál* . ITU.KAM [9] *áb-ša-ra-ni . li-mu-um A-šur-i-/mì-tí* [10] 2 GÍN.TA *i*-ITU.KAM *a-ma-na-im* [11] *ú-ṣa-áb* .
2 KIŠIB *kà-ri-im . Kà-ni-iš* [12] 6 *ma-na* KB . *ṣa-ru-pá-am* [13] *ší-hi-tám . ša A-šur-e-mu-qí* [14] DUMU *Bu-zu-ta-a . Ištar-ba-ni* [15] *me-ra-šu . a-na kà-ri-im* [16] *Kà-ni-iš . iš-qú-ul . iš-tù* [17] *ha-mu-uš-tim . ša Ku-lu-ma-a* [18] ITU.KAM *ku-zal-li . li-mu-um* [19] *E-na-ma-nim* 1 GÍN.TA [20] *i*-ITU.KAM *a-ma-na-im* [21] [*ṣí-ib-tá*]*m . ú-ṣú-bu*
3 [22] [KIŠIB *kà-ri-i*]*m . Kà-ni-iš* [23] [x *ma-na* K]B *ṣa-ru-pá-am* [24] [*ni*]*-pí-il*$_5$*-tám . ša ni-kà-sí* [25] [*š*]*a* 23 *ma-na*.TA [26] [*š*]*a A-šùr-e-mu-qí* DUMU *Bu-zu-ta-/a* [27] [A]*-šùr-ta-ak-lá-ku a-na kà-ri-/im* [28] [*K*]*à-ni-iš . iš-qú-ul iš-tù* [29] [*h*]*a-muš-tim . ša Ku-lu-ma-a* [30] ITU.KAM . *ku-zal-li . li-mu-um* [31] *E-na-ma-nim* . 2 GÍN.TA *i*-ITU.KAM [32] *a-ma-na-im . ṣí-ib-tám* [33] *ú-ṣa-ab* (...)
4 [36] ... 3 *ma-na* KB *ṣa-ru-pá-am . i-ṣé-er* [38] *A-šur-e-mu-qí* DUMU *Bu-zu-ta-/a* [39] *Puzur*$_4$*-A-na* DUMU *En-nam-A-/šur* [40] *i-šu . iš-tù ha-muš-tim* [41] *ša En-na-nim* DUMU *Am-ri-a* [42] ITU.KAM *áb-ša-ra-ni* [43] *li-mu-um A-šur-ma-lik* [44] DUMU *A-lá-hu-um*. Lines 1-11 are duplicated by kt f/k 13.1-8, where the amount in line 2 is 660 1/2 minas (by courtesy of L. Umur).

552 With Larsen 1976, p.311. BIN 6, 91.2-6: [2] ... *a-šu-mì* [3] [*ša n*]*i-kà-sí i-ša-tim* [4] [*ša a*]*-bu-ni i-mu-tù* [5] [*a-na n*]*i-kà-sí a-na-ku* [6] [*a-pu*]*-ùh a-bi*$_4$*-ni a-*[*z*]*i-iz*. The text was edited by C. Michel, 1991a II, pp.107-108 (text 77). In KTS 55a.29-34, 2 talents 40 minas of copper are (paid) for the *nikkassū*. Ilī-Alum

As stated above, texts about *šitapkum* usually express the value of the deposited goods in silver, but there are a number of texts where the value of the deposits is given in copper, and the amount to be paid or balanced as well, as in kt n/k 1464 (**20**; see below). This may be a particular feature in certain cities where much copper was traded, so that for instance rating in silver was mainly typical of the *kārum*-office in Kanish, but deposits valued in copper were typical of for instance Purušhattum. The deposits were recorded on special tablets in the *kārum*-office. In that of Purušhattum there were at least six special ledgers according to TMH 1, 27b.[553] A 'large tablet' (*ṭuppum rabium*) in the *kārum*-office on which traders were credited for amounts of silver is attested in several texts, but it is unclear whether the same ledger is meant, and in which city, though it was presumably kept in Kanish.[554]

In kt n/k 1464 (**20**), Uṣur-ša-Ištar is informed by Uzubiškum about the accounting in the *kārum*-office, presumably in Purušhattum, of his deposits made during *šitapkum*. The amounts are all given in copper:

3-6: when he left, he was credited with 560 minas of copper and 8 textiles as his share in the *kārum*-office;

7-10: after leaving the city, when one textile was deposited per shareholder, Uzubiškum deposited one textile for him and credited him with another 40 minas;

11-16: furthermore, when 1 mina of silver was deposited per shareholder, Uzubiškum deposited this amount; as an equivalent Uṣur-ša-Ištar could collect 200 minas from the *suhuppum*;

18-20: the accounting took place and the *dātum* per shareholder was fixed at 920 minas (of copper);

21-22: the 8 textiles were valued at 40 minas each, totalling 320 minas (of copper).

The book value of Uṣur-ša-Ištar's deposits was thus 560 + 320 (the 8 textiles) + 40 (1 textile) + 200 = 1,120 minas of copper. But the 200 minas for the silver were paid out separately from the *suhuppum* so that 920 minas remained as the amount credited. The *dātum* was 920 minas, and Uṣur-ša-Ištar had fulfilled his obligations.

I assume that the *dātum* amount was disbursed to the shareholder, perhaps during the *nikkassū* of the *kārum*. In the text the equivalent of the deposited silver was in fact paid in copper to the shareholder, and it is likely that the textiles were also paid for in copper. The reason why the equivalent of the silver was paid separately might be that at that time it was already clear that Uṣur-ša-Ištar's share had been sufficient, or that the silver had been used for a different purpose than the textiles which caused a separate payment.

The equivalents of copper used in this text are 200:1 for silver and 40 minas per textile, but the quality of the copper is not mentioned. The exchange ratio between copper and silver

---

charges the anonymous writer of the text with 16 minas 27 shekels of silver for 35 talents 42 minas of copper, including the (transportation) costs, which he owed him (= an exchange ratio of 130.2:1).

553 TMH 1, 27b: [1] 2 *me-at* 60 *ma-na* [2] *i-ṭup-pì-im* [3] *ša-li-ší-im* [4] 2 *me-at* URUDU [5] *i-ṭup-pì-im* [6] *ša-da-ší-im* [7] É *kà-ri-im* [8] *i-Pu-ru-uš-ha-tim* [9] *na-dá-ku*, 'I have deposited (and have been credited with) 260 minas on the third tablet and 200 (minas of) copper on the sixth tablet at the *kārum*-office in Purušhattum.' A 'first tablet or large *iṣurtum*' is mentioned in kt c/k 459.13-15, see above.

554 BIN 6, 156.21-22; Ka 430.6'-7'; VS 26, 46.4.

is low, and points to poor quality copper, while 40 minas for a textile is a relatively high price (see Appendix 1).

The payment for deposits made for one's share demonstrates that the *dātum* was not a tax, but rather a charge that was levied by the *kārum* organization and paid for by the merchants. The same structure is found with the 'tithe' paid to the palace, for which the trader was credited in the *kārum*-office.

5.6 *The 'five per cent' or* mētum hamšat *in Purušhattum*

In kt c/k 450 and c/k 452, a charge called *mētum hamšat* on textiles entering Purušhattum is mentioned.[555] Five per cent of the load was handed over to the *kārum*-office *ana mētum hamšat*, 'as the five per cent'. In the same city disbursements of copper occur which were made in the *kārum*-office, from a fund called *mētum hamšat* or 'five per cent' as well. Because in texts concerned with this fund the word *qātum* or 'share' is used for the amount to be disbursed, a comparison with *šitapkum* imposes itself. With 'five per cent' presumably a fund administered in Purušhattum is meant, in which merchants had a share (which might differ per participant). By means of payments by merchants to this fund, merchandise was made available to the *kārum* as an organization, which sold the goods for copper and disbursed this among the participants. The disbursement to each participant was not the same, because the quantities received were dependent on the number of textiles imported by each merchant. This consideration enables the equation of the levy called 'five per cent' with the fund by the same name. It would have been confusing if two different concepts bore the same name, one denoting a charge, the other a fund. As a hypothesis I assume that the levying of five per cent on textiles created the fund called 'five per cent'. The charge levied on goods which entered Purušhattum was thus returned in the form of copper to the merchant. This interpretation differs from the one given by M.T. Larsen who suggested that the levying of the 'five per cent' as mentioned in TC 3, 165.37 might be the local form of the *nishātum* taxation, which was also five per cent.[556] The *mētum hamšat* did not replace *šitapkum* in Purušhattum, but existed alongside it, as is shown by CCT 2, 8.30-36, where textiles are deposited as a share in this city for *šitapkum*.[557]

The collecting of copper 'of the five per cent' from the *kārum*-office occurs in the following texts:

BIN 4, 72 (second page of a letter): 'You collected 81 minas of copper as the "five per cent" in the *kārum*-office; out of this amount you deposited two textiles as my share for the

---

555 Kt c/k 450.8-12: [8] ... 1 *me-at* 54 1/2 TÚG.HI.A [9] *iz-ku-ú-nim* 8 LÁ 1/4 TÚG *i-na* [10] *Pu-ru-uš-ha-tim a-na me-at* [11] *ha-am-ša-at a-na* É *kà-ri-im* [12] *a-dí-in*, '154 1/2 textiles became available (after clearance) and I gave 7 3/4 textiles for the "5%" to the *kārum*-office in Purušhattum';
kt c/k 452.23-28: [23] 75 TÚG.HI-*tù-kà* [24] *iz-ku-ú-nim a-na* [25] *Pu-ru-uš-ha-tim ú-šé-ri-ib-ma* [26] *ša me-at ha-am-ša-at* [27] 4 LÁ 1/4 TÚG *a-na* É *kà-ri-im* [28] *a-dí-in*, 'Your 75 textiles became available (after clearance) and I brought (them) to Purušhattum where I gave 3 3/4 textiles as "5%" to the *kārum*-office.'

556 Larsen 1967, p.161.

557 CCT 2, 8.30-36: [30] 14 TÚG *ku-ta-ni i-na Pu-ru-uš-ha-tim* [31] *a-na* É *kà-ri-im qá-tí A-šùr*-DÙG *i-dí* [32] 6 *ku-ta-ni i-na Pu-ru-uš-ha-tim* [33] *a-na* É *kà-ri-im qá-tí E-nu-Be-li[m]* [34] *i-dí* 1 *ku-ta-nam i-na Pu-ru-uš-ha-tim* [35] *[qá]-tí A-du-ú a-na* É *kà-ri-im* [36] *[i]-dí*.

kārum-office. Add the remaining silver to my silver. My dear brother, should we fight among ourselves?! Buzāzu has been credited with 81 minas of copper *per year*. He demanded (it) here from me, saying (as explanation): "I did not collect it." He went there, but the *kārum* replied (to me): "Let Iddi(n)-Kūbum himself, or his representatives, give you the silver. The *kārum* will not give it you (or: or the *kārum* will give it to you)."'[558]

CCT 1, 43 (letter to Imdilum from Ennam-Aššur, Usānum, and Adu): 'I collected 130 minas of copper of the "five per cent" in the *kārum*-office as Puzur-Aššur's share, and 40? minas of copper of the "five per cent" as your share. The proceeds (in) fine copper are 180 minas each, with which Amur-Ištar has been credited in the *kārum*-office. I took from this 78 minas 15 shekels of copper, and Usānum took 100 (minas of) copper in the *kārum*-office as proceeds (in) fine copper; you have been fully paid with your copper. I gave 150 minas of copper - as much as I collected on different occasions as your share in the *kārum*-office - to Usānum and Adu. As much (copper) as ... as share in the *suhuppum* ... in the *kārum*-office, Aššur-taklāku son of Abu-šalim [will bring to] you. You have no claims on me. Adu took 18 from the 250 minas (for his expenses).'[559]

kt a/k 410: '(As for) the 120 minas of copper for Aššur-nādā, (and) the 75 minas of copper for Aššur-iddi(n), which Šū-Nunu had collected in the *kārum*-office as their share of the "five per cent", the representatives of Aššur-ṭāb son of Alāhum, Ennum-Aššur son of Zuzu, and of Uṣur-ša-Aššur son of Akua have collected (this) copper from Šū-Nunu.'[560]

In all three texts an amount of copper can be collected from the *kārum*-office as disbursement for the 'five per cent'. In BIN 4, 72, the disbursement per share for the 'five per cent' was 81 minas, and in both instances mentioned in this text the copper was exchanged for silver. The word *šattum* in line 11 remains problematic; it is here interpreted as suffixed with the locative-ending *-um*, and unnecessarily preceded by *ina*.[561] The

---

558 BIN 4, 72.3-18: [3] ... ⸢81⸣ *ma-/na** [4] URUDU *ša me-tum* 5 ŠÈ [5] É *kà-ri-im tal-qé* ŠÀ.BA [6] 2 TÚG *qá-tí a-na* É *kà-ri-/im* [7] *ta-dí ší-tí* KB *a-li-bi$_4$* [8] KB-*pì-a i-dí a-hi a-ta* [9] *i-ba-ri-ni ni-ta-ki-ir* [10] 81 *ma-na* URUDU *i-na* [11] *ša-tù-um Bu-za-zu* [12] *lá-pì-it a-na-kam e-ri-/ší-ma* [13] *um-ma šu-ut-ma lá al-qé* [14] *a-mì-ša-am i-lik-ma* [15] *um-ma kà-ru-um-ma I-dí-Ku-bi$_4$* [16] [*šu*]-*ut ù-lá ša ki-ma I-dí-Ku-bi$_4$* [17] KB *li-dí-nu-ni-kum* [18] *ù-lá kà-ru-um li-dí-na-/kum* (collation by K.R. Veenhof).

559 CCT 1, 43.3'-28': [3'] 1 *me-at* 30 *ma-na* URUDU [4'] *ša me-tum ha-am-ša-at* [5'] É *kà-ri-im qá-ti* [6'] *Puzur$_4$-A-šur al-qé* [7'] ⸢40⸣ *ma-na* URUDU *ša me-tum* [8'] *ha-am-ša-at qá-at-kà* [9'] *al-qé* 1 *me-at* 80 *ma-na*.TA [10'] *ší-im* URUDU SIG$_5$ *ša* É *kà-ri-im* [11'] *A-mur-Ištar lá-áp-tù* [12'] ŠÀ.BA 78 *ma-na* 15 GÍN [13'] *a-na-ku al-qé* [14'] 1 *me-at* URUDU [15'] É *kà-ri-im ší-im* [16'] URUDU SIG$_5$ *Ú-sà-num il$_5$-qé* [17'] URUDU-*kà ša-bu-a-tí* [18'] 1 *me-at* 50 *ma-na* URUDU [19'] *ma-lá* É *kà-ri-im* [20'] *qá-at-kà al-ta-qé-ú* [21'] *a-na Ú-sà-nim ú A-du* [22'] *a-dí-in . ma-lá* [23'] *qá-tám i-sú-hu-pì-im* [24'] [*a-hu*]-*ru*?-*kà* É *kà*-⸢*ri*⸣-[*im*] [25'] [x x *A-šur*]-*ta-ak-lá-ku* DUMU [*A-b*]*u-ša-lim* [26'] [*ú-ba-lá*]-*kum mì-ma i-na li-bi$_4$-a* [27'] [*lá*] *tí-šu i-na* 2 *me-at* 50 *ma-na* 18 *ma-n*[*a*] [28'] *A-du il$_5$-qé*.

560 Kt a/k 410.1-13: [1] 1 *me-at* 20 *ma-na* URUDU [2] *ša A-šùr-na-da* [3] 75 *ma-na* URUDU [4] *ša A-šùr-i-dí* [5] *ša i-na* É *kà-ri-im* [6] *ša me-tum ha-am-ša-/at* [7] URUDU *qá-sú-nu . Šu-Nu-nu* [8] *ša il$_5$-qé-ú . ša ki-ma* [9] *A-šùr*-DÙG DUMU *A-lá-hi-im* [10] *En-um-A-šùr* DUMU *Zu-zu* [11] *ù Ú-ṣur-ša-A-šùr* [12] DUMU *A-ku-a* . URUDU *iš-tí* [13] *Šu-Nu-nu . il$_5$-qé-ú*.

561 The word *šattum* does not occur in CCT 5, 39a.6 or 7 as suggested in Veenhof 1972, p.94, note 152. CCT 5, 39a.6-8 read: [6] 2/3 *ma-na* 3 GÍN *ša-*[*al*]*-/ša-tum* [7] 1 1/3 *ma-na* 6! GÍN [8] *a-na e-lá-tim a-lá-pá-at*, (following the levying of *nishātum* and 'tithe':) '46 shekels are the one-third; I will credit the enterprise with 1 mina 26 shekels.' The booking refers to the payment for the 'tithe' which was done according to a distributive code as in kt c/k 454 (see n.540), where the *mišit ellatim* had been credited for

suggested meaning 'per year' is a conjecture. The seemingly insignificant amount of 81 minas is parallelled by 80 minas as (disbursed) share in CCT 5, 42b, translated in Larsen 1976, pp.339-340.[562] The same text offers an example of a merchant credited with silver from the 'five per cent' of Purušhattum in Kanish.[563] The passage might be interpreted '(to) 3 shekels of silver (from the amount disbursed) from the "five per cent" of Purušhattum, (and to) 4 shekels (of silver) from the "tithe"[564] Aššur-taklāku is credited in Kanish, originating from Gazia's enterprise; I can collect it for the assets of (my) enterprise.' Both amounts were due to Aššur-taklāku; the second originated from a share Aššur-taklāku had in the payment for the 'tithe' which had been levied in Kanish on Gazia's textiles, because a number of these belonged to him. It remains uncertain whether the first amount was a credit to Gazia as well. It is more likely that Aššur-taklāku had sold copper obtained as 'five per cent' for silver, or had received silver instead of copper, which he had not yet paid to its owner.

The interpretation of CCT 1, 43 is complicated by the uncertainty about whether the amount which can be collected as Imdilum's share in the 'five per cent' was the same as the 'proceeds in fine copper' of 180 minas per shareholder. Amur-Ištar's share in the fine copper was collected by Ennum-Aššur (78 minas 15 shekels) and Usānum (100 minas); the deficit, 1 3/4 mina, could have been caused by losses originating from the breaking up of the ingots. The figures for Imdilum cause some problems: if in line 7' '50' is to be read and added to the 130 in line 3', Ennum-Aššur collected in all 180 minas and this would mean that the 'five per cent' disbursement is the same as the '180 minas of fine copper'. But according to line 18' the amount collected for Imdilum was 150 minas. Whatever the cause for this difference may be, it is interesting that the copper was not exchanged for silver after it had been collected.

Kt a/k 410 (like CCT 5, 42b) is concerned with the affairs of Aššur-iddi(n) and his family. Father and son have different amounts as shares if indeed the text records the full amount was disbursed: 75 for Aššur-iddi(n) and 120 minas for Aššur-nādā.

It is not known how many times during the year such disbursements were made out of the results of the 'five per cent' or exactly how the copper was obtained. Unpublished texts may provide more details which will enable a better understanding of this system.

---

two-thirds, and the *šalšātum*-fund for one-third of the 'tithe'. In CCT 5, 39a, line 7 contains the amount which will be booked on the enterprise's account, with *ellatim* elliptic for *mišit ellatim*. The total amount is 2 minas 9 shekels.

562 CCT 5, 42b.1-7: [1] *i-na* 80 *ma-na qá-tí* [2] *a-bi*$_4$*-ni ù* 80 *ma-na* URUDU [3] *qá-tí A-šùr-na-da* [4] *ša* 15 GÍN.TA KÙ «*ù*» TÚG *ku-ta-ni* / «*ku-ta-ni*» [5] 12 *ma-na ša A-šur-na-da* «*da*» [6] 6 *ma-na ša A-šur-ta-ak-lá-ku* [7] *i-na áb-ni-ni a-hu-ur*. This text is badly written as a comparison with the parallel shows: Cole 9.17-22: [17] ... 12 *ma-na* [18] URUDU *i-na* 80 *ma-na*.TA [19] *i-Pu-ru-uš-ha-tim* [20] *ša A-šùr-na-da* [21] 6 *ma-na ša a-bi*$_4$*-ni* [22] *i-ri-ha-am* (by courtesy of K.R. Veenhof).

563 CCT 5, 42b.22-29: [22] 3 GÍN KB *i-na me-tum* [23] 5 ŠÈ *ša Pu-ru-uš-ha-tim* [24] 4 GÍN *ša* 10-*tum* [25] *iš-té-en*$_6$ *i-na Kà-ni-ìš* [26] *iš-tí A-šùr-ta-ak-lá-ku* [27] *i-Kà-ni-ìš lá-áp-tù* [28] ILLAT-*at* <*Ga-zi*>-a . *a-mì-ší-it* [29] «*a*» ILLAT-*tim a-lá-qé*,
Cole 9.35-38: [35] 3 GÍN KB *i-na* 1 *me-tum* [36] 5 ŠÈ *ša Pu-ru-uš-ha-tim* [37] 4 GÍN *ša* 10 ŠÈ-*tum* 1 TÚG.TA *i-na* [38] ILLAT-*at Ga-zi-a* KI *A-šùr-ta-/ak-lá-ku*

564 The *išrātum* was levied in the *kārum*-office in CTMMA 1, 74, but in the palace in ATHE 62 and ATHE 61.9, where it is followed by another lot of textiles which the palace took as 'tithe'.

CONCLUSION

The aim of this study of the Old Assyrian copper trade in Anatolia during the 19th century B.C. was three-fold:

- to establish the origin of the copper;
- to determine in which qualities and shapes the metal was traded and to understand the terminology used by the Assyrians;
- to discover the motives for and mechanisms of this type of inner Anatolian trade.

The origin of the copper

The study of archaeological and archaeo-metallurgical publications combined with a fresh investigation into the geography of Central Anatolia on the basis of the Assyrian documents, permitted only an approximate location of the area where the copper traded by the Assyrian merchants in Central Anatolia was mined. Although Turkey possesses many deposits of copper, remains of ancient mining activities are scarce. This is to a large extent due to later metallurgical activies which destroyed the traces of earlier workings. Moreover, the frequent absence of objects, e.g. pottery or pieces of wood, seriously hinders efforts to date those old workings which have been identified. Such mines can at best be assigned to a specific period covering several centuries. There is at present no old working the date of which can be reliably attributed to the period of the Assyrian involvement.

The results of efforts to trace the origin of the metal used in the manufacture of an artefact to the ore deposit have been unsatisfactory. This so-called sourcing of artefacts, based upon a comparison of trace elements and lead isotope ratios in ore deposits and in copper objects, is only possible in cases where the metal originated from a single source. Even if the copper objects excavated at Kültepe would in the main have been made from first-use metal (which is not necessarily the case, as metal frequently was recycled), establishing its origin is seriously hindered by the absence of an adequate database containing all the necessary information.

From a re-examination of the geography of Central Anatolia on the basis of Old Assyrian textual evidence which recently became available for study, the conclusion could be drawn that the copper in general originated from the area around Çankırı and Ankara. There are no indications that copper mined east of Kültepe was traded in Central Anatolia. It follows that Ergani did not supply Central Anatolia with copper during this period. The role which ore deposits north of Kanish, near Çorum and Amasya, played in the Assyrian copper trade appears to have been very small.

The qualities and shapes of copper and the Old Assyrian terminology

Archaeological and archaeo-metallurgical research has demonstrated that during the second millennium B.C. copper was traded almost exclusively in the shape of ingots, and not as lumps of ore. The copper mined and traded by the Assyrian merchants apparently derived largely from oxide ore deposits. An iron-containing flux was added to facilitate the smelting of the ore. The purity of the copper often depended on the subsequent remelting or refining of the smelted metal.

The Old Assyrian terminology can be divided into two main categories: copper of high and that of low purity. The first often consisted of refined metal, while the latter contained a high level of impurities, usually iron, and can be considered a primary smelting product or the result of an unsuccessful refining operation. The adjectives employed in the texts usually plainly describe the metal's (im)purity. However, there are also a number of qualifications by means of a place name, which refer to a specific quality or perhaps ingot characteristic of the copper from that town. Most of such qualifications will refer to copper mined near, or refined in the town mentioned. Four words can be identified which were used to denote a type of ingot. They were rarely used, presumably because the ingots were usually broken up. There are no words referring to copper ore; native copper was called *ša šaduīšu*, lit. '(copper) of its mountain'. A small quantity of copper occurred as scrap, consisting of old implements and other objects. The scrap was not an object of trade, but was used to pay expenses en route. The copper was conveyed by means of donkeys, but also by porter and by wagon. The last were rented from the palace, which apparently also organized a transport system (*sikkum*) between towns which Assyrian merchants were allowed to use.

The copper was used in the production of copper and bronze objects, such as weaponry, agricultural implements, and household utensils. The range of artefacts made of copper and bronze is apparent from the excavated objects, workshops, and from several texts mentioning metal objects.

The trade

Studying documents from five selected firms in their archival context proved to be an essential tool for establishing the mechanisms of the copper trade. Especially the archive of Uṣur-ša-Ištar's firm, consisting largely of unpublished material, yielded important information. The imported textiles and tin were sold mainly to the local palace and dignitaries in exchange for silver, gold, and copper. This apparently occurred in all towns. It usually took some weeks before the metal was handed over. The palace transferred the copper to the local *kārum*-office, the building housing the administrative centre of the Assyrian settlement which had its own storage facilities, where the merchants could collect the copper that was due to them. Occasionally part of the copper was paid out by the *kārum*-office in the form of silver or gold.

Most copper was obtained in or near the towns of Turhumit and Tišmurna, which were located closest to the major mining areas. It was exchanged for textiles and tin, but also for wool, agricultural products, and even for precious metals and valuable stones. A large part of the copper obtained in that area appears to have been unrefined or poorly so. As the demand for poor copper was small, it was often exchanged for refined copper in Turhumit and Tišmurna. These towns were metallurgical centres, where a considerable amount of copper was refined, probably controlled by the local palace. There are no indications that the Assyrian *kārum*-office ordered the refining of copper. From the major copper markets of Turhumit and Tišmurna the metal was transported in a south-eastern direction where it was exchanged for silver or gold. Poor copper apparently was transported only as far as Wahšušana.

The indirect exchange of merchandise for copper increased the profit, but meant that it took longer before silver and gold arrived in Kanish, and that the merchandise or its proceeds were longer exposed to commercial risks, notably losses caused by war and default. The availability of personnel and of transport facilities was a precondition for this type of trade. The archives of Imdilum and Uṣur-ša-Ištar demonstrate the importance of having an agent living in Turhumit. It follows that obtaining copper on a regular scale and in considerable amounts in specific areas, e.g. Turhumit and Tišmurna, was only possible within the structure of a firm. The available evidence does not allow us to determine whether every firm might in theory embark on the copper trade. Each merchant may have observed that it was profitable to first obtain copper in the town where this was cheapest and then bring it to another town where he could acquire more silver for it, but possibilities and preferences seem to have dictated whether this was in fact done.

For most copper qualities a number of prices are attested, expressing the copper's relative value in silver. However, it is often difficult to determine in which town a price was in force. In addition, prices expressed in silver for some of the commodities exchanged for copper are scarce. Notwithstanding these limitations, it is possible to demonstrate the geographic differentiation of prices as well as to provide an indication of the profit (at least 25-100%) which could be made by prior exchange of merchandise for copper. A chronological development could not be observed because most prices cannot be dated or placed geographically. Moreover, all major archives stem from the same period of a few decades, which represents the hey-day of Assyrian trade in Anatolia. That there were price fluctuations is evident from some texts referring to the effects of war and scarcity of money.

The metals gold, silver, tin, and copper were used as money. The tin was imported by the Assyrians with the aim of exchanging it for silver. The other three metals were mined in Anatolia itself and circulated in various qualities. Making payments in copper was preferred to payments in silver or gold, in accordance with Gresham's law that 'bad money drives out good'. However, this did not preclude merchants from using occasionally precious metals to obtain copper. Copper may have been more widely accepted than silver, because for many people its purity and hence value was easier to assess than that of silver. Considerable quantities of gold and silver left Anatolia and were brought to Assur, but no attempt has been made in this study to give an estimate. We must realize that part of it will have found its way to Syria by the activities of non-Assyrian traders. Copper remained in Anatolia, and, with the exception of small quantities, was not exported. The reasons for this were its high price and substantial transport costs, which made export to Mesopotamia unprofitable.

The powerful position of the palace, represented by the ruler and his dignitaries, made it the chief economic power in the Anatolian city-states. Bronze was the main alloy from which weapons were made, and this made copper of strategic importance. Although the palace therefore had a vested interest in the trade in copper and tin, the commercial activities of the Assyrian merchants demonstrate that it did not excercise a monopoly on the trade in copper. On the contrary, it encouraged trade, buying this metal from the Assyrians and, depending on the situation, selling it to them as well. The palace also provided transport facilities for the merchants. At times, this positive attitude might change and a ruler could block all traffic on the roads passing through his territory.

The role of the Anatolian traders still remains obscure. It is likely that they too brought copper from e.g. Turhumit to other towns, but it must remain unclear at the present state of the epigraphic documentation (with only part of a single Anatolian's archive having been published) whether this was done in the service of the palace.

The Assyrian *kārum*-institution excercised some control over the copper trade of the Assyrian merchants by stipulating interest percentages on loans in copper and by occasionally intervening in the trade itself. More importantly, it organized collective enterprises (*šitapkum* and the 'five per cent in Purušhattum') which enabled the merchants to obtain copper at low risk. Its office functioned as transfer point for copper which the palace credited to the office's *suhuppum*-fund.

Since, as mentioned above, only small amounts were sent to Assur, mainly to relatives for private use, copper was not exported to Assur from Kanish. The copper used in Assur must have had a different origin, presumably Oman and Iran, and perhaps from Ergani as well.

Many data about copper occur only a limited number of times in the texts which could be studied. In view of the huge number of documents still awaiting publication, it is likely that our understanding of the mechanisms of the copper trade will be enlarged, and that features which are still unclear will be better understood in the future.

SELECTED TEXTS

1 = KTS 54d	**6** = BIN 4, 35	**11** = CCT 2, 29	**16** = KTH 14
2 = ICK 2, 54	**7** = BIN 4, 31	**12** = CCT 4, 12b	**17** = TC 3, 25
3 = AKT 1, 82	**8** = BIN 4, 148	**13** = TC 2, 18	**18** = VS 26, 6
4 = kt c/k 263	**9** = BIN 4, 160	**14** = ATHE 32	**19** = AKT 2, 39
5 = TC 2, 33	**10** = BIN 6, 76	**15** = kt a/k 583	**20** = kt n/k 1464

AKT 1, 82 = **3**	BIN 4, 148 = **8**	ICK 2, 54 = **2**	KTS 54d = **1**
AKT 2, 39 = **19**	BIN 4, 160 = **9**	kt a/k 583 = **15**	TC 2, 18 = **13**
ATHE 32 = **14**	BIN 6, 76 = **10**	kt c/k 263 = **4**	TC 2, 33 = **5**
BIN 4, 31 = **7**	CCT 2, 29 = **11**	kt n/k 1464 = **20**	TC 3, 25 = **17**
BIN 4, 35 = **6**	CCT 4, 12b = **12**	KTH 14 = **16**	VS 26, 6 = **18**

1 KTS 54d

See p.33.

1 *me-at* 95
GÚ *we-ri*
ŠÀ.BA 3 GÚ
a-na DUMU *A-hi-a*
as-ni-iq
[IGI *Puzur*$_4$]-*A-na*
R D[UMU *Pu*]*zur*$_4$-*A-šur*

'195 talents of copper, of which I handed over 3 talents to the son of Ahia in the presence of Puzur-Ana son of Puzur-Aššur.'

2 ICK 2, 54

See pp.39, 130, 155, and 157.

A transcription and translation of the whole text can be found in Ichisar 1981, pp.67-69. Lines 1-20 were discussed in Larsen 1976, p.92, with transcription and translation in n.25; discussion of lines 1-5 in Muhly 1983, pp.35-36.

[4 GÚ] URUDU SIG$_5$
[*ša T*]*a-ri-ta-a*[*r ú*]
[2] GÚ 30 *ma-na* URUDU
[*š*]*a ša-ad-wa-na-am*
la ú-kà-lu
ŠUNIGIN 6 GÚ 30 *ma-*[*na* URUDU]
i-ṣé-er
Id-na-a-a
DUMU *Sà-li-a*
*Im-dì-*DINGIR *i-šu*
iš-tù ha-muš-tim
ša A-la-hi-im
DUMU *I-dí-Sú-in*
ù E-na-ma-nim
Im-dì-lúm a-na
Tù-ur-hu-mì-it
ṣú-ha-ar-šu . i-ša-pár
a-na Kà-ni-iš . a-na
ni-ga-lim . URUDU
ú-ša-ak-ša-dam šu-ma
[*lá*] *uš-ta-ak-ší-dam*
[*ki-m*]*a a-wa-at*
[*kà-ri-im* 1 1/2 *m*]*a-na.*TA
[*i-*ITU.KAM *a-na*]
1 GÚ*-tim ú-*[*ṣa-ab*]
IGI *Lá-qé-*[*pí-im*]
IGI ^{d}I[M-]

'4 talents of fine copper from Taritar and 2 1/2 talents of copper that does not contain haematite, in total 6 1/2 talents of copper, Imdilum has as claim on Idnāya son of Sallia. From the week of Alāhum son of Iddi(n)-Suen and Ennam-Anum, Imdilum will send his servant to Turhumit, (and) he will see to it that the copper reaches Kanish before harvest-time. If it does not reach (it) in time, he will add according to the *kārum* rule each month 1 1/2 mina per talent. Witnessed by Lā-qē[pum] and Adad-[ ].'

3 AKT 1, 82

Transcription and translation in Turkish of this text (excavation number kt y/k 14) in AKT 1, p.101, hand copy on p.185.

a-na Im-dí-lim qí-bi$_4$*-ma*
um-ma Šu-mì-a-bi$_4$*-a-ma*
*A-al-*DÙG *ù Ku-zi-zi-a-ma*
1 GÚ AN.NA 10 *ku-ta-nu*
7 TÚG *šu-ru-tim Kà-ki*
ùb-lam AN.NA-*kà*
ni-is-<*ni*>*-iq-ma* 1 *ma-na*
im-ṭí mì-šu ša ta-áš-p[*u-ra-ni*]
um-ma a-ta-ma
a-na AN.NA-*ki-a*
R 7 GÍN.TA *a-n*[*a*]
ku-ta-nim 1/3 [*ma-na*.TA]
a-na šu-ru-tim
12 GÍN.TA . *lá ta-áš-ta-*[*na-me*]
ki$^!$*-ma* KB-*pu* . *wa-aq-ru-ni*
ù AN.NA 9 GÍN.TA
i-za-zu-ni KB *a-na*
URUDU-*im* 1 1/2 *ma-na* ⸢5 GÍN⸣.TA
i-za-zu-šum AN.NA-*kà*
ù TÚG.HI-*kà a-na*
ma-lá té-er-tí-kà
ni-il$_5$*-qé* KB 1 GÍN.TA
ra-ma-ni ni-ša-gi$_5$*-iš-ma*
ma-lá šé-bu-lam nu-šé-ba-lá-kum

'[1-3] Speak to Imdilum, thus (say) Šumi-abia, Al-ṭāb, and Kuzizia: [4-6a] Kaki brought us 1 talent of tin, 10 *kutānum*-textiles, and 7 *šurum*-textiles. [6b-8a] We checked (the weight) of your tin and there was a deficit of 1 mina. [8b-14a] Why did you write: "(Fetch a price) for my tin of (1 shekel of silver) per 7 shekels of tin, for each *kutānum*-textile 20 shekels (of silver), and for each *šurum*-textile 12 shekels"? [14b-17a] Do you not hear that silver is expensive? Hence, the exchange ratio between tin and silver is 9:1. [17b-19a] The ratio between copper and silver is 95:1. [19b-24] (Nevertheless,) we have accepted your tin and textiles on your conditions. We will exert ourselves to the utmost for the silver, each shekel of it, and we will send you as much as possible.'

4 Kt c/k 263.1-45

See pp.113-114, 155, and 156.
Unpublished text in Ankara (inv. no. 154-262-64); duplicated by kt c/k 735. Lines 18-29 were quoted by K. Balkan (1979, pp.57-58).

um-ma Ta-ha-ša-i-li-ma . a-na
*A-šur-be-el-*SA.TU *ù A-šur-ma-lik*
qí-bi$_4$-ma ta-áš-pu-ra-ni (erasure) *um-ma*
a-tù-nu-ma 2 *me-at* 77 TÚG.HI.A
ù 2 GÚ AN.NA . *ša a-bi$_4$-ni*
*A-šur-*DÙG *e-zi-ba-kum . i-na lu-qú-/tim*
*a-ni-tim . i-nu-mì A-šur-*DÙG
i-li-kà-ni-ma . AN.NA *il$_5$-qé-ú*
9 TÚG.HI.A . *a-li-wi-tim . il$_5$-qé*
ŠUNIGIN 2 *me-at* 68 TÚG.HI.A 25 *ma-na.*TA
ta-ad-nu 2 GÚ AN.NA 18 *ma-na.*TA
ta-dí-in ŠÀ.BA 2 GÚ URUDU *ma-sí-am*
i-nu-mì . AN.NA . *il$_5$-qé-*<*ú*> . *a-na A-šur-*DÙG
a-dí-in . i-nu-mì . A-lá-hu-um . iš-tù
pá-zu-ur-tim . i-li-kà-ni
5 GÚ *ma-sí-am . iš-tí* AN.NA-*ki-/šu*
A-lá-hu-um ú-kà-il$_5$-šum
ur-kà-num 6 GÚ *ma-sí-am*
ù 14 GÚ *ší-kam* 10 GÍN-*tum*
E *a-ṣé-er* 10 *ma-na-e . ta-ur*
i-na e-re-qí-im . a-na
R *Pu-ru-uš-ha-tim A-lá-hu-um*
ù Wa-li-a . ub-lu-šum
21 *ma-na ig-ri e-re-qí-im*
32 1/2 *ma-na . a-qá-tí-šu-nu*
a-dí-in 17 GÚ 40 *ma-na* URUDU
ma-sí-am ù 3 ANŠE *ṣa-lá-me*
*A-šur-be-el-*SA.TU . *il$_5$-qé*
2 1/2 MA.NA KÙ.KI . *pá-ša-lam*
A-lu-lá-a ù Ta-ki-lá-a . a-ṣé-er
*A-šur-*DÙG *a-Kà-ni-iš . ub-lu-nim*
ší-im KÙ.KI 58 *ma-na.*TA
ší-im-šu 53 GÚ 10 *ma-na*
*i-nu-mì A-šur-*DÙG . *i-li-kà-ni-ma*
AN.NA *a-na Pu-ru-uš-ha-tim*
ú-šé-ri-bu 20 *ma-na* URUDU *a-qá-tí-šu*
a-dí-šum 2 GÚ 30 *ma-na* URUDU
ša-du-a-tám . a-kà-ri-im

a-dí-i 18 *a-na* TÚG . *iš-tù*
Kà-ni-iš . *a-dí* . *Wa-ah-šu-ša-na*
3 *ma-na*.TA . *da-tum* . *ik-šu-/dí*
LoE ŠUNIGIN 54 *ma-na* 5 *ma-na* KB
A-šur-be-el-SA.TU . *i-li-bi*$_4$
lu-qú-tim a-ni-tim
LE *il*$_5$*-qé*

'1-3 Thus (says) Tahašaili, speak to Aššur-bēl-šadue and Aššur-mālik: 3-6a you wrote me thus: "Aššur-ṭāb left 277 textiles and 2 talents of tin belonging to our father behind with you." 6b-9 Out of this merchandise Aššur-ṭāb took 9 textiles for wrapping when he came to take the tin. 10-12a In all 268 textiles were sold at 25 minas (of copper) each; the 2 talents of tin were sold for 18 minas (of copper) per (mina). 12b-14a Out of this, I gave 2 talents of refined copper to Aššur-ṭāb when he took the tin. 14b-17 When Alāhum arrived from smuggling, Alāhum offered him 5 talents of refined (copper) together with his tin (for sale). 18-23 Subsequently, Alāhum and Walia brought him by wagon in Purušhattum 6 talents of refined (copper) and 14 talents of *šikkum* (copper), a 10 shekel weight added to (every) 10 mina weight. 24-26a 21 minas (I paid for) the rent of the wagon, 32 1/2 minas I gave (them) for their expenses. 26b-28 Aššur-bēl-šadue took 17 talents 40 minas of refined copper and 3 black donkeys. 29-33 2 1/2 minas of auriferous ore Alulāya and Takilāya brought to Aššur-ṭāb in Kanish; the price of the gold at 58 minas per (...), was 53 talents 10 minas. 34-37a When Aššur-ṭāb came to bring the tin into Purušhattum, I gave him 20 minas of copper for his expenses. 37b-39a I paid 2 talents 30 minas of copper as *šaddu'atum*-tax to the *kārum*. 39b-42a 18 (textiles): the road-tax (*dātum*) from Kanish to Wahšušana cost me 3 minas per textile, in total 54 minas (of copper). 42b-45a Aššur-bēl-šadue took 5 minas of silver out of this merchandise.'

5 TC 2, 33

See pp.42, 135, and 137.
K.R. Veenhof discussed lines 4-5 (1972, p.380 n.507), and lines 8-10 (1972, pp.449 and 451).

a-na Puzur$_4$*-A-šur qí-bi-/ma*
um-ma I-ku-pí-a-ma
3 *hi-dim* KÙ.⸢GI⸣ *ù*
1 5/6 *ma-na* 3 1/2 GÍN KB
da-na-⸢*am*⸣ *ša tù-šé-bi-lá-ni*
a-na ša-lá-ší-⸢*šu*⸣ . *ni-im-ha-sú-/ma*
iš-té-et qá-tám KB
a-na-ku al-qé-ma 1 GÍN *a-na*
3 *ma-na ší-ki-im ù*
KÙ.GI 1 GÚ *ší-ki-im*

[a-na] u$_4$-me a-na Ib-ni-lí
[a-dí-i]n ki-ma ša-qá-al
[URUDU] *a-da-gu$_5$-lu*
[a-ma-l]a u$_4$-me-e ir-ku-sú
R [ ] *i-ṣa-ba-sú ki-ma*
[ší]-ki-im URUDU *lá-mu-lnàm*
[a]l-té-qé ŠÀ.BA 4 GÚ
[URUDU] *ší-kam ù a-ha-ma*
20 *ma-na ší-kam ší-im*
ar-ša-tim Ta-as-ku-ul
na-áš-a-kum ší-tí
URUDU-*i-kà* 53 *ma-na* / 15 GÍN *ší-ku-um*
3 ANŠE.HI.A *ṣa-lá-me*
da-nu-tim šé-ri-a-ma
URUDU *lu-šé-bi-lá-kum*
šu-ma li-bi-kà
URUDU-*a-kà i-pá-ni-a*
LeE *a-Kà-ni-iš lá-dí-šu a* x [ ]
i-pá-ni-a a-Kà-ni-[iš]
a-na-ší

‘[1-2] Speak to Puzur-Aššur, thus (says) Ikuppia: [3-13a] we have divided into three parts the 3 gold beads and the 1 mina 53 1/2 shekels of “strong silver” that you sent and I have taken one part of the silver, and gave (the rest) for a term to Ibni-ilī at 1 shekel for 3 minas of *šikkum* copper and (the bead of) gold for 1 talent of *šikkum* copper. [13b-15a] Because I am expecting the payment of the copper, ... he will seize it (the copper?) not later than the term for which he contracted him. [15b-17] Instead of *šikkum* copper I have collected poor copper. [17b-21a] Out of this Taskul is bringing you (worth) 4 talents of *šikkum* copper and furthermore 20 minas of *šikkum* as the proceeds of the wheat. [21b-22] The rest of your copper is 53 minas 15 shekels of *šikkum.* [23-30] Sent 3 strong black donkeys and I will send the copper to you. If you agree, I will personally give your copper (to somebody to bring) to Kanish. I will bring the ... personally to Kanish.’

6 BIN 4, 35

See pp.133, 136, 138, and 152.
J. Lewy discussed lines 36-42 in EL II, p.121 n.*c*. See Garelli 1963, p.124 n.1, for lines 8-14 and 36-38. Lines 47-48 were discussed in Veenhof 1972, pp.450-451.

a-na Puzur$_4$-A-šùr qí-bi-ma
um-ma Bu-za-zu-ma . i-nu-mì
a-na Kà-ni-iš . ša-sú-ha-ku-ma
a-li-kà-ni . AN.NA *lu i-a-am*

lu ku-a-am a-na Ili$_5$*-we-da-ku*
e-zi-ib um-ma a-na-ku-ma a-dí
a-tù-ra-ni . URUDU SIG$_5$. *e-pu-uš-ma*
li-im-hu-ra-ni . *iš-tù*
Kà-ni-iš . *a-tù-ra-ni* . *lá li-bi*
DINGIR-*ma* . *e-er-šu-um i-ṣa-áb-ta-ni*
iš-tù . *at-bi-ú* . *um-ma a-na-ku-ma*
URUDU *lu i-a-am lu ša Puzur*$_4$*-A-šùr*
dí-nam-ma a-na Pu-ru-uš-ha-tim
lá-li-ik-ma . KB 10 *ma-na*
lu a-na Puzur$_2$*-A-šùr lu i-a-tí le-pu-/uš*
i-ta-pá-lá-ni-ma . *sú-ku-ur-tum*
i-ta-áš-kán-ma . ITU.5.KAM *ù* ITU.6.KAM
E *a-sú-hu-ur* . *ù i-nu-mì*
iš-tù . *Kà-ni-iš* . *a-na* URUDU
ta-áš-ta-na-pá-ra-ni
R *a-dí* 5-*ší-šu ù šé-ší-šu* «*šu*»
aq-bi-šu-um . *um-ma a-na-ku-ma*
URUDU *lu nu-šé-bi-il*$_5$*-šu-um ù i-a-tí*
URUDU-*i* . *ta-da-nam* . *lá i-mu-a*
ù ku-a-tí . *šé-bu-lam* . *lá i-mu-a*
i-ṣé-er . URUDU . *ir-ta-na-bi*$_5$*-iṣ*
ù mì-li-ik . URUDU *i-ma-li-/ik*
lá na-ṭù-ma . *ri-ig-ma-am*
ú-lá a-na-dí-i . *um-ma a-na-ku-ma*
pá-ni-ša-am . *bu-lá-tí* . *qá-tí*
Puzur$_2$*-A-šùr* . *ša ṣa-ba-tim* . *li-iṣ-/ba-at*
a-hi-dí-ir-ma . *a-na ší-ip-ri-im*
ra-mì-ni áš-ku-un-ma
um-ma a-na-ku-ma . URUDU *lu i-a-am*
lu ša Puzur$_2$*-A-šùr ša a-ba-ki-im*
i-pá-ni-a . *lá-bu-uk* URUDU . *a-na*
Ša-lá-tù-a-ar i-pá-ni-a
a-bu-kà-ma . *i-ba-áb* . *ha-ra-ni-a*
LoE DAM.GÀR . *ú-ta-dí-a-am* . *a-dí* . *qá-tí*
URUDU *li-iṣ-ba-at* . *um-ma a-na-ku-ma*
e-zi-ib . DAM.GÀR . *ša li-bi*$_4$*-šu*
lu-wa-dí-a-am . *ki-ma a-Ša-lá-tù-a-ar*
e-ru-bu . *ur-ki-a-ma*
2 *ša-zu-úz-tám* . *uš-ta-zi-za-am* . *ú ku-a-tí* / *ù i-a-tí*
a-na ra-pu-dí-ni . *i-tí-zi-iz-ma* . / URUDU *a-na ša* DAM.GÀR
LE *a-he-e-em* . *iš-ta-kán ù u*$_4$*-ma-am a-na dí-nim* . *i-šé-e-a-ni*
ú a-na tár-ki-is-tim . *iš-pu-ra-am um-ma šu-ut-ma* 1/2 *ma-na* KB

ù 1 GÍN.TA . *ra-ki-sà-ni-šu . a-wi-lam . ṭé-em-šu* <*ša-ni*>

'[1-2a] Speak to Puzur-Aššur, thus (says) Buzāzu: [2b-9] When I had been summoned to move to Kanish, and I was about to leave, I left tin, both mine and yours, behind with Ilī-wēdāku, saying: "Acquire before I return fine copper so that it falls to my share." [9b-15] Unfortunately, I was suddenly confined to bed after I returned from Kanish. After I had recovered, I said: "Give me the copper, both mine and that of Puzur-Aššur, so that I can go to Purušhattum and earn silver, about 10 minas, both for Puzur-Aššur and for myself." [16-25] While he kept arguing with me, a blockade came in force and I was delayed for 5 or 6 months. And since you repeatedly write me from Kanish for the copper, I said to him 5 or 6 times: "Let us send him the copper," but as for me, he refuses to give me my copper, and as for you, he refuses to send (it) to you. [26-27] He keeps sitting on the copper and is still making up his mind what to do with it. [28-31] It is not the right moment for me to lodge a complaint. I said: "First, let Puzur-Aššur take from the merchandise I have in trust (*būlātu*) whatever he can." [32-36a] But I became anxious and decided to act as an envoy, thinking: "I will personally transport as much copper as possible, both mine and that of Puzur-Aššur." [36b-46a] I personally brought the copper to Šalatuar, and when I was about to leave, he made known to me the merchant('s name) (who was the owner in Ilī-wēdāku's view), lest I myself would seize the copper. I thought: "Let it be, he can indicate to me (as the copper's owner) whichever merchant he wants." As soon as I entered Šalatuar, he appointed 2 persons as agents and took action to cause problems to you and me, and has made the copper the property of a strange merchant! [46b-48] And today he is suing me (for the copper)! And he wrote for a contract, asking: "Engage me for the copper at 1 shekel of silver per 30 shekels of copper." The man has gone mad.'

7 BIN 4, 31

See pp.132, 133, 135, 139, and 156.
See for this text Veenhof 1972, p.52. Cf. Michel 1992, p.144 n.38 for lines 19-20. Lines 43-46 were discussed by J. Lewy in KTH, p.7, cf. Veenhof 1972, p.249.

um-ma Pu-šu-ke-en$_6$ *ša ki-ma*
Puzur$_4$*-A-šur ú Sú-e-a-ma*
a-na Puzur$_4$*-A-šur qí-bi*$_4$*-ma*
19 GÚ URUDU *ma-sí-am*
ku-nu-ki ša I-a-šar nu-sà-ni-iq-ma
13 2/3 *ma-na im-ṭí* ŠÀ.BA
1 GÚ 40 *ma-na* URUDU *ší-ku-um*
i-na URUDU *ma-sí-im pá-ni-im*
ú ur-ki-im ub-lu-nim
7 1/2 GÚ URUDU *ší-kam*
ku-nu-ki ša I-a-šar nu-sà-ni-iq-/ma
10 *ma-na im-ṭí mì-ma a-nim*

I-a-šar uš-té-bi-lam
3 GÚ URUDU *ṣa-lá-mu-um*
i-na URUDU-*i-šu na-dí*
2 GÚ URUDU *ma-sí-am*
ša i-na Wa-lá-ma i-ni-du
ub-lu-ni-ma nu-sà-ni-iq-ma
1 *ma-na im-ṭí* 20 *ma-na* URUDU *<ší>-ku-um*
i-pá-ni i-lá-tim
E *na-dí* 10 GÚ 30 *ma-na*
URUDU *ší-kam ku-nu-ki ša Ku-ra*
R *ú Šu-Be-lim nu:sà-ni-iq-ma*
12 *ma-na im-ṭí* 52 *ku-ta-ni*
iš-tù Wa-ah-šu-ša-na
ša ki-ma ku-a-tí ú-šé-bi$_4$*-lu-nim*
ŠÀ.BA 30 *ku-ta-ni* 9 1/4 GÍN.TA
a-Pu-šu-ke-en$_6$ *ni-dí-ma*
KB-*áp-šu-nu a-dí-ni ú-lá iš-qú-ul*
48 *ku-ta-ni I-dí-A-šur*
ir-dí-a-am ŠUNIGIN TÚG-*tí-kà*!
1 *me-at* 48 TÚG-*ba-tù*
ša šé-ep I-dí-A-šur i-ba-ší-lú
20 *ma-na* KB *ku-nu-ki-ni*
En-na-Sú-in DUMU *E-lá-lí*
na-áš-a-ku-um lu URUDU *ú* TÚG-*tí*
lu ba-a-<ab>-tám ša té-zi-bu
a-ma-lá a-wi-lu-tí-ni
a-KB *nu-ta-ar-m[a]*
i-šé-ep Dan-A-šur [KB-*k*]*à*
nu-za-kà-ma KB-*kà*
LoE *nu-šé-ba-lá-kum*
a-ma-lá té-er-tí-kà
ANŠE-*ru-kà*
LE *a-na-áb-ri-tim na-du ša-am-kà-kà*
ša-li-im

‘[1-3] Thus (say) Pūšu-kēn, Puzur-Aššur's representative, and Suea, speak to Puzur-Aššur: [4-9] we checked 19 talents of refined copper, sealed (and sent) by Ea-šar, and there is a deficit of 13 2/3 minas. 1 talent 40 minas of it are *šikkum* copper; they brought (it) among the first and the second (consignment) of refined copper. [10-12a] We checked 7 1/2 talents of *šikkum* copper sealed by Ea-šar, and there is a deficit of 10 minas. [12b-13] Ea-šar sent all of this consecutively. [14-15] 3 talents of black copper have been deposited among his copper. [16-19a] They brought 2 talents of refined copper which had been stored in Ulama, and we checked it: there is a deficit of 1 mina. [19b-21a] 20 minas of *šikkum* copper is placed in front of the bags. 21b-24a We checked 10 talents 30 minas of *šikkum* copper sealed by Kura and Šū-Bēlim: there is a deficit of 12 minas. 24b-29 Your representatives sent from Wahšušana 52 *kutānum*-textiles; of these we sold 30 *kutānum*-textiles to Pūšu-kēn at a price of 9 1/4 shekels (of silver) a piece, but up till now he has not paid their silver. 30-33 Iddi(n)-Aššur led (donkeys) here (carrying) 48 *kutānum*-textiles. The total of your textiles, 148 pieces, brought by Iddi(n)-Aššur, is on hand. 34-36a Enna-Suen son of Elāli is bringing you 20 minas of silver under our seals. 36b-42 We will exchange for silver, in a honest way, both the copper and the textiles, as well as the outstanding claims which you left behind. And when your silver is freely available, we will send it you with Dān-Aššur. 43-46 In accordance with your instructions your donkeys have been put to pasture (to recover). Your children are well.'

8 BIN 4, 148

See p.155.
A transcription and translation of lines 1-4 and 23-30 was given by J. Lewy in EL II, p.96 n.*b*. See Veenhof 1972, pp.433 and 437-438 for lines 26-29.

i-lu-qú-tim pá-ni-⸢tim⸣
5 GÚ URUDU 9 *ma-na*
ša qá-tí-im 8 *ma-na*
ša-du-ú-tum e-ku-ul
ša 1 *me-at* 17 TÚG.HI.A
56 1/2 GÚ 3 *ma-na*
ší-<im>-šu-nu 23 TÚG.HI.A
ša Ša-lá-tù-ar
4! *ma-na* 1/2 GÍN KB
ší-im-šu-nu 10 TÚG *ku-ta-nu*
1 *šu-lu-up-ki-ú-um*
É *kà-ri-im*
ŠUNIGIN URUDU 74 GÚ 20 *ma-/na*
ŠÀ.BA «*a*» 5 GÚ 15 *ma-/na*
ší-im ANŠE *ṣa-lá-mì-im*
E 2 1/2 *ma-na* LÁ 1 1/2 GÍN
R KB *iz-ku-am*
ší-im 10 GÚ 20 *ma-na* URUDU
ša šé-ep Ša-lim-Be-lí
5 5/6 *ma-na* 1 1/2 GÍN KB-*/áp-šu*
ŠÀ.BA 2 *ma-na* KB
a-na Ur-šu ú-šé-bi-lam
ší-im 1/3 *ma-na* 6 GÍN KÙ.GI
2 1/3 *ma-na* 6 GÍN KB-*áp-šu*
A-šur-na-da e-ku-ul

12 GÚ 28 *ma-na* URUDU *ša Ì-lí-A-lúm*
a-na Ša-lá-tí-ME-ar ú-šé-ri-bu
5 1/2 *ma-na* 2 GÍN KB-*áp-šu*
iš-ku-nam ŠÀ.BA 5 *ma-na* LÁ 4 GÍN
KB *A-šùr-na-da e-ku-ul ší-im*
2 GÚ URUDU SIG$_5$ 4 GÚ 10 *ma-na* URUDU$^{!}$-*šu*
LoE 3 *ma-na* 4 GÍN KÙ *A-mur-A-š*[*ùr*]
kà-ṣa-ar Dan-A-šur ub-lá-ku[*m*]
ŠÀ.BA 5/6 *ma-na* LÁ 1 <GÍN> KB
LE *ší-im* KÙ.GI *ṭá-hi* 1 *ma-na* KB
KI *Puzur*$_4$-*A-na ṭup-pu-šu* É
Am-ri-a i-ba-ší

'[1-4] (Proceeds) from the previous consignment of mechandise: 5 talents of copper; 9 minas (paid) for expenses, 8 minas consumed by the *šaddu'atum*-tax. [5-7a] The proceeds of (the sale of) 117 textiles are 56 talents 33 minas (of copper). [7b-10a] The 23 textiles from Šalatuar: the proceeds (of their sale) are 4 minas 1/2 shekel of silver. [10b-12] 10 *kutānum*-textiles and 1 *šulupkium*-textile (are deposited) in the *kārum*-office. [13] In all, (this represents a total valued in) copper (of) 74 talents 20 minas. [14-15] (Paid) out of this: 5 talents 15 minas, price of a black donkey. [16-17] 2 minas 28 1/2 shekels of silver became available for me. [18-22] The proceeds of 10 talents 20 minas of copper transported under the responsibility of Šalim-Bēlī are 5 minas 51 1/2 shekels in silver. Out of this he sent 2 minas of silver to the city of Uršu. [23-25] The proceeds of 26 shekels of gold, 2 minas 26 shekels in silver, Aššur-nādā consumed. [26-30a] (For) the 12 talents 28 minas of copper which Ilī-Alum transported to Šalatuar I was charged for 5 minas 32 shekels of silver. Out of this, Aššur-nādā consumed 4 minas 56 shekels of silver. [30b-33] As proceeds of 2 talents of fine copper (and of) his 4 talents 10 minas of copper, Amur-Aššur, the transporter of Dān-Aššur, brought 3 minas 4 shekels of silver to you. [34-35a] The proceeds of the sale of gold, (being) 49 shekels of silver, are included in this. [35b-37] Puzur-Ana owes 1 mina of silver, his debt-note is in Amria's house.'

9 BIN 4, 160

See n.413.
See RGTC 4, p.128, for the personal name in line 1; Lewy 1955, p.158 n.19, for line 36. P. Garelli discussed lines 2-3 (1963, p.177 n.1), 4 and 6 (1963, p.296 n.3), and 15-18 (1963, p.166 n.2).

4 *li-me* URUDU *ší-kum* KI *Ú-lá-mì-il*$_5$
45 GÚ URUDU *ší-kam* É *kà-ri-im*
i-na sú-hu-pí-im a-lá-qé
2 *me-at* URUDU *ší-kam ší-im a-bar-ni-im*
KI *Zu-li-a* GAL *ma-ṣa-ar-tim*

2 *me-at* 20 *ma-na* URUDU *ší-kum*
KI *Sà-ba-sí-a* 2 *me-at* 60 *ma-na* URUDU *ší-kam*
KI GAL *ki-ta-a-tim* . ŠÀ.BA *me-at* 20 *ma-na*
URUDU *al-qé* 2 1/2 GÚ URUDU *ma-sí-um*
1 *ma-na* 9 GÍN KB KI *Ili*$_5$*-we-da-ku*
11 *ma-na* URUDU SIG$_5$ KI *Šu-ma-li-ba-A-šùr*
1 TÚG *bu-ra-am* SIG$_5$ DIRI *ú-lá-bi*$_4$*-šu*
1/3 *ma-na* 1/2 GÍN KB *iš-tí*
Nu-ur-ki-li . ÌR *ša kà-ri-im*
1 *me-at* 50 *ma-na* URUDU *ma-sí-um*
KI *Nu-wa-nu-wa ù tap-pá-i-šu*
me-eh-ra-sú . KI *A-šùr-ták-lá-ku*
al-qé URUDU *šu-a-tí* . *i-lá-ak*
10 *ku-ta-ni* . *i-na* É *kà-ri-im*
E *na-ad-a*$^{!}$*-ku* 30 *ma-na*.TA
3 *me-at* URUDU *ší-im-šu-nu*
ŠÀ.BA 1 *me-at* 95 *ma-na* URUDU
R *a-na Be-lúm-ba-*[*ni*] *i-lá-ak*
17 *ma-na* URUDU [*ú* x+] 2 GÍN KB
KI *A-lu-lá-a x* [ ] *i-dí-in*
5/6 *ma-na* 7 [+x GÍN] KB
75 *ma-na* URUDU *ší-kum iš-tí*
Ni-mar-Ištar DUMU *Puzur*$_4$*-S*[*ú*]*-in*
1/3 *ma-na* 2 GÍN KB KI . *Pì-lá-ha-nim*
DUMU *Da-num*-DINGIR *iš-tù ha-muš-/tim*
ša Bu-za-zu ITU.KAM *hu-bur*
li-mu-um A-ku-tum . *a-na*
[ ] *ha-am-š*[*a-tim*] *i-ša-qal Ni-mar-/Ištar*
[*m*]*a-na* URUDU SI[G$_5$]*-áb-lá-lá*
[ ] *ma-na* URUDU *ší-kam* [*ša*] *ar-ša-/tim*
[ ] *x x a-hu-*(erasure)*-sà-ri-im*
[ ] 4 *ma-na* URUDU *a-šu-mì*
[*ku-nu-k*]*i-im ša hu-sà-ri-im*
[*a-na A*]*-mur-Ištar a-dí-in*
[ ] GIG *i-na* 10 *ma-na-e*
[*ma-n*]*a* URUDU 1 DUG *ar-ša-tim*
[*i-nu-mì*] *a-na Kà-ni-iš*
LoE [*i-t*]*al-ku a-na am-tim*
[*ša*] *Ú-ṣur-ša-A-šùr a-dí-in*
[ ] 2 DUG *a-lá-ni* . *iš-tí*
I-ku-pí-a DUMU *A-da-làl*
LE 5/6 *ma-na* KB KI *Šu-Ištar* DUMU *Ì-lí-iš-tí-kál*
[*x*] *ma-na* URUDU SIG$_5$ KI *A-šùr-i-dí-in* 5 *ku-ta-/ni*

[ ]-*ib* 1 *ta-ma-lá-kum ša ṭup-pì*
[KI *a*]-*hi ša Ú-ṣur-ša-A-šur*

'[1] 4,000 (minas of) *šikkum* copper are owed by Ulamil. [2-3] 45 talents of *šikkum* copper I can take from the *suhuppum* at the *kārum*-office. [4-5] 200 (minas of) *šikkum* copper, the price of an *abarnium*-textile, is owed by Zulia, the commander of the guard. [6-7a] 220 minas of *šikkum* copper is owed by Sabāsia. [7b-9a] 260 minas of *šikkum* copper is owed by the official in charge of linen, of this I have received 120 minas. [9b-10] 2 1/2 talents of refined copper - 1 mina 9 shekels of silver - is owed by Ilī-wēdāku. [11-12] 11 minas of fine copper are owed by Šumma-libb-Aššur (because) I provided him with a *burā'um*-garment of excellent quality. [13-14] 20 1/2 shekels of silver are owed by Nūr-ki-ili, the servant of the *kārum*. [15-18] 150 minas of refined copper are owed by Nuwanuwa and his partners - I have received its equivalent from Aššur-taklāku, the copper is (now) due to him. [19-23] I have deposited 10 *kutānum*-textiles at the *kārum*-office, at a price of 30 minas (of copper) a piece, their value is 300 (minas of) copper. Out of this 195 minas of copper are due to Bēlum-bānī. [24-25] 17 minas of copper [and/or x+] 2 shekels of silver are owed by Alulāya x [ ]-iddin. [26-28] 57 +[x] shekels of silver (and) 75 minas of *šikkum* copper are owed by Nimar-Ištar son of Puzur-Suen. [29-33a] 22 shekels of silver are owed by Pilah-Anum son of Dannum-ilum; from the "week" of Buzāzu, the sixth month of the year eponym of Akutum, he will pay within [x] "weeks". [33b-34] Nimar-Ištar [x] minas of fine copper [are owed by]ablala. [35] [x] minas of *šikkum* copper [from (the sale of)] wheat. [36] [ ] x x for *husārum*-stone. [37-39] [x]+ 4 minas of copper I gave to Amur-Ištar for a seal made of *husārum*. [40] [x sacks] of wheat with the 10 mina weight ... [41-44] [x] minas of copper. 1 jar with wheat I gave to the wife of Uṣur-ša-Aššur when [he/I] left for Kanish. [45-46] [ ] 2 jars of acorns are owed by Ikūppia son of Adallal. [47] 5/6 mina of silver are owed by Šū-Ištar son of Iliš-tikal. [48a] [x] minas of fine copper are owed by Aššur-iddin. [48b-50] 5 *kutānum*-textiles [ ] 1 container with tablets [is with PN,] brother of Uṣur-ša-Aššur.'

10 BIN 6, 76

See p.125.

A transcription and translation of this text was given in Ichisar 1981,pp.299-301. See for lines 10-12 Garelli 1963, p.177 n.4.

um-ma A-mur-Ištar-ma a-na
«a-na» Im-dì-DINGIR *qí-bi-ma*
A-šùr-ta-ak-lá-ku kà-ṣa-ar
Hu-ra-ṣa-nim ṭup-pí ub-lá-kum
i-na ṭup-pí-a a-na SÍG.HI.A
ú-lá a-ṭá-hi i-na tù-a-ri-a
[š]a-pá-at-kà a-da-ma KB 1 *ma-na*
[ú]-šé-lá-a-kum . ki-ma Ú-zu-a
[a-n]i-e [i]k-ta-áš-dì-ni um-ma
[a]-na-ku-ma lá-ma a-tal-ku SÍG.HI.A
a-na URUDU *li-tur$_4$-ma* GEME *ù*
Ú-zu-a a-na URUDU SIG$_5$ *lu-ta-e-ru*
a-na 60 GÚ SÍG.HI.A *lu iš-tí*
ni-a-ú-tim lu iš-tí pá-ší$^{!}$-/ri
a-mì-gi$_5$-ir-ma ŠÀ.BA 40 GÚ
a-ni e-ru-ba-ma qá-tí-ma
E (erasure) *i-ta-am-ša-ah*
ší-[tí SÍG.HI.A] *i-na*
R *T[í-iš-mu-ur]-na na-ad-/a-at*
SÍG.HI.[A *wa-ar]-ki-tám*
ša [ú-bu-lu]-ni-ni
[a-ni] a-na Tí-iš-mu-ur-na
[ta-ú-ra]-at um-ma a-na-ku-ma
[a-tal-k]am-ma ANŠE.HI.A *ù*
[URUDU] *a-na Kà-ni-[i]š ṭù-ru-ud*
[ší-im SÍG.HI].A *ša a-na ší-[mì-i]m a-dí-nu*
[URUDU] *mì-ma lá ú-ba-lá-am*
[ ]- *x -lu-tù-ni* SIG$_5$
[ ]*-ar šu-ma li-bi-kà*
[lu pá]-ni-tum lu wa-ar-ki-/tum
[a-dí] tù-a-ri-a i-na
[Tí-iš]-mu-ur-na-ma li-bi-ší
[ ]*-li a-li ta-ma-li-ku*
[SÍG.H]I.A *mì-lik a-na a-wa-tim*
LoE *[ša š]a-pá-tim ša a-ma-kam*
ni-iš-ta-na-me-ú i-na
ṭup-pí-im ša a-ṣé-ri-a
LE *[t]ù-šé-ba-lá-[ni (x) la]-pí-tám*

tù A-šùr-[ ]*-um-kà*
[SÍG].HI[.A]

'[1-2] Thus (says) Amur-Ištar, speak to Imdilum: [3-4] Aššur-taklāku, Huraṣānum's transporter, has brought my tablet to you. [5-8a] In my letter (I wrote): "I cannot get at the wool. When I return I shall sell your wool and produce for you silver, (every) single mina (of it)." [8b-12] Because Uzua has just reached me, I said: "Before I leave, the wool should be converted into copper and then let the servant girl and Uzua convert (this copper) into copper of good quality." [13-17] For 60 talents of wool, I have reached an agreement (as to its delivery) both with our people and with the retail sellers, and now 40 talents of this have arrived (here), but my own share has been taken away by force. [18-19] The remainder [of the wool] is stored in the city of Tišmurna. [20-23a] The [subsequent consignment of] wool which they [brought] to me, [has now been returned] to Tišmurna. [23b-25] I said: "[Leave], send donkeys and [the copper] to Kanish." [26-27] He does not (want to) bring at all the proceeds [in copper] of the wool which I sold. [28-29a] ... [29b-32] If you agree, let both the previous and the later (consignments of wool) remain in store in Tišmurna itself until I return. [33-34a] [ ] wherever you decide (to bring it) ... take a decision. [34b-40] For the rumours about wool which we hear many times here, write me in the tablet that you will send me ... .'

11 CCT 2, 29

See p.133.
Cf. Veenhof 1972, p.430 for lines 33-35.

um-ma Ah-ša-lim-ma a-na
Puzur$_4$*-A-šur qí-bi*$_4$*-ma*
i-na 20 GÚ 15 *ma-na*
URUDU *ša i-na Wa-ah-šu-ša-na*
a-šu-mì-a i-ša-qú-lu-ni
10 GÚ 55 *ma-na* URUDU SIG$_5$
a-na A-šur-ma-lik a-ma-lá
té-er-tí-kà ú-sà-ni-iq
10 *ma-na ú-ma-am*
ku-nu-ki-a ak-nu-uk-ma
na-áš-a-kum ší-tí URUDU*-i-kà*
9 GÚ 20 *ma-na* URUDU
ṣú-ha-ar-kà
ta-e-ra-ma lu-šé-bi$_4$*-lá-kum*
a-ša A-šur-ma-lik
R *ú-ša-as-hi-ru-ú*
a-dí té-er-tum iš-tù
Wa-ah-šu-ša-na

i-tù-ra-ni a-ší-a-tí
ú-ša-as-hi-ir-šu
1 GÚ 10 *ma-na* URUDU SIG$_5$
i-na ší-im ṣú-ba-tí-kà
Šu-Ištar ub-lá-kum
20 *ma-na* URUDU *a-na*
Šu-Ištar kà-ṣa-ri-kà
20 *ma-na* URUDU *a-na*
Hi-ma-a ù ṣú-ha-ri-kà
i-nu-mì maš-ki ù
ša-nam iš-du-du-ni
ni-dí-in ší-tí
KB-*pì-kà iš-tí*
URUDU-*i-kà-ma*
LE *ú-šé-ba-lá-kum* ⸢3⸣ GÍN KB
Bu-za-zu il$_5$-qé um-ma šu-ut-ma
a-na Puzur$_4$-A-šur ṣa-he-er
URUDU *a-na A-li-a-ša*
a-bu-kam

'[1-2] Thus (says) Ah-šalim, speak to Puzur-Aššur: [3-8] In accordance with your instructions, I have handed over to Aššur-mālik 10 talents 55 minas of fine copper from the 20 talents 15 minas of copper which they have to pay me in Wahšušana to my name. [9-14] 10 minas (are for his expenses). Today, I sealed (this copper) with my own seals and (Aššur-mālik) is bringing it to you. The rest of your copper is 9 talents 20 minas. Send your servant back so that he can bring it to you. [15-20] As for the fact that I have kept Aššur-mālik here: I retained him till (your?) message returned to me from Wahšušana. [21-23] Šū-Ištar is bringing you 1 talent 10 minas of fine copper from the sale of your textiles. [24-30a] We gave 20 minas of copper to your transporter Šū-Ištar and 20 minas to Himāya and your servant when they carried the hides and the kettle. [30b-33a] I will send the rest of your silver together with your copper. [33b-37] Buzāzu took 3 shekels of silver, saying: "It is nothing to Puzur-Aššur!" I will transport the copper to (the town of) Aliaša.'

12 CCT 4, 12b

See pp.134 and 138.
Lines 15-21 were discussed in Veenhof 1972, p.11.

a-na Puzur$_4$-A-šùr qí-bi$_4$-ma
um-ma Ili$_5$-we-da-ku-ma
a-dí ša A-šùr-ba-ni ša
ta-áš-pu-ra-ni a-ma-lá

té-er-tí-kà 50 GÚ SIG$_5$
ú-kà-il$_5$-šu-ma um-ma šu-ut-ma
a-na Pu-ru-uš-ha-tim
lá e-ra-áb a-bi$_4$ a-ta
a-na Pu-ru-uš-ha-tim
a-na KB *a-da-gal*
KB *i-ma-qú-tám-ma*
40 *ma-na* KB
a-kà-na-ak-ma
ú-šé-ba-lá-kum
R 10 ANŠE-*ri kà-aṣ-ru-/tim*
ša bi$_4$-il$_5$-tim šé-ri-a-/ma
URUDU-*a-kà . ša i-ba-ší-ú*
lu a-na Kà-ni-ìš
lu a-na Pu-ru-uš-ha-tim
a-li . té-er-ta-kà
i-lá-kà-ni lu-ub-lu
a-dí ša Da-da-a DUMU
Tí-li-a ša ta-áš-pu-ra-ni
a-dí iš-ri-šu ù iš-ra
té-er-tí i-li-ik
a-ma-lá té-er-tí-kà ṣú-ha-ra-<*am*>
a-ṭá-ra-ad-ma KB
ú-ša-áš-qú-lu-šu-ma
E *ú-ša-ak-na-ak-šu-ma*
ú-šé-ba-lá-kum

'1-2 Speak to Puzur-Aššur, thus (says) Ilī-wēdāku: 3-8a as for Aššur-bānī about whom you wrote me, I offered him 50 talents of fine (copper) in accordance with your instructions, but he said: "I will not enter Purušhattum." 8b-14 My dear father, I am expecting silver from Purušhattum. When the silver arrives here, I will seal 40 minas of silver and send it to you. 15-21 Bring 10 harnessed pack donkeys here, so that they can bring your copper which is available to Kanish or to Purušhattum, wherever you order. 22-30 As for Dadāya son of Tilia about whom you wrote me, I already sent 10, even 20 times a message to him! In accordance with your instructions I will now dispatch a servant, (with the message) that they should collect the silver from him, and I will send it you under seal.'

13 TC 2, 18

See pp.106, 133, 135, 138, and 156.
See Veenhof 1972, p.384, for lines 7-12.

a-na Pu-šu-ke-en$_6$
qí-bi-ma um-ma A-šur-ma-lik-ma
I-a-šar ù ^{d}IM-*ṣú-lu-li-ma*
URUDU *ša Puzur*$_4$*-A-šur*
i-li-bi I-a-šar i-šu-ú
URUDU *ta-áš-bi-ta-am*
i-dí-ni-a-tí ki-ma URUDU-*i-ú*
i-na Pu-ru-uš-ha-tim
ma-ah-ṣú-ni i-re-qám
lá ni-gu$_5$*-ur-ma*
a-na Pu-ru-uš-ha-tim
R *lá nu-šé-ri-ib*
e-ma-ri-i šé-ri-a-ma
URUDU-*ú-šu a-na Pu-ru-uš-ha-tim*
lu nu-šé-ri-ib
šu-ma ṭup-pu-šu
ša Puzur$_4$*-A-šur a-ṣé-ri-ni*
i-li-kam ší-ta-me-šu-ma
a-ni-ša-am šé-bi$_4$*-lam*
i-ša-am-ší ṭup-pì-ni
ta-ša-me-ú
LE *e-ma-ri . šé-ri-am*

'[1-3] Speak to Pūšu-kēn, thus (say) Aššur-mālik, Ea-šar, and Adad-ṣulūlī: [4-7a] as for the copper which Puzur-Aššur owns from Ea-šar, he gave us copper, the final instalment. [7b-15] Because (the market for) copper is paralyzed in Purušhattum we did not rent a wagon to bring (it) to Purušhattum. Send us donkeys and we will bring his copper to Purušhattum. [16-19] If a letter from Puzur-Aššur arrived for us, you must read it carefully and then send it here. [20-22] Send donkeys on the very day you read our letter!'

14 ATHE 32

See pp.151, 153, 160-161, and 164.
Edited in transcription and translation by B. Kienast in ATHE, p.47, with hand copy on plates 14-15. Discussed in Veenhof 1988, p.260. Cf. Larsen 1976, p.44 n.69, for line 22.

a-na Pu-šu-ke-en$_6$
qí-bi-ma um-ma A-mur-Ištar-/ma
a-šu-mì AN.NA *ša Hi-na-a*
ša ta-áš-pu-ra-ni
DUMU *a-wi-lim ú kà-ṣa-ar-šu*
a-na-kam wa-áš-bu
ú-šé-ra-áb-šu-nu-ma
ma-ah-ri-šu-nu AN.NA
ku-nu-ki ša a-lim^{ki}
ša a-bi$_4$*-šu lu-sà-ni-iq-ma*
E *ba-tí-iq ú wa-tù-ur*
R *a-da-an-ma* KB
ú-šé-ba-^{lá}*lam*
AN.NA *ša* É *A-šur-ma-lik*
a-na [UR]UDU ⸢*ší*⸣*-ki-im*
15 *ma-na.*TA *a-d*[*í-m*]*a*
URUDU *i-ba-ší Eb-lá-i-ú*
ma-<*du*>*-tum i-li-ku-ni-ma*
URUDU *ma-dum i-na* É.GAL-*/lim*
i-ší-qí-il$_5$*-šu-nu-tí-ma*
LoE KB 2 1/3 *ma-na.*TA
a-mu-ra-am ú-ta-e-ru
a-dí 10 *u*$_4$*-me* URUDU-*šu*
LE *i-ga-mu-ru* KB *a-ša-a-/ma*
ú-šé-ba-lam

'1-2 Speak to Pūšu-kēn, thus (says) Amur-Ištar: 3-13 Concerning the tin of Hināya about which you wrote me, the son of the gentleman and his transporter are staying here. I will bring them (into my house) and check in front of them the tin of his "father" which is sealed by the City; then I will sell it at any price and send the (proceeds in) silver. 14-17a I sold the tin from the house of Aššur-mālik for *šikkum* copper at 15 minas (of copper) each (mina of tin); the copper is in store. 17b-25 Many Eblaites have come here and at the palace a lot of copper has been weighed out to them for which they paid with *ammurum* silver at a rate of 2 1/3 minas (of copper for) each (shekel of silver). Within 10 days, when its stock of copper is exhausted, I will purchase silver and send it.'

15 Kt a/k 583/b.5-31(end)

See p.168.
Unpublished text in Ankara (inv. no. 109-239-64).

a-dí 10 *u$_4$-me-e ha-ra-ni ú$^!$-ṣí-a-am*
a-na ša u$_4$-me-e ma-du-tí-im
a-na-kam uš-bu-ni 24 *li-me*
URUDU É.GAL-*lúm ù té-ra-tù-šu*
i-hi-ib-lu-nim-ma lá li-/bi$_4$
i-li-ma i-na sí-kà-tí-im
[*uš-bu-m*]*a e-ša-ni-ma ša* KB
[x] *ma-na er-ba-am a-na*
E É.GAL-*lim ag-mur-ma*
R [1+] 1 *ší-ta ša-na-tí-im*
É.GAL-*lam ú-ni-ih-ma*
URUDU-*i* . *kà-lá-šu ar-ta-kà-/as*
iš-tù ITU.KAM *eq-lá-tim*
ù ki-ri-a-tim ša 20 *ma-na*
KB *ù e-li-iš* É.GAL-*lúm*
i-š[*a$^?$-kà*]*-nam e-zi-ib-ší-na-ma*
a-ta-bi-a-ma a-ta-lá-kam A-šur
ù DINGIR *il$_5$-kà li-iš-a-lá-ni*
iš-tù 3 *ša-na-tim* 10 *ma-na* KB
ù KÙ.KI 1 *ma-na ù* 2 *ma-na*
kà-ni-ik-ma a-pá-ṭá-ri-im
LoE *pá-al-ha-ku*
a-hi a-ta be-lí a-ta a-dí
10 *u$_4$-me* . *a-ma-kam uš-ba-ku*
LE *ki-ma ha-ra-ni qur-ba-at-n*[*i*]
mì-ma KB *ú* K[Ù.G]I
ú-lá uš-té-bi$_4$-lam

(To Asānum from Uṣur-ša-Aššur:) '[5] Within 10 days my caravan will leave. [6-11a] As for the fact that they have been staying here for a long time, (that is because) the palace and its officers (= for *bēl tērātim*) became indebted to me for 24,000 (minas = 400 talents) of copper, and now, unfortunately, they are present at the *sikkatum* festival. [11b-16] Once more, I incurred an expense (by presenting) to the palace a goodwill gift worth about [x] minas of silver, and (in this way) I could appease the palace for 2 years, but I have now fixed by contract the payment of all my copper. [17-21a] For one month, the palace has been offering me/putting at my disposal fields and orchards worth 20 minas of silver and more; I gave them up, I want to leave. [21b-27] Aššur and your god may question me (if I lie): for 3 years, 10 minas of silver and gold, about 1 or 2 minas, have been lying packed and under seal, but

I did not dare to open (the packages)! 27-31 My dear brother and lord! Within 10 days I will be staying over there (with you). I did not sent any silver or gold because my departure is near.'

16 KTH 14

See pp.155 and 167

Published in transcription and translation by J. Lewy in KTH, pp.22-25, hand copy on plate 9. Cf. for 'Harana' Lewy 1956, p.22 n.95. Lines 6-8 were discussed in Garelli 1963, p.303 n.1. See also Larsen 1976, pp.91-92 nn.23 and 28.

a-na Ha-na-na-ri-im
I-tur$_4$-DINGIR . *ù Puzur*$_4$*-A-šùr*
qí-bi-ma um-ma Šu-Ištar-ma
a-na-kam 10 *li-me-e* URUDU GAL *sí-ki-tim*
iq-bi-am . URUDU *iq-bi-am ú-lá*
ú-šé-pí-am . *um-ma šu-ut-ma ra-dí-e*
ù ṣú-ha-ar-kà . URUDU *a-na* KB
ù KÙ.GI . *li-dí-nu-ma*
ù lu-ub-lu-nim-ma . *a-na-kam*
a-da-na-kum . KB *ù* KÙ.GI
ra-dí-šu . *ù ṣú-ha-ri* . *e-pu-šu-ma*
i-na ṣí-li-a-ni-šu ra-dí-um
ik-nu-uk-šu-ma is-kam iṣ-ba-at-ma
a-na Wa-ah-šu-ša-na a-ṣé-er
GAL *sí-ki-tim* . *ú-ba-lá-šu*
i-na ha-ra-nim . *iš-pu-ru-ma*
KB *uš-té-li-ú* . *ṣí-li-a-ni*
ra-qú-tim . *Ša-ha-ra-na-a-ma*
a-ṣé-er . GAL *sí-ki-tim*
a-šé-lu-im . *i-zi-iz-ma*
E *a-na-kam* . *kà-ru-um*
ip-nu-a-ma um-ma kà-ru-ma
R *ṣí-li-a-ni-šu* . *ṣí-li-a-ni-šu-ma*
e-ṭé-er-šu . *Ša-ha-ra-na*
as-li-ma ṣí-li-a-ni
a-ṣé-er GAL *<sí>-ki-tim*
ú-lá uš-té-li a-na É *be-tí-a*
uš-té-ri-ib-šu-nu um-ma a-na-ku-ma
a-li-ik ṣú-ha-ri ša iš-tí-kà
KB *iš-ú-mu* . *ru-a-ma*
zu-ku-a . *šu-ma-me-en* . *ṣí-li-a-ni*
ra-qú-tim . *a-ṣé-er* GAL *sí-ki-tim*

ú-šé-li lu i-na Pu-ru-uš-ha-tim
lu a-na-kam a-na qá-qá-da-tí-ni
ni-ip-lá-ah-me-en
lá ta-áš-ta-na-me-a ki-ma
a-wi-lu-um na-ba-al-kà-tim
ma-al-ú . ki-am i-da-an
ki-am e-ṭé-er KB *lá šé-pu-am*
2 *li-im* URUDU *e-ri-ší-ma*
a-dí-šu-um . a-na-kam A-hu-qar
ší-ip-ru-um ša-bu-ú
LoE *a-ba-ú-a be-lu-ú-a*
a-tù-nu KB *lá-qá-ta*
kà-ra-am mu-ùh-ra-ma
LE KB *lu-šé-ru-ni-ma e-ri-iš-tí a-wi-lim*
lá a-kà-lá-ma a-na a-wa-tim lá i-tù-ar

'[1-3] Speak to Hananarum, Itūr-ilum and Puzur-Aššur, thus (says) Šū-Ištar: [4-5a] Here, the general promised me (to pay) 10,000 (minas of) copper. [5b-10a] Although he had promised (it), he did not produce (it) for me! He said: "Let my escort and your servant sell the copper for silver or for gold and let them then bring (it), then I will give (that) here to you." [10b-15] His escort and my servant did obtain silver and gold and the escort put it under seal in his boxes and seized the *iskum* in order to bring it to the general in Wahšušana. [16-20] When they sent (him) a message during the trip, they removed the silver. Šaharana himself took steps to send those empty boxes to the general. [21-31a] Here, the *kārum* turned to me, saying: "His boxes are indeed his boxes, save him!" I argued with Šaharana so that he did not send the boxes to the general. I brought them (= the escort Š. and members of his party) into my house, saying: "Go, bring me my servant who bought the silver together with you, and finish the affair." [31b-41a] Would we not have to fear for our life, in Purušhattum as well as here, if I had sent empty boxes to the general? Have you not heard repeatedly that the man is capricious? At times he gives, at others he takes. The silver was not produced for me! Still he asked me for 2,000 (minas of) copper and I gave (it) him! [41b-47] Here, Ahu-waqar, the envoy, has been satisfied. My dear fathers and lords, collect the silver! Approach the *kārum* so that they release the silver to me, then I shall not further delay the request of the man, lest he comes back to this case.'

17 TC 3, 25

a-na Pu-šu-ke-en$_6$ qí-bi$_4$-ma
um-ma Dan-A-šur-ma
a-bi a-ta be-lí a-ta
a-na ša-al-ṭim
lá ta-ša-ki-ni-a-tí
URUDU *ša* É.GAL-*lim a-dí*
u$_4$-mì-im a-ni-im
i-ta-na-pu-lu-ni-a-tí
ù i-a-tí mu-ur-ṣú-um
im-qú-tám-ma
a-dí a-ni a-sú-hu-ur
E *ša-al-ma-ku*
R *mì-ma li-ba-kà*
lá i-pá-ri-id
a-dí 5 *u$_4$-me ú-ta-qá-ma*
šu-ma URUDU *ša*
É.GAL-*lim lá i-ta-ṣa-am*
kà-ki a-lá-qé-ma
a-Pu-ru-uš-ha-tim
a-ta-lá-ak

'[1-2] Speak to Pūšu-kēn, thus (says) Dān-Aššur: [3-5] My dear father and lord! Do not treat us harshly! [6-8] Concerning the copper due from the palace: they keep us dangling until this very day. [9-14] As for me, an illness befell me so that I was confined until now. (But) I am well (again), do not worry. [15-20] I will wait for 5 days and if the copper of the palace has then not become available, I will take up my weapon (= prepare for my departure) and leave for Purušhattum.'

18 VS 26, 6

See p.158.
Museum number is VAT 9255. Cf. K.R. Veenhof, VS 26, p.18.

a-na Pu-šu-ke-en$_6$ qí-bi-ma
um-ma A-mur-Ištar
ù Dan-A-šùr-ma i-na
1 GÚ 2 1/2 *ma-na* 9 GÍN AN.NA
ša É.GAL-*lúm il$_5$-qé-ú*

3 *me-at* 43 *ma-na* URUDU
a-1 GÚ-*tim ú-ṣa-ma*
4 *me-at* 10 *ma-na*
ik-šu-ud-ni-a-tí ŠÀ.BA
2! *me-at* 44 *ma-na*
ṣa-lá-ma i-ba-ší ší-tí
1 *me-at* 68 *ma-na*
R 1 2/3 *ma-na* 6 1/2 GÍN.TA
KB-*ap-šu* ⌜2⌝ *ma-na* 10 1/2 GÍN
ŠÀ 15 GÍN *ša-du-a-tám*
ni-dí ší-tí 1 5/6 *ma-na* 6 1/2 GÍN
KB *ku-nu-ki-ni Dan-A-šùr*
na-áš-a-ku-um ší-tí AN.NA
a-dí-ni i-É.GAL-*lim*
ga-ma-ar-šu mì-ma
ú-lá a-dí-šu-um

'1-3a Speak to Pūšu-kēn, thus (say) Amur-Ištar and Dān-Aššur: 3b-9a By means of the 1 talent 2 minas 39 shekels of tin which the palace bought - 343 minas of copper came out for one talent (of tin) -, 410 minas fell to our share. 9b-11a Out of this, 244 minas (of copper) is black; it is in store. 11b-14 The remainder is 168 minas - at a rate of 1 mina 46 1/2 shekels (of copper for) each (shekel of silver), its value in silver is 2 minas 10 1/2 shekels. 15-16a Out of this, we paid 15 shekels as *šaddu'atum*-tax. 16b-21 The remainder, 1 mina 56 1/2 shekels of silver under our seals, Dān-Aššur is bringing to you. The rest of the tin is still in the palace. I did not give him (= Dān-Aššur) anything for his expense(s).'

19 AKT 2, 39

See p.57.
A transcription and a translation in Turkish of this text (excavation number kt n/k 577) were given in AKT 2, p.74-75.

um-ma Ú-ṣur-ša-Ištar
ù Ir-ma-A-šùr-ma
a-na Puzur$_4$*-ì-lí qí-bi-ma*
10 GÚ URUDU *Zu-zu-wa ub-lam*
ŠÀ.BA 7 1/2 GÚ *ma-sí-um*
20 LÁ 2/3 *ma-na* URUDU *ma-sí-a-am*
a-na É *kà-ri-im a-na*
kà-ú-nim ni-dí-in
ší-tí URUDU-*i-kà* 7 GÚ 10 2/3 *ma-na*
5/6 *ma-na* 7 GÍN.TA
KÙ.BE 7 1/2 *ma-na* 3 1/3 GÍN

E 1 GÚ *ša Ha-bu-ra-at*
1 *ma-na* 5 GÍN.TA KÙ.BE
5/6 *ma-na* 5 1/6 GÍN
1 GÚ 30 *ma-na* URUDU *ší-ku-um*
ù 18 *ma-na ša*
É *kà-ri-im ù-kà-i-nu-ni-a-*[*t*]*í-ni*
ŠÀ.BA 9 *ma-na*
ša-du-a-tám ni-dí-[*i*]*n*
11! *ma-na i-na áb-ni-ni*
a-na Zu-zu-wa ni-dí-in
a-ma-kam lu-ba-li-ṭá-kum
ší-tí URUDU 1 GÚ 28 *ma-na*
2 *ma-na* 10 GÍN.TA KB
2/3 *ma-na* 1/2 GÍN ŠUNIGIN 9 *ma-na* 9 GÍN
E ŠÀ.BA
LE 7 *ma-na* 9 GÍN KB
ku-nu-ki-ni En-um-A-šur
ub-lá-kum

'[1-3] Thus (say) Uṣur-ša-Ištar and Ir(i)m-Aššur, speak to Puzur-ili: [4-8] Zuzuwa brought (us) 10 talents of copper, of which 7 1/2 talents were refined. We gave 19 1/3 minas of refined copper to the *kārum*-office for a *ka''unum* charge. [9-11] The remainder of your copper - 7 talents 10 2/3 minas - (was sold) at a ratio of 57:1, and its proceeds in silver amounted to 7 minas 33 1/3 shekels. [12-14] (As for) the talent of copper from Haburata, (it was sold) at a ratio of 65:1, and its proceeds in silver amounted to 55 1/6 shekels. [15-19] (As for) the 1 1/2 talent of *šikkum* copper and the 18 minas which they "secured" for us at the *kārum*-office: we paid 9 minas of it as *šaddu'atum*-tax. [20-22] We gave 11 minas - according to our weights - to Zuzuwa. He should pay (it) back to you there. [23-29] The remainder of the copper was 1 talent 28 minas. At a ratio of 130:1, the proceeds in silver amounted to 40 1/2 shekels. The total amount of silver (obtained by selling the copper) was 9 minas 9 shekels. Of it, Ennum-Aššur brought you 7 minas 9 shekels under our seals.'

20 Kt n/k 1464

See pp.145, 156, 171-172, 175-176.
A hand copy of this text made by H. Sever and S. Çeçen was distributed during the XIth Congress of the Turkish Historical Society in Ankara, 1990.

a-na Ú-ṣú-ur-ša-Ištar
qí-bi-ma um-ma Ú-zu-bi$_4$-iš-ku-/ma
i-nu-mì . a-na-kam . ta-ta-al-ku
5 *me-at* 60 *ma-na* URUDU
ú 8 TÚG.HI.A . *qá-at-kà*

É kà-ri-im . na-dá-tí
iš-tù ta-ta-al-ku
1 TÚG.TA . *qá-tám i-dí-ú-ma*
qá-at-kà . a-dí-ma 40 *ma-na* URUDU
al-pu-ut-kà a-ha-ma
E 1 *ma-na*.TA KB
qá-tám . i-dí-ú-ma
R *qá-at-kà . a-dí-ma*
2 *me-at* URUDU *i-na ab-na-/tí-ni*
i-na wa-ṣa-a . URUDU *i-na*
sú-hu-pì-im . ta-lá-qé
É *kà-ri-im al-pu-ut-kà*
ni-kà-sú . i-ší-sí-ú-ma
9 *me-at* 20 *ma-na*.TA URUDU
da-tum ik-šu-ud
40 *ma-na*.TA *a-na* TÚG-*tí-/šu-nu*
i-na ni-kà-sí . iš-ku-nu
LoE *mì-šu . ša ki-a-ma*
té-er-ta-kà . lá i-lá-kà-ni
LE *ma-tum sá-hi-at . ki-ma*
iš-li-mu . ú-ṣí-am
a-ma-kam e-kà . i-na
ṣé-er ṣú-ùh-ri-im
li-li-ik

'[1-2] Speak to Uṣur-ša-Ištar, thus (says) Uzubiškum: [3-6] When you left here, you had deposited at the *kārum*-office 560 minas of copper and 8 textiles as your share. [7-10a] When, after you left, they deposited 1 textile each, I deposited (one as) your share and credited you for 40 minas of copper. [10b-17] Furthermore, when they deposited 1 mina of silver each as share, I deposited your share and you can take 200 (minas of) copper, according to our weight stones, from the *suhuppum* when the copper becomes available; I credited you at the *kārum*-office. [18-20] The accounts have been settled and as a result the *dātum* charge (of the *kārum*-office) amounted to 920 minas of copper each. [21-22] They valued their textiles at the accounting at 40 minas (of copper) a piece. [23-26] Why do you not just send me a message? The country is in uproar; as soon as it becomes peaceful again, I will leave. [27-29] Keep a watchful eye on the children/servants there!'

CATALOGUE OF COPPER IN OLD ASSYRIAN TEXTS

Evidence for each of the qualifications discussed in chapter 3 is presented in this catalogue. This involves the use of the qualification, the prices in silver attested for it (sale, purchase, and equivalent), and the quantities. In the case of rare qualifications, all evidence is listed, for the other qualifications, a selection has been made.

werium (§ 3.1.1)

Prices:

purchase

140: KB 2 1/3 *ma-na*.TA *a-mu-ra-am* (purchase by Eblaites probably of *šikkum*), ATHE 32.21-22

120: URUDU *ša-na-um iš-ta-ak-nu*, kt a/k 259.27

80: 7 GÚ 30 *ma-na* URUDU *qá-ta-am ša Pu-šu-ke-en*$_6$ *iš-tí En-nam-A-šur* 1 1/3 *ma-na*.TA *I-dí-Sú-in iš-am*, CCT 1, 17a.1-5

equivalent

240: 7 LÁ 1/3 GÍN KB KI *Ì-lí-dan* 4 *ma-na*.TA URUDU *i-ša-qal*, CCT 5, 30d.1-4

172: *a*-2 GÚ 9 *ma-na* URUDU 2/3 *ma-na* 5 GÍN KB, kt 90/k 137.10-12 (by courtesy of C. Michel)

150: 20 *ma-na*.TA TÚG.HI-*tí qá-tám a-na* É *kà-ri-im ni-ta-na-dí* 8 GÍN.TA *a-na* KB *ú-lá kà-ší-id*, AKT 3, 63.21-24

120: ... 2 *ma-na*.TA URUDU *a-na* KÙ-*pì-kà* , RA 60, Dessallien.12-13 (in Turhumit)

108: 7 5/6 *ma-na* KB (...) ŠÀ.BA 1 GÚ 30 *ma-na* URU[DU] *a-ha-sú tal-qé ší-tí* KB 7 *ma-na i-ší-tám*, kt b/k 117.1-7

105: ... *ší-tí* URUDU-*kà iš-tí Lu-lu i-ba-š*[*í*] *ú* 18 GÚ 45 *ma-na iz-ku-am* 1 2/3 *ma-na* 5 GÍN.TA 10 2/3 *ma-na* 2 5/6 GÍN KB-*áp-šu*, CCT 4, 23a.38-42

100: ... ŠUNIGIN URUDU 32 *ma-na* LÁ 5 GÍN KB-*áp-šu* 1/3 *ma* LÁ 5/6 GÍN, KUG 27.42-43

97: ... 4 GÚ 51 *ma-na* URUDU *ša* 3 *ma-na* KB (...) 2 GÚ 9 1/3 *ma-na* URUDU *ša* 1 1/3 *ma-na* KB, TC 3, 173.9-14

90: ... *a-na* URUDU 1 1/2 *ma-na*.TA KB (...) *a-pu-ul-ma*, CCT 5, 45a.11-14

72: 7 1/2 GÍN KB *ší-im* 9 *ma-na* URUDU, RA 58, Goudchaux 1.19-20

60: 5 GÍN KB 5 *ma-na* URUDU, BIN 4, 88.26

60: ... *a-ša*[*r*] URUDU *i-na* É *be-tí-ki e-zi-bu* 1 *ma-na*.TA 1 GÍN KB *lá* [*t*]*a-ša-qí-li-ni*, I 581.13-16 (by courtesy of K. Hecker)

60: 1 *ma-na* KB *ù* 1 GÚ URUDU, kt n/k 534.10-11 (by courtesy of C. Günbattı)
60: 1 GÚ URUDU *ù* 1 *ma-na* [K]B, LB 1290 rev.7'

sale

105: URUDU *i-na Wa-ah-šu-ša-na a-na* KB 1 2/3 *ma-na* 5 GÍN.TA *ni-dí-in-ma*, I 507.10-13 (by courtesy of K. Hecker)
95: ... K[B *a*]-*na* URUDU-*im* 1 1/2 *ma-na* 5 GÍN.TA *i-za-zu-šum*, AKT 1, 82.17-19
80: 1 1/3 *ma-na*.TA URUDU-*um a-na* KB *i-za-az*, kt v/k 128.3-4 (by courtesy of C. Günbattı)
77: [1 *ma*]-*na* 17 GÍN.TA [ ] *Ú-lá-ma* URUDU *ù* [ ] x *i-na* [*Ša-la-t*]*ù-ar* URUDU-*um* [*da-mu*]-*qú-um* 1 *ma-na* 13 GÍN.TA *i-za-az*, kt v/k 128.13-18 (by courtesy of C. Günbattı)
72 1/2: ... URUDU-*a-kà* 1 *ma-na* 12 1/2 GÍN.TA *a-na* KB *ta-dí-in*, kt a/k 259.3-5
60: 7 *ma-na* KB *ší-im* URUDU 7 GÚ, RA 81, 1.14-15
40: 1 1/2 GÍN KB *ší-im* 1 *ma-na* URUDU *a-na Wa-qar-A-šùr áš-qúl*, I 773.14'-16' (by courtesy of K. Hecker)

masium **- 'refined'** (§ 3.1.2.1)

Qualities attested:
URUDU *masium*, passim
10 GÚ URUDU *ma-sí-am* SIG$_5$, CTMMA 94.1-2
30 *ma-na* URUDU SIG$_5$ *ma-sí-am*, ICK 2, 99.3-4
55 *ma-na* URUDU *ma-sí-am da-mu-qám ša-bu-ra-am*, OIP 27, 56.33-34
6 GÚ URUDU *masiam* (...) URUDU *ša-bu-ra-am*, kt c/k 671.23-25
18 GÚ 45 *ma-na* URUDU *ma-sí-am ša-bu-ra-am*, TC 3, 10.10-11
35 *ma-na* URUDU *ma-sí-am ša Tišmurna*, KTB 5.10-11

Prices:
equivalent
30: 1/2 *ma-na*.TA ù 1 GÍN.TA KB, kt a/k 486.14-15 (punitive ratio)

sale
71: *ší-im* 21 GÚ 9 *ma-na* URUDU *ma-sí-im* (...) 17 5/6 *ma-na* 2 GÍN KB, kt 91/k 305.6-14 (by courtesy of K.R. Veenhof)
57: 5/6 *ma-na* 7 GÍN.TA, AKT 2, 39.10
55: 5/6 *ma-na* 5 GÍN.TA, Berytus 3, Pl. 17.20 (special ratio for a *gimillum*-loan)
46: 4 GÚ 8 *ma-na* 15 GÍN *ma-sí-a-am* 2/3 *ma-na* 6 GÍN.TA, TC 1, 55.14-16
45: 2 1/2 GÚ URUDU-*a-kà* (= *masium* in line 8) 2/3 *ma-na* 5 GÍN.TA *a-na* KB *a-dí-šu-ma*, kt a/k 1056.23-25

Notable quantities:
1 *ma-na* URUDU *ma-sí-a-am*, BIN 4, 145.6 (gift)

1 1/3 *ma-na* URUDU *ma-sí-a-am*, kt 91/k 156.1-2 (gift) (by courtesy of K.R. Veenhof)
2 1/3 *ma-na* URUDU *ma-as-am*, VS 26, 27.10-11
3 *ma-na* URUDU *ma-as-am*, BIN 6, 94.13
13 1/2 *ma-na* URUDU *ma-sí-ú-um*, CCT 1, 17b.17-18
30 *ma-na* URUDU *ma-sí-am*, BIN 4, 79.6' (for expenses)
40 *ma-na* URUDU *ma-sí-am*, BIN 4, 62.5
URUDU 50 GÚ *ma-sí-am*, BIN 4, 43.5-6 (in line 19: SIG_5)
me-at 5 GÚ URUDU *ma-sí-a-am*, kt u/k 4.13

***dammuqum* - 'fine'** (§ 3.1.2.2)

broken up:
30 *ma-na* URUDU SIG_5 [*ša-b*]*u-ra-am*, kt n/k 426.6-7 (by courtesy of S. Bayram)
1 GÚ 18 *ma-na* URUDU SIG_5 *ša-bu-ru-um*, ATHE 37.18
1 GÚ 45 *ma-na* URUDU SIG_5 *ša-bu-ra-am*, BIN 4, 197.1-2
2 GÚ 50 *ma-na* URUDU SIG_5 *ša-bu-ra-am*, KUG 2.1-2
3 GÚ 20 *ma-na* URUDU SIG_5 *ša-bu-ra-am*, AKT 1, 66.8-9 (to be paid in Turhumit)
30 *ma-na* URUDU SIG_5 *Ha-bu-ra-ta-i-um*, kt 86/k 176.3
4 *ma-na* URUDU SIG_5 *ša-bu-ru-um Ha-bu-ra-ta-i-a-am*, Kayseri 71/1.3-5 (by courtesy of K.R. Veenhof)
[ ] GÚ 16 *ma-na* URUDU SIG_5 *Ha-bu-ra-ta-i-um ša-bu-ru-um*, Pa 36.1-3

with masium:
30 *ma-na* URUDU SIG_5 *ma-sí-am*, ICK 2, 99.3-4
15 GÚ URUDU SIG_5 *mì-iš-lu-um ma-sí-um mì-iš-lu-um Ha-bu-ra-ta-i-um*, VS 26, 26.5-7

massuhum:
12 *ma-na* URUDU SIG_5 *ma-sú-ha-am*, ATHE 38.25-26

from Haburata:
URUDU SIG_5 *Ha-bu-ra-ta-i-am*, JCS 14, 11.6
15 GÚ URUDU SIG_5 *mì-iš-lu-um ma-sí-um mì-iš-lu-um Ha-bu-ra-ta-i-um*, VS 26, 26.5-7
URUDU SIG_5 *ša-bu-ru-um Ha-bu-ra-ta-i-a-am*, Kayseri 71/1.3-5 (by courtesy of K.R. Veenhof)
[ ] GÚ 16 *ma-na* URUDU SIG_5 *Ha-bu-ra-ta-i-um ša-bu-ru-um*, Pa 36.1-3
URUDU *Ha-bu-ra-ta-am* SIG_5, Ka 276.3-4

from Taritar:
10 *ma-na* URUDU SIG_5 *Ta-ri-ta-ra-i-am*, kt n/k 1831.4-5 (by courtesy of V. Donbaz)
1 GÚ 43 *ma-na* URUDU SIG_5 *ša Ta-ri-ta-ar*, TC 3, 211.35-36
[4 GÚ] URUDU SIG_5 [*ša T*]*a-ri-ta-a*[*r*], ICK 2, 54.1-2

from Tišmurna:
5 1/2 *ma-na* URUDU SIG_5 *Tí-iš-mu-ur-[na-i-am]*, ICK 2, 45.1-2
30 *ma-na* URUDU SIG_5 *Tí-iš-mu-ur-na-i-a-am*, JCS 14, 11.34-35
URUDU SIG_5 20 *ma-[na ...] Tí-iš-mar-na*, kt a/k 955.8-9

Prices:
purchase
80: *ša* 5/6 *ma-na* 5 GÍN KB 1 1/3 *ma-na*.TA 1 GÚ 13 1/3 *ma-na* URUDU SIG_5, kt a/k 294.16-19
70: *ša* 1 2/3 *ma-na* 5 GÍN KB 1 *ma-na* 10 GÍN.TA-*ma* 2 GÚ 2 1/2 *ma-na* URUDU SIG_5, CCT 3, 1.16-18 (in Turhumit)

equivalent
120: 2 *ma-na* URUDU SIG_5 1 GÍN KB, TC 3, 208.13-14
90: 30 *ma-na* URUDU SIG_5 *Ha-bu-ra-ta-i-um ù* 1/3 *ma-na* KB *ṣa-ru-pu-um*, kt 86/k 176.3-4
88 1/4: 2 5/6 *ma-na* KB 4 GÚ 10 *ma-na* URUDU SIG_5, CCT 6, 46b.1-2
82 1/2: 2 1/2 GÚ URUDU SIG_5 = 1 2/3 *ma-na* 9 GÍN KB, AKT 1, 65.1, 11-12
80: ... *ša* 5/6 *ma-na* 5 GÍN KB 1 1/3 *ma-na*.TA 1 GÚ 13 1/3 *ma-na* URUDU SIG_5, kt a/k 294.16-19
60: 5 GÍN KB *ší-im* 5 *ma-na* URUDU SIG_5, CCT 1, 33b.5-6
60: 3 *ma-na* URUDU SIG_5 3 GÍN KB *ší-im-šu*, CCT 1, 33b.14-15
60: 3 *ma-na* URUDU SIG_5 3 GÍN KB *a-na qá-tí En-na-nim a-dí-in*, ICK 1, 66.16-18
60: 30 GÚ URUDU SIG_5 *li-dí-nu-ni-ku-ma* 30 *ma-na* KB *ṣa-ru-pá-am šu-qú-ul-šu-nu-tí*, kt a/k 308.10-13
60: 1 GÚ URUDU SIG_5 1 *ma-na* KB, kt c/k 839.70-71

sale
73: ... *i-na [Ša-la-t]ù-ar* URUDU-*um [da-mu]-qú-um* 1 *ma-na* 13 GÍN.TA *i-za-az*, kt v/k 128.15-18 (by courtesy of C. Günbattı)
70: 1 *ma-na* 10 GÍN.TA *ta-dí-in*, TC 1, 23.15-16
68: 1 *ma-na* 8 GÍN.TA URUDU SIG_5 *ta-dí-in*, TC 1, 23.11-12
60: 1 *ma-na*.TA *ù* 1 GÍN.TA *a-na a-mu-ri-im ta-dí-in*, kt 87/k 462.10-12 (in Kanish; by courtesy of K. Hecker)
60: 1 *ma-na*.TA *ù* <1 GÍN.TA>, KTS 6.38 (cf. C 13.26-27)
50: *a-na* URUDU *a-nim* 5/6 *ma-na*.TA *a-na-kam da-na-sú iq-bi₄-a-am*, kt a/k 294.21-23

Notable quantities
1 *ma-na* URUDU SIG_5, kt a/k 431.8
5 *ma-na* URUDU SIG_5, CCT 3, 12a.14
10 *ma-na* URUDU SIG_5, CCT 3, 44b.22
1/2 GÚ URUDU SIG_5*, Pa 26.27 (collation by K.R. Veenhof)

44 *ma-na* URUDU *da-mu-qám*, TC 3, 200.1-2
URUDU 50 GÚ SIG_5, BIN 4, 43.19

watrum **- 'excellent'** (§ 3.1.2.3)

Occurrences:
1/2 *ma-na* 4 GÍN AN.NA *ù* 1/2 *ma-na* URUDU *ba*[!]*-at-ra-am*, CCT 1, 27a.9-11
3 *ma-na* URUDU *wa-at-ra-am*, I 548.9-10 (by courtesy of K. Hecker)
2 *ma-na* URUDU DIRI (...) *áš-qúl*, kt a/k 259.10-13
1 GÚ 40 *ma-na* URUDU *wa-at-ra-am*, kt c/k 51.5

zaku'um **- 'clean'** (§ 3.1.2.4)

Occurrences:
URUDU *za-ku-ú-um ša ma-at Ša-wi-it*, AKT 3, 56.37
ša 2 *ma-na-im* KB URUDU *za-ku-tum iš-tù a-lim*[ki] *e-li-a-ni-ma*, RA 58, Sch. 2.7-9
URUDU *Ta-pá-ša-ta-i-am za-ak-wa-am*, kt 91/k 390.6-7 (by courtesy of K.R. Veenhof)

lammunum **- 'poor'** (§ 3.1.3.1)

Prices:
sale
80: 1 1/3 *ma-na*.TA, kt 87/k 436.18 (probably in Wahšušana; by courtesy of K. Hecker)
72: 1 GÚ 12 [*ma-n*]*a la-*[*mu-nam*] *a-na* 1 *ma-*[*na*] KB, KTS 2, 8.8-9

Notable quantities:
4 *ma-na* URUDU *lá-mu-nam*, BIN 4, 172.1-2
40 LÁ 1/2 *ma-na* URUDU *lá-mu-nam*, TC 3, 5.21
40 *ma-na* URUDU *lá-mu-nam*, TC 3, 200.11
45 [*ma-na*] *la-*[*mu-num*], KTS 2, 8.10
1 GÚ URUDU *la-mu-nu-um*, kt n/k 1385.22'-23' (by courtesy of C. Günbattı)
1 GÚ 12 [*ma-n*]*a la-*[*mu-nam*], KTS 2, 8.8
1 GÚ 20 *ma-na* URUDU *lá-mu-nu-um*, kt 91/k 346.2-3 (by courtesy of K.R. Veenhof)
2 GÚ URUDU *lá-mu-nim*, Pa 26.26
3 GÚ 37 1/2 *ma-na* URUDU *lá-mu-nam*, kt a/k 1030.14-15
4 GÚ URUDU *lá-mu-num*, kt a/k 625.4
4 GÚ 14 *ma-na* URUDU *lá-mu-nàm*, RA 58, Sch. 2.1
2 *me-at* 60 *ma-na lá-mu-nu-um*, KUG 7.27-28
5 GÚ 20 *ma-na* [UR]UDU *lá-mu-num*, CCT 5, 46a.18-19
6 GÚ URUDU *lá-mu-nam*, kt a/k 977.13-14
6 GÚ 17 *ma-na* URUDU *lá-mu-nam*, Kayseri 1834.6-7
6 1/2 GÚ URUDU *lá-mu-nam*, KUG 49.3
7[?] GÚ 15 *ma-na* URUDU [*lá*]*-mu-nam*, kt a/k 1030.18-19

8 1/2 GÚ URUDU *lá-mu-nam*, kt a/k 976.5-6
10 GÚ [x+] 30 *ma-na* URUDU *lá-mu-nam*, kt a/k 1030.6-7
13 GÚ 20 *ma-na* URUDU *la-mu-nam*, kt a/k 1030.2-3
15 GÚ 52 *ma-na* URUDU *lá-mu-nim*, kt a/k 300.5-6, 14-15
23 GÚ 5 *ma-na*, kt 87/k 436.11-12 (by courtesy of K. Hecker)

***massuhum* - 'dirty'** (§ 3.1.3.2)

Used as qualification in:
URUDU SIG_5 *ma-sú-ha-am*, ATHE 38.25-26

Quantity:
12 *ma-na* URUDU SIG_5 *ma-sú-ha-am*, ATHE 38.25-26 (cf. K. Hecker, GKT, p.48 n.1, § 29b)

***ṣallāmum* - 'black'** (§ 3.1.3.3)

Price:
sale
110: *ší-tí* URUDU 18 [*ma-na* LÁ 10 GÍN] *ṣa-lá-mu-um* KB-[*áp-šu*] 10 LÁ 1/4 GÍN, TC 3, 107.17-19

Occurrences:
URUDU *ṣa-al-ma-am ki-iṣ-ra-am*, kt 91/k 422.22-23 (by courtesy of K.R. Veenhof)
15 *ma-na* URUDU *ṣa-la-mì-im*, kt 91/k 371.12 (by courtesy of K.R. Veenhof)
17 *ma-na* URUDU *ṣa-lá-ma-am*, TC 3, 107.3
18 [*ma-na* LÁ 10 GÍN] *ṣa-lá-mu-um*, TC 3, 107.17-18
50 *ma-na* URU[DU] *ṣa-lá-mì-im*, VS 26, 152.3'-4'
2 GÚ 40 *ma-na* URUDU *ṣa-lá-mì-im*, kt 91/k 514.5-6 (by courtesy of K.R. Veenhof)
3 GÚ URUDU *ṣa-lá-mu-um*, BIN 4, 31.14
2[!] *me-at* 44 *ma-na ṣa-lá-ma i-ba-ší*, VS 26, 6.9-11

***ša masā'im* - 'which has to be refined'** (§ 3.1.3.4)

Occurrences:
24 *pá-tá-li* URUDU *ša ma-sá-im*, I 792.3' (by courtesy of K. Hecker)
30 *ma-na* URUDU *ša ma-sá-im*, AKT 3, 48.1
11 GÚ URUDU *ša ma-sá-e-im*, Dalley 13.6

***hassu'um* - '?'** (§ 3.1.4.1)

Occurrence:
10 *ma-na* URUDU *ha-sú-a-am iš-tù A-lim*[ki] *ú-ša-ak-ší-/dam*, CCT 1, 42a.21-23

kiṣrum **- 'lump'** (§ 3.1.4.2)

Occurrence:

URUDU *ṣa-al-ma-am ki-iṣ-ra-am*, kt 91/k 422.22-23 (by courtesy of K.R. Veenhof)

ṣahhirum **- 'small-sized'** (§ 3.1.4.3)

Occurrences:

10 *ma-na* URUDU *ṣa-hi-ru-um*, TC 3, 178.8-9
[x+] 20 *ma-na* URUDU *ṣa-hi-ra-am*, kt m/k 147.87 (by courtesy of K. Hecker)
40 *ma-na* URUDU *ṣa-hi-ru-um*, CCT 1, 35.19-20
1 GÚ URUDU *ṣa-hi-ra-am*, kt 90/k 150.14 (by courtesy of C. Michel)
1 GÚ 1 *ma-na* URUDU *ṣa-hi-ri-im*, Pa 8.41-42
16 GÚ 30 *ma-na* URUDU *ṣa-hi-ru-um*, LB 1220.4-5
i-na ma-hi-ri-im ša-na-ú-um a-na URUDU *ṣa-hi-ri-im ip-šu-⌜ur⌝-ma*, Ka 435 r.3'-5'

šabburum **- 'broken'** (§ 3.1.4.4)

Used as qualification in:

57 *ma-na* URUDU *ša-bu-ra-am*, VS 26, 60.21-22
(URUDU SIG$_5$) *ša-bu-ur*, kt c/k 254.6
1 GÚ 18 *ma-na* URUDU SIG$_5$ *ša-bu-ru-um*, ATHE 37.18
<URUDU> *ša-bu-ra-am* SIG$_5$, kt 75/k 84.19 (by courtesy of C. Michel)
18 GÚ 45 *ma-na* URUDU *ma-sí-am ša-bu-ra-am*, TC 3, 10.10-11
55 *ma-na* URUDU *ma-sí-am da-mu-qám ša-bu-ra-am*, OIP 27, 56.33-34
4 *ma-na* URUDU SIG$_5$ *ša-bu-ru-um Ha-bu-ra-ta-i-a-am*, Kayseri 71/1.3-5 (by courtesy of K.R. Veenhof)
[x] GÚ ⌜16⌝ *ma-na* URUDU SIG$_5$ *Ha-bu-ra-ta-i-um ša-bu-ru-um*, Pa 36.1-3

ša šaduīšu **- 'native copper'** (§ 3.1.4.5)

Occurrences:

URUDU *ša ša-du-i*[!]*-/šu*, kt t/k 76+79 v.7 (Hecker 1993, p.287. The last sign but one is partly written on the edge of this writing exercise. It is given as 'BI' on the hand-copy, and Hecker read accordingly *ša-du-pí-šu*, but on the photo on plate 48 this sign looks more like 'I'.)
⌜2⌝ *ma-na* URUDU *ša ša-du-/i-šu ù* 10 *ma-na lu-lá-am a-na Kà-ni-iš šé-bi-lam*, kt a/k 265.30-32
3 *ma-na ša ša-du-i-šu*, OIP 27, 62.42

šikkum **- '?'** (§ 3.1.4.6)

Quality attested:

12 GÚ URUDU *ší-kam* SIG5, CCT 1, 7b.6

Prices:

purchase

140: KB 2 1/3 *ma-na*.TA *a-mu-ra-am ú-ta-e-ru*, ATHE 32.21-22 (probably *šikkum*)

equivalent

150 1/3: *ša* 1 GÚ 54 *ma-na* URUDU *ší-ki-im* (...) 2/3 *ma-na* 5 1/2 GÍN KB-*ap-šu*, kt c/k 47.14-16

136 1/3: 2 *me-at* URUDU *ší-ku-um* 1 1/3 *ma-na* 8 GÍN KB, Berytus 3, Pl. 17.1-2

In the following passages the copper probably was of *šikkum* quality:

c. 370: 40 *ma-na* URUDU *a-ba-an ma-tim* É *kà-ri-im ni-il5-qé* KB-*ap-šu* 6 1/2 GÍN, I 520.18-20 (by courtesy of K. Hecker)

285: 2 *me-at* 16 *ma-na i-na ṭup-pì-im pá-ni-im i-ṣú-ur-tim ra-bi-tim* 2/3 *ma-na* 5 1/2 GÍN KB É *kà-ri-im a-lá-qé*, kt c/k 459.12-17

266 2/3: textiles sold for 40 minas *aban mātim* a piece and 9 shekels of silver a piece, CCT 2, 24.21-29

sale

270: 1 GÚ 10 *ma-na* URUDU *ší-kam i-na* 4 1/2 *ma-na*.TA *a-dí-na-kum*, kt a/k 245.5-7

180: 1 GÍN (silver) *a-na* 3 *ma-na ší-ki-im*, TC 2, 33.8-9

150-140: *šu-ma* 2 1/3 *ma-na*.TA *dí-na šu-ma* 2 1/3 *ma-na*.TA *lá-šu* 2 1/2 *ma-na*.TA *dí-na*, kt n/k 98.22-26 (by courtesy of S. Bayram)

140: 2 1/3 *ma-na*.TA URUDU-*i-a-kà ší-kam ni-dí-in*, TC 1, 23.16-18

130: 2 *ma-na* 10 GÍN.TA, AKT 2, 39.24

130: 2 *ma-na* 10 GÍN.TA *ù* 1 GÍN.TA KB, kt m/k 5.16-17 (by courtesy of K. Hecker)

Notable quantities:

1 1/3 *ma-na* URUDU *ší-kam*, ATHE 18.14

10 *ma-na* URUDU *ší-kam*, AKT 1, 53.1

15 *ma-na* URUDU *ší-kum*, ICK 2, 125.8-9

29 *ma-na* URUDU *ší-kam a-ba-ma-tim*, kt c/k 459.7-8

51 1/2 *ma-na* URUDU *ší-kam*, Archaeology 24 (1978), p.181.6

1 GÚ 40 *ma-na* URUDU *ší-ku-um*, BIN 4, 31.7

1 *me-at* URUDU *ší-kum*, TC 1, 72.28

1 *me-at* 85 *ma-na* URUDU *ší-kam*, TC 3, 257.1-2

2 *me-at* URUDU *ší-ku-um*, Berytus 3, Pl. 17.1

6 GÚ URUDU *ší-kum*, CCT 5, 28b.8
12 GÚ URUDU *ší-kam* SIG5, CCT 1, 7b.6
8 *me-at* URUDU *ší-kam*, CCT 5, 28b.11-12
50 GÚ URUDU *ší-kam*, BIN 6, 94.4
4 *li-me* URUDU *ší-kum*, BIN 4, 160.1
73 GÚ 10 *ma-na* URUDU *ší-ki-im*, BIN 4, 199.10-11
98 GÚ 50 *ma-na* URUDU *ší-kam*, I 471.1-2 (Lewy 1964, p.183 n.3)

***Haburatāyum* - 'from Haburata'** (§ 3.1.5.1)

Qualities attested:
URUDU *Ha-bu-ra-ta-i*, TC 1, 55.4-5
URUDU *Ha-bu-ra-ta-i-am*, kt a/k 1254.2
URUDU [*Ha-bu*]-*ra-ta-i-am*, kt 75/k 89.1-2; in line 14 called [URU]DU SIG5 (by courtesy of C. Michel)
URUDU *Ha-bu-ra-ta-i-am*, LB 1295.5'-6'
URUDU SIG5 *Ha-bu-ra-ta-i-a-am*, JCS 14, 11.6
URUDU SIG5 *Ha-bu-ra-ta-i-um*, kt 86/k 176.3
15 GÚ URUDU SIG5 *mì-iš-lu-um ma-sí-um mì-iš-lu-um Ha-bu-ra-ta-i-um*, VS 26, 26.5-7
URUDU SIG5 *ša-bu-ru-um Ha-bu-ra-ta-i-a-am*, Kayseri 71/1.3-5
URUDU SIG5 *Ha-bu-ra-ta-i-um ša-bu-ru-um*, Pa 36.1-3
URUDU *Ha-bu-ra-ta-am* SIG5 *wa-at-ru-um*, Ka 276.3-5

Prices:
purchase
120: URUDU *Ha-bu-ra-ta-am* SIG5 *wa-at-ru-um ša-na-um i-za-az*, Ka 276.5

sale
90: 30 *ma-na* URUDU SIG5 *Ha-bu-ra-ta-i-um ù* 1/3 *ma-na* KB *ṣa-ru-pu-um*, kt 86/k 176.3-4
65: 1 GÚ *ša Ha-bu-ra-at* 1 *ma-na* 5 GÍN.TA KÙ.BE 5/6 *ma-na* 5 1/6 GÍN, AKT 2, 39.12-14
55: [ ] x *Ha-bu-ra-ta-i-am* 1 *ma-na* LÁ 5 GÍN.TA *ta-dí-in*, LB 1276.2'-3'
55: 1 GÚ *Ha-bu-ra-ta-i* KB-*áp*-<*šu*> 1 *ma-na* 5 1/2 GÍN, TC 1, 55.19-20

Quantities attested:
4 *ma-na* URUDU SIG5 *ša-bu-ru-um Ha-bu-ra-ta-i-a-am*, Kayseri 71/1.3-5 (by courtesy of K.R. Veenhof)
30 *ma-na* URUDU *Ha-bu-ra-ta-i-am*, LB 1295.5'-6'
30 *ma-na* URUDU SIG5 *Ha-bu-ra-ta-i-um*, kt 86/k 176.3
1 GÚ URUDU *Ha-bu-ra-ta-i*, TC 1, 55.4-5
1 GÚ *ša Ha-bu-ra-at*, AKT 2, 39.12

[ ] GÚ ⌜16⌝ *ma-na* URUDU SIG_5 *Ha-bu-ra-ta-i-um ša-bu-ru-um*, Pa 36.1-3
[2 GÚ 10 *ma*]-*na* URUDU [*Ha-bu*]-*ra-ta-i-am*, kt 75/k 89.1-2 (by courtesy of C. Michel)
3 GÚ LÁ 7 *ma-na* URUDU *Ha-bu-ra-ta-i-am*, kt a/k 1254.1-2
9 GÚ URUDU SIG_5 *Ha-bu-ra-ta-i-am*, kt 91/k 474.7-8 (by courtesy of K.R. Veenhof)
15 GÚ URUDU SIG_5 *mì-iš-lu-um ma-sí-um mì-iš-lu-um Ha-bu-ra-ta-i-um*, VS 26, 26.5-7
36 GÚ URUDU-*kà* SIG_5 [*Ha*]-*bu-ra-ta-i-um*, kt 87/k 453.6-7 (by courtesy of K. Hecker)

Ištunuwāyum **- 'from Ištanuwa'** (§ 3.1.5.2)

Attested in:
40 *ma-na* URUDU *Iš-tù-nu-a-i-am*, kt n/k 1131.1-2 (by courtesy of S. Çeçen)

Kunanamāyum **- 'from Kunanamit'** (§ 3.1.5.3)

Quantities attested:
30 *ma-na Ku-na-na-ma-i-um*, I 471.7 (Lewy 1964, p.183 n.3)
40 *ma-na Ku-na-na-ma*[!]-*i-ú-um*, TC 3, 209.23-24
1 GÚ 11 *ma-na*, AKT 3, 57.4-5
5 GÚ 40 *ma-na* URUDU *Ku-na-na-ma-i-am*, CCT 2, 23.10-11
10 GÚ *Ku-na-na-ma-i-um*, AKT 1, 17.25-26 (= URUDU 10 GÚ (...) *ina Kunanamit*, CCT 2, 23.36-37)
4 *li-mì-im* URUDU *Ku-na-na-ma-i-u*[*m*], kt u/k 4.18

ša Ninašša **- 'from Ninašša'** (§ 3.1.5.4)

Attested in:
53 *ma-na* URUDU *ša Ni-na-ša-a-a*, kt a/k 461/b.1-2 (the sign NI is written with 4 vertical wedges)

ša Purušhattim - 'from Purušhattum' (§ 3.1.5.5)

Attested in:
i-nu-mì 10 *li-me* URUDU *ša Pu-ru-uš-ha-tim* É.GAL-*lúm i-dí-nu-šu-ni*, CCT 2, 13.5-7
cf. *a-ša-me-ma* URUDU *i-na Pu-ru-uš-ha-tim ṭá-áb*, BIN 4, 64.12-13

ša māt Sawit - 'from the country of Sawit' (§ 3.1.5.6)

Attested in:
URUDU *za-ku-ú-um ša ma-at Ša-wi-it*, AKT 3, 56.37

Tapašattāyum - 'from Tapašatta' (§ 3.1.5.7)

Qualities attested:
URUDU *Ta-pá-ša-ta-i-am*, kt n/k 1952.3-4
URUDU *Ta-pá-ša-ta-i-am za-ak-wa-am*, kt 91/k 390.6-7 (by courtesy of K.R. Veenhof)

Quantity:
50 *ma-na* URUDU *Ta-pá-ša-ta-i-am za-ak-wa-am*, kt 91/k 390.6-7

Ta/iritarāyum, ša Ta/iritar - **'from Ta/iritar'** (§ 3.1.5.8)

Qualities attested:
URUDU *ša Ta-ri-tár*, BIN 6, 13.5-6
URUDU SIG_5 [*ša T*]*a-ri-ta-a*[*r*], ICK 2, 54.1
URUDU SIG_5 *ša Ta-ri-ta-ar*, TC 3, 211.35-36
URUDU SIG_5 *Ta-ri-ta-ra-i-am*, kt n/k 1385.18'

Prices:
purchase
150: *Ti-ri-ta-a-am* 2 1/2 *ma-na i-za-az* (...) *šu-ma* URUDU *Ti-ri-ta-ú-a-am* te_9*-pá-áš*, Ka 276.6-7

sale
45: *Ta-ri-ta-ra-i-um* 2/3 *ma-na* 5 GÍN.TA *ta-dí-in*, LB 1276.4'-6'

Quantities attested:
2 *ma-na* URUDU *Ta-ri-ta-ra-i-im*, kt n/k 1289.23-24 (by courtesy of S. Çeçen)
4 2/3 *ma-na* URUDU *ša Ta-ri-tár*, BIN 6, 13.5-6
10 *ma-na* URUDU SIG_5 *ša Ta-ri-ta-ar*, AKT 2, 26.4
10 *ma-na* URUDU SIG_5 *Ta-ri-ta-ra-i-am*, kt n/k 1831.4-5 (by courtesy of V. Donbaz)
16 *ma-na* URUDU SIG_5 *ša Ta-ri-tár*, AKT 2, 27.14-15
43 1/3 *ma-na* URUDU *Ta-ri-ta-ra-i-am*, kt a/k 822.6-7
1 GÚ 43 *ma-na* URUDU SIG_5 *ša Ta-ri-ta-ar*, TC 3, 211.35-36
2 GÚ URUDU *ša Ti-ri-ta-ar*, CCT 6, 19b.18
9 GÚ URUDU [*Ti-ri-ta-ra-i*]*-a-am*, CCT 6, 37a.4-5

***Tišmurnāyum, ša Tišmurna* - 'from Tišmurna'** (§ 3.1.5.9)

Qualities attested:
URUDU SIG_5, CCT 4, 27a.31
URUDU SIG_5 *Tí-iš-mu-ur-[na-i-am]*, ICK 2, 45.1-2
URUDU SIG_5 *Tí-iš-mu-ur-na-i-a-am*, JCS 14, 11.34-35
URUDU SIG_5 (...) (obtained) [*ina*] *Tí-iš-mar-na*, kt a/k 955.8-9
URUDU *ma-sí-am ša Tí-iš-mu-ur-na*, KTB 5.10-11

Prices:
purchase
164: 1 GÚ URUDU SIG_5 *i-na Tí-iš-mur-na* 2 2/3 *ma-na* 4 [GÍN.TA] *ú-kà-in*, CCT 4, 27a.31-33

Quantities attested:
URUDU SIG_5 20 *ma-[na i-na] Tí-iš-mar-na* (obtained), kt a/k 955.8-9
30 *ma-na* URUDU SIG_5 *Tí-iš-mu-ur-na-i-a-am*, JCS 14, 11.34-35
35 *ma-na* URUDU *ma-sí-am ša Tí-iš-mu-ur-⌜na⌝*, KTB 5.10-11
5 1/2 *ma-na* URUDU SIG_5 *Tí-iš-mu-ur-[na-i-am]*, ICK 2, 45.1-2
1 GÚ URUDU SIG_5 *i-na Tí-iš-mur-na* (obtained), CCT 4, 27a.31-32

***ša Tuhpia* - 'from Tuhpia'** (§ 3.1.5.10)

Quantities attested:
50 *ma-na* URUDU *ša Tù-ùh-pì-a*, kt 85/k 27.28
2 [GÚ] URUDU <*ša*> *Tù-ùh-pí-[a]*, KTS 2, 8.3
1 *me-at* URUDU *ša Tù-ùh-pì-a*, kt 85/k 27.24

Context:
kt 85/k 27.22-30: 1 TÚG *ku-ta-nam er-ba-am i-ší-a-am . a-na er-bi₄-šu* 1 *me-at* URUDU *ša Tù-ùh-pì-a a-dí-šu-um-ma . lá-qá-am lá i-mu-a-ma* É *Zu-me-a e-zi-ib-šu-ma . Šu-Be-lúm i-li-kam-/ma* 50 *ma-na* URUDU *ša Tù-ùh-pì-a* 30 *ma-na* URUDU SIG_5 . 2 *kà-ar-pá-/tim ša ša-am-ni-im a-dí-šu-um*, 'He brought me 1 *kutānum*-textile as a goodwill-gift and I gave him 100 (minas) of copper from Tuhpia which he refused to accept, and I left it behind in Zumea's house. When Šū-Bēlim came I gave him 50 minas of Tuhpia copper, 30 minas of fine copper, and 2 jars with oil' (Günbattı 1994).

***ša Wašhania* - 'of Wašhania'** (§ 3.1.5.11)

Quantities attested:
32 GÚ 30 *ma-na* URUDU *ša Wa-áš-ha-ni-a*, BIN 4, 113.6-7
[x GÚ URUDU *ša Wa-á*]*š-ha-ni-a*, kt m/k 131.1 (by courtesy of K. Hecker)

Scrap (§ 3.1.6)

hušā'ū

Quality attested:

šu-ma hu-ša-ú lu 30 *ma-na lu* 20 *ma-na* SIG5 *i-ša-qú-lu-ni-kum*, 'if they want to pay you either 30 or 20 minas of fine scrap,' I 611.6-7 (by courtesy of K. Hecker)

Price:

equivalent

75: 5 *ma-na hu-ša-e i-ša-am-ší tù-uṣ-ú ta-al-qé* KB-*áp-šu-nu* 4 GÍN, KUG 27.28-30

Quantities attested:

15 shekels, VS 26, 148.4
45 shekels, TC 1, 108.11
1 mina, I 785.7
2 minas, BIN 6, 175.16
3 1/2 minas, TC 3, 157.11
5 minas, AKT 3, 20.9; BIN 4, 133.1; I 705.22; ICK 2, 321.9; KUG 27.28; TTC 30.10
6 minas, VS 26, 129.3
10 minas, ICK 1, 7.13
11 minas, TC 1, 108.2
20 minas, I 434.15, 36; I 611.7; ICK 1, 55.7; kt n/k 524.11; OIP 27, 62.37
30 minas, I 611.7; kt n/k 524.34; TTC 30.18

Selected examples:

BIN 4, 133.1-6: 5 *ma-na hu-ša-e ù* 6 1/2 GÍN KB *a-qá-tí Š. a-dí-in*, 'I gave 5 minas of scrap (copper) and 6 1/2 shekels of silver for Š.'s expenses'

I 434.14-16: ... 40 *ma-na* URUDU SIG5 *ù* 20 *ma-na hu-sa-e* (by courtesy of K. Hecker)

ICK 1, 55.7-8: 40 *ma-na* URUDU SIG5 *ù* 20 *ma-na hu-sa-e ub-lu-šu-um*, 'they brought him 40 minas of fine copper and 20 minas of scrap (copper)'

ICK 2, 321.7-10: ... 5 *ma-na* AN.NA-*ak qá-tim* 5 *ma-na hu-ša-e* 11 GÍN NA4.GUG, '(I gave) 5 minas of hand-tin (and) 5 minas of scrap (copper) (and) 11 shekels of carnelian (to I.)'

kt n/k 524.11-12, 34-35: ... 20 *ma-na hu-ša*[!]*-e e-li-a-tim* 4 *ma-na ta-bu-ú-tim* (...) 30 *ma-na* [*hu-ša-e*] *e-li-a-tim*, '20 minas of scrap (copper) in the shape of *e*.-s, 4 minas of *t*.-ed (copper) 30 minas of scrap (copper) in the shape of *e*.-s' (by courtesy of C. Günbattı)

TC 1, 108.1-18: 2 GÚ 56 *ma-na* URUDU ŠÀ.BA 11 *ma-na hu-sa-ú* 1 ANŠE *ṣa-la-mu-um* 1/2 *ma-na* KB *a-na qá-ti-šu a-din* (...) 1 1/2 *ma-na* URUDU 2/3 *ma-na* 5 GÍN *hu-sa-ú* 2 *a-ni-qé-en ša-hi-re-en a-na ṣú-ha-ri-im . mì-ma a-nim A. a-na A-lim*[ki] *ú-bíl*, '2 talents 56 minas of copper, of which 11 minas are scrap. One black donkey. I gave him 1/2 mina of silver for his expenses (...) 1 1/2 minas of copper (and) 45 shekels of scrap (copper). Two rings (with matching?) *š.*-s for a boy, all this A. brought to the City.'

TC 3, 157.9-13: ... *i-na* [x] *ma-na hu-ša-e ša ta-dí-nu-šu-ni* 3 1/3 *ma-na hu-ša-e a-na ta-ší-a-tim il$_5$-qé-ma*, 'he took 3 1/2 minas from the x minas of scrap (copper) you had given him to pay for transport costs.'

VS 26, 129.1-6: 1/3 *ma-na* KB *ṣa-ru-pá-am* 6 *ma-na hu-ša-ú a-na qá-tí E. a-dí-in*, 'I gave 20 shekels of refined silver (and) 6 minas of scrap (copper) to E.'

ZI.IR

Attested in (both texts are a letter from Ilia and Lā-qēp to Iddi(n)-Ištar):

URUDU ZI.IR, BIN 4, 95.3

URUDU ZI.IR 2/3 *ma-na*.TA, JCS 41, 3.14

Price:

sale

40: URUDU ZI.IR 2/3 *ma-na*.TA, JCS 41, 3.14

tabbu'um (§ 3.5.1)

Attested in:

kt n/k 524.11-13: 2/3 *ma-na* KB 20 *ma-na hu-ša*[!]*-e e-li-a-tim* 4 *ma-na ta-bu-ú-tim a-na A-bi-tí-pá-an E. ú-bi-il$_5$-šum*, 'E. brought him in Abitipan (...) 40 shekels of silver, 20 minas of scrap (copper) in the shape of *e.*-s (and) 4 minas of *t.*-ed (copper)' (by courtesy of C. Günbattı)

the *verb* is attested in:

TC 3, 163.7-8: 1/3 GÍN KB *a-na* URUDU *ta-bu-e-im*

elītum

I 804.5-6: 2 *e-li-te$_9$-en a-na ta-ší-a-tim a-dì-in*, 'I gave 2 *e.*-s as payment for costs of transportation' (by courtesy of K. Hecker)

kt n/k 524.11-13, 31, 34-36: 2/3 *ma-na* KB 20 *ma-na hu-ša*[!]*-e e-li-a-tim* 4 *ma-na ta-bu-ú-tim a-na A-bi-tí-pá-an E. ú-bi-il$_5$-šum* (...) 30 *ma-na* AN.NA *qá-tim* (...) 30 *ma-na* [*hu-ša-e*] *e-li-a-tim* 1 *me-at* 20 *sà-a*[*m-ru-a*]*-t*[*im*] *a-na qá-tí-šu . a-dí-in*, 'E. brought him in Abitipan (...) 40 shekels of silver, 20 minas of scrap (copper) in the shape of *e.*-s, 4 minas of *t.*-ed (copper). (...) I gave him 30 minas of hand-tin (...) 30 minas of scrap (copper) in the shape of *e.*-s and 120 nails (of tin/copper)' (letter from Aššur-imittī and Imdilum to Pūšu-kēn and Uṣur-ša-Ištar; by courtesy of C. Günbattı)

TTC 30.9-12, 16-20, 22: ... 5 *ma-na* AN.NA 5 *ma-na hu-ša-e ù* 4 ANŠE.HI *ša* 15 GÍN.T[A] *a-na qá-tí I.* (...) 35 *ma-na* AN.NA*-ak qá-tim* 30 *ma-na hu-ša-e e-li-*

a-tim [x+] 4 *ma-na sà-am-ru-a-tim* (...) *a-na A. áp-qí-id*, '5 minas of tin, 5 minas of scrap (copper), and 4 donkeys (costing) 15 shekels (of silver) each, (I entrusted) to I. (...) I entrusted to A. (...) 35 minas of hand-tin, 30 minas of scrap (copper in the shape of) *e.*-s, x+4 minas of nails (...)' (letter to Imdilum, Ennam-Aššur, and Aššur-ṣulūlī, from Puzur-Aššur)

The word probably denotes a top-pack in KTH 18.17:

KTH 18.13-20: 2/3 *ma-na* KB *lu a-hu-ša-e lu a-na sà-he-er-tim a-dí-šu-nu-tí a-ha-ma* 1 GÍN KB *a-sà-he-er-tim* KÁ *ha-ra-nim a-dí-šu-nu-tí* 10 *ma-na sà-am-ru-a-tum* 1 *me-at mu-nu-sí-na* 10 *e-li-a-tim ù* 4 *mu-ta-tim a-ta-ší-a-tim i-dí-nu-ma a-dí Qá-ta-ra ta-ší-a-tim a-na-ku a-dí-in*, 'I gave them 40 shekels of silver (to buy) either scrap (copper) or small wares. I also gave them 1 shekel of silver for small wares when they were about to depart. They gave 10 minas of nails - 100 in number -, 10 top-packs and 4 half-packs for transportation, and I myself paid for the shipment as far as Qatara.'

***niggallum* - 'sickle'**

Quantities attested:

20 shekels, KUG 24.5
2 1/2 minas, KUG 24.1
4 minas, ICK 1, 120.8
5 minas, BIN 6, 227.11
15 minas, CCT 5, 28a.6
20 minas, CCT 1, 38a.9
30 minas, I 471.8; TC 3, 178.9
40 minas, VS 26, 19.10

Selected examples:

BIN 6, 145.6-7: 5 GÍN KB *pá-ni-um i-ší-im ni-ga-li i-ri-ih*, 'there remained the first? 5 shekels of silver from the proceeds of the sickles'

BIN 6, 227.11-13: 5 *ma-na ni-ga-li a-Ki-is-ha-nu-el a-dí-in*, 'I gave 5 minas (of copper in the shape of) sickles to Kishanuel'

I 471.6-9: 5 GÚ URUDU *ma-sí-um* 30 *ma-na Ku-na-na-ma-i-um* 30 *ma-na ni-ga-lu* 5 *ma-na ṣí-ba-ra-tim*, '5 talents of refined copper, 30 minas of copper from Kunanamit, 30 minas of copper) in the shape of sickles, 5 minas in the shape of *ṣ.*-s' (Lewy 1964, 183 n.3)

kt a/k 339.6: *ša* 2 GÍN KB *ni-ga-li a-dí-na-[kum]*, 'I gave you sickles for a value of 2 shekels of silver'

KTK 57.1-6: ⌜7⌝ GÍN *ni-ga-li* 1/3 *ma-na* UD!.KA!.BAR 30 *ṣí-ba-ra-tim I* DUMU *U ha-bu-ul*, 'I. son of U. owes me 7 shekels (of silver? for) sickles, 20 shekels (of silver? for) bronze(?) (and) 30 *ṣ.*-s'

TC 3, 178.8-10: 10 *ma-na* URUDU *ṣa-hi-ru-um* 30 *ma-na ni-ga-lu lá-bi-ru-tum*, '10 minas of small-sized copper and 30 minas of old sickles'

APPENDICES

APPENDIX 1
VALUE OF COMMODITIES EXPRESSED IN COPPER

A Metals and minerals
B Textiles
C Animals and slaves
D Agricultural products

m. = mina(s); sh. = shekel(s)

A Metals and minerals

silver

quality:	purchase:	equivalent:	sale:
Haburatāyum	120 (very fine)	–	90; 65; 55 (2x)
Taritarāyum	150	–	45
Tišmurnāyum	164 (fine)	–	–
masi'um	–	30 (punitive ratio)	71; 57; 55; 46; 45
dammuqum	80, 70	120; 90; 88 1/4; 82.5; 80; 60 (5x)	73; 70; 68; 60 (2x); 50
lammunum	–	–	80; 72(?)
ṣallāmum	–	–	110
hušā'ū	–	75	–
ZI.IR	–	–	40
šikkum	(140)	±370; (285); (266 2/3); 150 1/3; 136 1/3	270; 180; 150; 140 (2x); 130 (2x)

See for exchange rates of 'copper' the Catalogue, *werium*.

gold

1 *hiddum* bead = 1 talent of *šikkum* copper, TC 2, 33.10 (given to commissioner)
1 *hiddum* bead = 40 m. of copper, kt 91/k 490.20-21, 23 (by courtesy of K.R. Veenhof)
1 sh. of *pašallum* gold = 1276 sh. of (poor) copper, kt c/k 263.32 (purchase)
1 sh. of *pašallum* gold = 635 1/2 sh. of copper, kt n/k 506.14 (half of 1024 m. for 48 1/3 sh. of gold) (by courtesy of C. Günbattı)

tin

(x sh. of copper per sh. of tin)

30: 5 *ma-na* AN.NA *i-na li-bi* AN.NA-*ki-kà a-na Ha-na-nim a-na* 30 *ma-na*.T[A URU]DU *lá-mu-nim a-dí-šu-ma*, ATHE 38.13-18 (poor copper)

(16: *ší-im* 15 GÍN AN.NA 16 GÍN.TA, KUG 7.16-17 equivalent in silver?)

15: *a-na* [UR]UDU ⌜*ší*⌝-*ki-im* 15 *ma-na*.TA *a-d*[*í-m*]*a*, ATHE 32.15-16

13: 13 *ma-na*.TA *a-na a-ni-ki-kà*, kt n/k 234.3-4 (by courtesy of S. Bayram)

11: *i-nu-mì Sú-e-a i-li-kà-ni* AN.NA 9 *ma-na*.TA *ú* 10 *ma-na*.TA *i-ta-na-dí-nu a-na-ku* 11 *ma-na*.TA *ú-šé-bi-lá-kum*, 'when Suea came, they used to sell tin at 9:1 and 10:1. I have sent you (copper obtained by selling it) at 11:1' ATHE 38.6-11 (probably fine copper; in Turhumit)

11: *a-na* AN.NA-*ki-kà* 11 *ma-na*.TA, RA 60, 35.7

10 1/2: 10 1/2 *ma-na*.TA *a-na* SIG$_5$ *a-dí-in*, kt n/k 1484.7-8 (fine copper; by courtesy of S. Çeçen)

10: 10 *ma-na*.TA, kt n/k 99.7 (fine copper; by courtesy of S. Bayram)

10 or 9: ATHE 38.8, see above

7: *a-dí ša*! *Ha-ni-a-nim* 10 *ma-na* AN.NA *i-dí-na-ma* 7 *ma-na*.TA *a-na* AN.NA-*ki-šu ú-šé-bi*$_4$-*lá-šu-um*, AKT 2, 49.7-11

5 3/4: 3 *me-at* 43 *ma-na* URUDU *a*-1 GÚ-*tim ú-ṣa-ma*, VS 26, 6.6-7 (tin bought by a palace)

amūtum-mineral

a ring weighing 2 3/4 sh. = 2 talents of fine copper, ICK 1, 39.1-6

husārum - '*lapislazuli(?)*'

nearly 554 sh. of fine copper per shekel of *h.*:

head of a *tudittum* pin made of *h.* weighing 3 1/4 sh. is sold for 30 m. of fine copper, kt n/k 695.5 (by courtesy of C. Günbattı)

hiddum beads

1 *hiddum* bead of gold = 1 talent of *šikkum* copper, TC 2, 33.10 (given to commissioner)

1 *hiddum* bead of gold = 40 m. of copper, kt 91/k 490.20-21, 23 (by courtesy of K.R. Veenhof)

1 *hiddum* bead of silver = 12 m. of (*šikkum*) copper, kt u/k 3.17 (line 19: weighs 3 shekels of silver)

B Textiles

abarnium (Veenhof 1972, pp.156-158)

200 m. of *šikkum*: 2 *me-at* URUDU *ší-kam ší-im a-bar-ni-im* KI *Zu-li-a* GAL *ma-ṣa-ar-tim*, BIN 4, 160.4-5

išrum - 'woollen scarf' (Veenhof 1972, pp.176-178)

77 sh.: 20 1/2 *ma-na* 2 GÍN URUDU *a-na* 16 *iš-ra-tim a-Tal-ha-dí-e áš-qúl*, 'I paid 20 m. 32 sh. of copper to the men from Talhad for 16 *i.*'s' BIN 4, 180.6-8

kutānum - 'a textile' (Veenhof 1972, pp.145-151)

11 m. of fine copper: *ku-ta-nam a-na* 11 *ma-na* URUDU SIG_5 *a-na T* DUMU *Z A i-dí-in*, kt a/k 514.12-16

12 m. of refined copper: *ší-im* 50 *ku-ta-ni* 10 GÚ URUDU *ma-sí-um i-na li-bi E.* DUMU *A.*, kt n/k 87.1-4 (by courtesy of S. Bayram)

15 m. of refined copper: 15 *ma-na* [URUD]U *ma-sí-a-am ší-im* 1 *ku-ta-ni-a*, kt a/k 977.5-6

20 m. of fine copper: 1 GÚ URUDU SIG_5 *ší-im* 3 TÚG *ku-ta-ni-kà*, kt n/k 99.3-4 (by courtesy of S. Bayram)

30 m.: 10 *ku-ta-ni i-na* É *kà-ri-im na-ad-a-ku* 30 *ma-na*.TA 3 *me-at* URUDU *ší-im-šu-nu*, BIN 4, 160.19-21

30 m.: 33 TÚG *ku-t*[*a-ni*] (...) *a-na* 30 [+3] TÚG *ša* 30 *ma-na*.TA *tal-qé-ú* 10 GÍ[N.T]A KB *al-qé*, kt a/k 265.8-11

32 m.: 32 *ma-na a-na ku-ta-nim áš-*^{qú}*qúl-ma a-na qá-tí-kà a-na* É *kà-ri-im a-dí-i*, CCT 4, 23a.19-21

35 m.: kt n/k 506.11-13 (by courtesy of C. Günbattı)

40 m.: 32 *ku-ta-ni* 40 *ma-na*.TA *i-na Pu-ru-uš-ha-tim E-nu-Be-lúm a-na* É.GAL-*lim i-din*, CCT 2, 8.63-65

70 m. of fine copper: 3 GÚ 30 *ma-na* URUDU SIG_5 *ší-im* 3 TÚG *ku-ta-ni a-ha-ma*, I 432.25-26 (by courtesy of K. Hecker)

lubūšum paṣium - 'white garment' (Veenhof 1972, pp.163-165)

150 m.: 12 TÚG *lu-bu-ší-e pá-ṣí-ú-tim* 2 1/2 GÚ.TA *Š. i-dí-in-ma* URUDU *a-na Š.* DUMU *A. e-zi-ib*, ATHE 66.22-26

makūhum (Veenhof 1972, pp.169-170)

20 m.: 17 TÚG *ma-ku-hi ša P* 3 TÚG *ku-a-ú-tim* É.GAL-*lúm* 20 *ma-na*.TA *a-na ší-mì-im* il_5*-qé*, KTS 18.5-8

maškanum - 'tent' (here used for the packing of wool)

4 m.: 14 *maš-ku-nu* (...) *ší-im maš-ki-ni* 56 *ma-na* URUDU, kt n/k 463.5, 8-9 (by courtesy of C. Günbattı)

matlihšum - 'part of a pack-saddle' (Veenhof 1972, pp.5-6)

15 sh.: 2 *ma-na* 15 GÍN URUDU *ší-im* 9 *ma-at-li-ih-šu*, C 44.5-6

sapdinnum (Veenhof 1972, p.170)

5 m. 20 sh.: 5 1/3 *ma-na* URUDU *ša-ar-ša-ri-té-en*$_6$ *a-sà-ap-dí-nim*, BIN 6, 227.8-10

simply called *ṣubātum*, 'a textile'

15 m.: 1 TÚG *a-na* É *kà-ri-im a-na qá-tí-kà i-dí-ma* 15 *ma-na* URUDU *ú-ṣa-he-er*, kt a/k 303.9-11

18 m.: 18 *ma-na* URUDU *A. a-na B. i-da-an ú B.* TÚG *a-na A. i-dá-an*, AKT 3, 35.12-18

18 m. 30 sh.: 2 TÚG 37 *ma-na* URUDU *it-bu-lu*, BIN 4, 88.10-11 (purchase in Assur)

20 m.: *né-nu* 20 *ma-na*.TA TÚG.HI-*tí qá-tám a-na* É *kà-ri-im ni-ta-na-dí* 8 GÍN.TA *a-na* KB *ú-lá kà-ší-id*, AKT 3, 63.20-24

20 m.: *i-na* 70 TÚG.HI.A *ša i-na* É.GAL-*lim il*$_5$-*qé-ú-ni ša* 9 1/2 TÚG.HI.A 20 *ma-na*.TA URUDU É *kà-ri-im a-lá-qé*, BIN 4, 156.1-7

24 m. 42 1/2 sh.: 34 TÚG.HI.A (...) 6 GÚ 4 *ma-na* URUDU *it-bu-lu*, CCT 1, 39b.1-5 (purchase in Assur)

28 m.: 7 TÚG *a-na* 3 GÚ 16 1/2 *ma-na* URUDU *ma-sí-«a»-im ta-ad-nu*, LB 1235.6-9 (by courtesy of K.R. Veenhof)

30 m.: 30 *ma-na* URUDU *ší-im* TÚG-*tim ša* É *kà-ri-im i-Pu-ru-uš-ha-tim i-ni-id-ú a-na-ku al-qé* 10 1/3 GÍN KÙ-*ap-šu*, CCT 5, 38a.31-33

40 m.: 1 TÚG.TA *qá-tám i-dí-ú-ma qá-at-kà a-dí-ma* 40 *ma-na* URUDU *al-pu-ut-kà*, kt n/k 1464.8-10

ṣubātum dammuqum- 'a fine textile'

90 m.: *a-na* 1 TÚG SIG$_5$ *ša* 1 [G]Ú 30 *ma-na* URUDU *tal-qé-ú* 1/2 *ma-na* KB *al-qé*, kt a/k 265.11-13

ṣubātum ša qātim - 'a textile of normal quality'

12 and 13 m.: *i-na* 10 TÚG.HI-*tí ša qá-tim* (...) 9 TÚG.HI.A 13 *ma-na*.TA 1 TÚG *a-na* 12 *ma na ta-ad-nu*, kt 90/k 137.4-9 (poor copper?; by courtesy of C. Michel)

40 m.: 20 TÚG-*tù-kà ša qá-tim i-na* É.GAL-*lim i-le-qé*-⌈*ú*⌉ ⌈40⌉ *ma-na*.TA URUDU *i-na a-ba-an ma-tim*, CCT 2, 24.21-24 (probably poor copper)

šulupkā'um (Veenhof 1972, pp.165-166)

30 m. of refined copper: 30 *ma-na* URUDU *ma-sí-um ší-im* TÚG *šu-lu-up-kà-im*, kt n/k 118.12-14 (by courtesy of S. Bayram)

50 m. of fine copper: *šu-lu-up-kà-e-kà* 50 *ma-na*.TA URUDU SIG$_5$, RA 60, 35.8-9

šurum (Veenhof 1972, pp.154-156)

13 m.: 3 TÚG *šu-ru-tum a-na* 40 LÁ 1 *ma-na* URUDU *ta-ad-nu*, CCT 4, 23a.18-19

20 m.: 12 TÚG *šu-ru-tum a-na* 20 *ma-na*.TA *ni-dí-ma*, kt n/k 506.36-37 (by courtesy of C. Günbattı)

ukāpum - 'woollen saddle rug' (Veenhof 1972, p.7)

1 m.: 5 *ú-kà-pì té-zi-ba-ma a-na* 5 *ma-na* URUDU *a-dí-šu-nu-tí*, I 689.12-14 (by courtesy of K. Hecker)

C Animals and slaves

emārum - 'donkey'
7 m.: ANŠE 7 *ma-na ta-dí-in*, JCS 41, 3.33
7 m. 15+x sh.: 14 1/2 *ma-n[a] ší-im* 2 ANŠE, BIN 4, 214.8'-9'
15 m.: *a-na* 15 *ma-na* URUDU *ṣa-la-mì-im il$_5$-qé-šu*, kt 91/k 371.11-13 (by courtesy of K.R. Veenhof)
25 m.: 25 *ma-na* SIG$_5$ *ší-im* ANŠE *i-li-bi$_4$-a*, ATHE 37.22-23
26 m.: 1 GÚ 18 *ma-na* URUDU SIG$_5$ *ša-bu-ru-um ší-im* 3 ANŠE.HI.A *ša* DAM.GÀR *ta-qí-pu*, ATHE 37.18-20
27 m.: ANŠE-*kà a-na* 27 *ma-na* URUDU *ta-dí-in*, kt c/k 256 edge and left edge
35 m.: *ù* 35 *ma-na* URUDU *ší-im* ANŠE *i-li-bi$_4$-šu*, kt c/k 100.20-22
40 m.: 40 *ma-na* URUDU *ma-sí-am ší-im* ANŠE, kt 91/k 371.1-3 (by courtesy of K.R. Veenhof)

12 *sá-da-li* URUDU *ší-im* ANŠE, TC 2, 62.5-6

unclear: 5 GÚ 15 *ma-na ší-im* ANŠE *ṣa-lá-mì-im*, BIN 4, 148.14-15

emmerum - 'sheep'
(the exchange ratio between copper and silver was in this text 100:1)
1 m. 50 sh.: 1 UDU 1 5/6 *ma-na ša I-d[í-li]m*, KUG 27.35-36 (purchase)
2 m. 10 sh.: 1 UDU 2 *ma-na* 10? GÍN *ša A-šur*-SIPA, KUG 27.37-38 (purchase)
2 m. 20 sh.: 1 UDU 2 1/3 *ma-na ša Zu-a*, KUG 27.38 (purchase)
2 m. 25 sh.: 1 UDU 2 1/3 *ma-na* 5 GÍN, KUG 27.37 (purchase)
2 m. 40 sh.: 1 UDU 2 2/3 *ma-na ša Du-ni-a*, KUG 27.39 (purchase)
2 m. 45 sh.: 2 UDU *ša* DUMU *Pì-[d]a-a* 5 1/2 ma-na, KUG 27.35 (purchase)
3 m.: 1 UDU 3 *ma-na*, KUG 27.36 (purchase)
3 m. 20 sh.: 1 UDU 3 1/3 *ma-na* DUMU *Du-ni-a-ma* GAL, KUG 27.40 (purchase)
5 m. 30 sh.: 11 *ma-na* URUDU *ší-im* 2 UDU.HI.A, KUG 25.21-22 (purchase)

subrum - ± 'donkey'
12 m.: 12 *ma-na* URUDU *ší-im sú-ub-ri-im*, TC 3, 262.4-5
36 sh.: 2 *ma-na* URUDU *tù-šé-bi$_4$-lam* 5 *sú-bar-kà* 1/2 *ma-na* 6 GÍN.TA 3 *ma-na* URUDU *ta-ša-qal* 1 *ma-na* URUDU *a-na ša* ANŠE.HI.A *ša-bu-im . šé-bi$_4$-lam*, 'You sent 2 m. of copper, but your 5 *s.*-s are 36 sh. each, so you have to pay 3 m. of copper. Send 1 m. of copper to satisfy the donkey-seller,' BIN 4, 40.4-8
c. 47 minas: 8 *sú-ub-ra-am i-dí-nu-ma* 6 GÚ 17 *ma-na* URUDU *lá-mu-nam iš-ú-mu*, Kayseri 1834.5-7

amtum - 'slave girl'
43 *ma-na* URUDU *ší-im am-tim*, kt 91/k 490.37 (by courtesy of K.R. Veenhof)

wardum

2 slaves are given as a pledge for 2 talents 10 m. of *šikkum* copper in kt a/k 447

D Agricultural Products

aršātum - 'wheat'

1 DUG GIG = 1 m. of copper, L 39 (by courtesy of C. Zaccagnini). One jar (*karpatum*, DUG) contained 30 sila or litres, see Lewy 1964, p.193 note 4.

kirānum - 'wine'

1 *kukkubum*-jar with wine = 3 1/2 m. (of copper), kt 89/k 239.39-41 (by courtesy of Y. Kawasaki)

7 *kūtum*-jars sold for 200 m.: 7 *ku-a-tim ki-ra-na-am* (...) *ni-dí-šu-ma* 2 *me-at* URUDU *ni-il₅-qé*, kt a/k 1060.20-24

šāptum - 'wool'

1 m. of copper per 3 m. of wool:

Ka 435.5-7: *šu-lu-ša-ú-um a-na kà-ba-ri-im ú-kà-lu-ni*

1 m. of copper per 2 1/3 m. of wool:

kt a/k 230.8-10: *ší-im* 3 *me-at* SÍG.HI.A 1 *me-at* 37 *ma-na* URUDU *qá-sú al-qé*

1 m. of copper per 2 m. of wool:

CCT 4, 38a.12-14:*ú ša-áp-t[um a-na-kam]* 2 *ma-na*.TA [*a-na* URUDU] *ší-ki-im i-za-az*

CCT 6, 19b.14-16: ... *ša-pá-at-kà a-na i-ta-aṭ-lim ša-na-um ni-dí-in-ma* URUDU-*a-kà i-ba-ší*

Ka 435.13'-14': *ša-na-ú-um a-na* URUDU *ṣa-hi-ri-im ip-šu-⌈ur⌉-ma*

kt n/k 463.5-6: *ší-im* SÍG.HI.A *ša-na-ú-ma*, (by courtesy of C. Günbattı)

KTS 18.8-10: ... 4 *me-at* 20 *ma-na* URUDU *ma-as-am* É.GAL-*lúm a-na ší-mì-im ša-na-um il₅-qé-ma*

abazuhum - '?'

8 m.: 1 *a-pá-zu-ha a-na* 8 *ma-na* URUDU *a-dí-in*, KBo 28, 169.1-2 (in the subsequent lines *a*.-s were sold for 2 sh. of silver a piece)

APPENDIX 2
MOULDS FOUND AT KÜLTEPE

I For ingots
II For weapons
III For implements

L. = length; W. = width

I For ingots

for bars

- kt f/k 78; one broad side has an open mould for 5 bars; level Ib; Özgüç 1986, p.47, Pl. 92, 4.
- kt j/k 132; open mould for 4 bars; from a house; level Ib; Özgüç 1986, p.47, Pl. 91, 7b.
- kt k/k 169; one of the broad sides a bar, one of the narrow sides 7 short bars (?); from a house; Özgüç 1986, p.47, Pl. 91, 1b.
- kt t/k 73; open mould for 2 bars; from the house of an Assyrian merchant; level II; Özgüç 1986, p.47, Pl. 90, 5d.
- kt 81/k (no number); from a house; level Ib; Özgüç 1986, p.47.
- kt f/t 219; open mould with on one narrow side a bar; contemporary with level Ib; Özgüç 1986, p.47, Pl. 91, 4 (ingot).
- terracotta open mould on a stone base, L. 80 cm, W. 50 cm; from the level Ib workshop discovered in 1954, room 3; Özgüç 1986, p.41 and 1955b, Fig. 4.

Open moulds for the casting of bars were found in workshop L/14 (Özgüç 1986, p.42, Pl. 82, 1) and in workshop bb-cc/19-20 (Özgüç 1986, p.42, Pl. 83, 2 left of wall) as well.

for a disc-shaped ingot

- kt 82/k 157; open mould; room 1 of level II workshop; Özgüç 1986, p.43, Pl. 86, 3.

for an oval object

- kt 82/k 181; room 1 of level II workshop; Özgüç 1986, p.43.

for rectangular ingots

- kt 82/k 194; part of bivalve mould; room 2 of level II workshop; Özgüç 1986, p.44, Pl. 86, 4.
- kt 82/k 195; part of bivalve mould; room 2 of level II workshop; Özgüç 1986, p.44, Pl. 86, 5.

- kt 82/k 196; part of bivalve mould; room 2 of level II workshop; Özgüç 1986, p.44, Pl. 86, 6.
- kt 82/k 197; part of bivalve mould; room 2 of level II workshop; Özgüç 1986, p.44.
- kt 82/k 198; part of bivalve mould; room 2 of level II workshop; Özgüç 1986, p.44.
- kt 82/k 199; part of bivalve mould; room 2 of level II workshop; Özgüç 1986, p.44.

for ingot of uncertain shape

- kt f/t 219; open mould; contemporary with level Ib; Özgüç 1986, p.47, Pl. 92, 2 and 91, 6 (ingot).

II For weapons

for 'axes'

- kt 82/k 160; fragment of open mould with 4 sides; room 1 of level II workshop; Özgüç 1986, p.43.

for battle axes

- kt f/k 61; open mould; 1954 workshop from level Ib; Özgüç 1986, p.41, Pl. 80, 4.
- kt i/k 183; part of bivalve mould; from a house; level II; Özgüç 1986, p.47, Pl. 91, 2.

for fenestrated axes

- kt k/k 20; part of bivalve mould; 1959 workshop from level II; Özgüç 1986, p.46, Pl. 89, 4 and 90, 1 (ingot).
- kt 82/k 155; part of bivalve mould; room 1 of level II workshop; Özgüç 1986, p.43, Pl. 86, 2a-b.

for flat axes

- kt f/k 58; open mould; L. 25 cm; 1954 workshop from level Ib; Özgüç 1986, p.40, Pl. 79, 2 and 5 (ingot).
- kt f/k 58; open mould; 1954 workshop from level Ib; Özgüç 1986, p.40, Pl. A (mould with two ingots) and 79, 3 (on the left).

for lugged axes

- kt k/k 19; open mould; L. 7.5 cm; level II; Özgüç 1986, p.40, Pl. 79, 10a.
- kt f/k 58; open mould; 1954 workshop from level Ib; Özgüç 1986, p.40, Pl. 79, 3 and 7 (ingot).
- kt f/k 60; open mould; 1954 workshop from level Ib; Özgüç 1986, p.40, Pl. 79, 8 and 9.

for shaft-hole axes

- kt f/t 219; part of bivalve mould; found on the mound = level Ib; Özgüç 1986, p.45, Pl. 89, 1.
- kt t/k 75; part of bivalve mould; found in level II house of an Assyrian merchant; Özgüç 1986, p.45, Pl. 88, 8a-b.

- kt 82/k 154; part of bivalve mould; level II workshop; Özgüç 1986, p.43, Pl. 86, 1a-b.
- kt 82/k 200; part of bivalve mould; room 2 of level II workshop; Özgüç 1986, p.44, Pl. 87, 1a-b.
- kt 82/k 203 and 204; complete bivalve mould; room 2 of level II workshop; Özgüç 1986, p.44, Pl. 3a-b, c.
- kt 85/k 121; part of bivalve mould; level Ib; Özgüç 1986, p.92, Pl. 134, 2.

for daggers

- kt f/k 61; open mould; 1954 workshop from level Ib; Özgüç 1986, p.41, Pl. 80, 4.
- kt f/k 63; open mould; 1954 workshop from level Ib; Özgüç 1986, p.40, Pl. 80, 3.
- kt j/k 132; open mould; from a house; level Ib; Özgüç 1986, p.47, Pl. 91, 7a.
- kt 82/k 160; fragment of open mould with 4 sides; room 1 of level II workshop; Özgüç 1986, p.43.
- kt 82/k 201 and 202; complete bivalve mould; room 2 of level II workshop; Özgüç 1986, p.44, Pl. 87, 2a-b.

for a spearhead

- kt f/k 58; open mould; 1954 workshop from level Ib; Özgüç 1986, p.40, Pl. 79, 4 and 6 (ingot).

III For implements

for a chisel

- kt 81/k (no number); open mould; from a house; level Ib; Özgüç 1986, p.47, Pl. 91, 3.

for files

- kt k/k 19; open mould; L. 8 cm; level II; Özgüç 1986, p.40, Pl. 79, 10b.
- kt k/k 169; open mould; from a house; Özgüç 1986, p.47, Pl. 91, 1a.
- kt t/k 73; open mould with 3 out of 4 sides for files; from level II house of an Assyrian merchant; on one of the broad sides a rectangular rasp with lozenge grid (Özgüç 1986, p.47, Pl. 90, 5b); on one narrow side a rasp with zigzags (Pl. 90, 5a); on other narrow side rasp with lozenge grooves (Pl. 90, 5c).
- kt 82/k 97; open mould; from a house; Özgüç 1986, p.47, Pl. 90, 6.

for a 'hammer' (T-shaped object)

- kt f/k 63; open mould; L. c. 7.5 cm; 1954 workshop from level Ib; Özgüç 1986, p.41, Pl. 80, 8 and 81, 1 (ingot).
- kt ş/k 9; open mould; level Ib; Özgüç 1986, p.47, Pl. 92, 3.

for rings

- kt f/k 64; part of bivalve mould(?) for 2 rings; Özgüç 1986, Pl. 80, 5.
- kt i/k 185; open mould for 3 rings; from a house; level Ib; Özgüç 1986, p.47, Pl. 92, 1.

APPENDIX 3
COPPER AND BRONZE OBJECTS: TEXTUAL EVIDENCE

Objects listed in kt c/k 1517 and kt h/k 87 (see pp.76-77) and summarized there as 'bronze' are considered to consist of this alloy.

agannum ša qātim - 'hand bowl'; bronze; CAD A/I, p.142b (*agannu*, 'a large bowl')
  kt h/k 87.21-22: 1 *a-ga-nu-um . ša qá-tim* 1 *it-qú-ru-um ša qá-tim*
allum - 'hoe'; copper; CAD A/I, p.356b
  KTS 57c.9-11: 2/3 *ma-na* 5 GÍN AN.NA *a-na-pá-hi-im ša a-le-e ú-pu-šu*, '45 shekels of tin (as payment) for the smith who made the hoes'
allunātum - 'a tableware item (a knive?)'; bronze; CAD A/I, p.360b ('tongs(?)'); AHw 38a 2) 'Krebsfigürchen')
  CCT 4, 20a.4: 3 *a-lu-na-tum ša sí-pá-ri*, '3 bronze *a.*-s'
apšuhum - 'a knive'; bronze
  kt c/k 1517.2: 1 GÍR *ap-šu-hu-um*
arzallum - '?'; metal?; CAD A/II, p.324b (an implement)
  CCT 4, 20a.10: 4 *ar-za-lu*
  kt m/k 1.32-33: ... 1 *ma-na pá-šu-um ar-za-lu-um* (by courtesy of K. Hecker)
ašhalum - 'a vessel'; bronze; CAD A/II, p.428a (a box made of metal, wood or reed)
  kt h/k 87.18: 2 *áš-hu-lu* 2 *mu-lá-/tum*
haburrum - 'a vessel'; bronze; Balkan 1965, p.160: = *hubūru*
  kt h/k 87.9-10: 9 *ha-bu-ra-a-tum* ŠÀ.BA *sà-pu-um ša na-aṣ-bi4-tim*, '9 *h.*-s including a *s.*-bowl with handle'
haṣṣinnum - 'axe'; copper?; CAD H, p.133a
  CCT 4, 20a.9: 3 *ha-ṣí-nu*
  kt m/k 1.26: 2 *ha-ṣí-nu . ú sà-sú-šu-nu* (by courtesy of K. Hecker)
hubkum - 'a vessel?'; bronze
  kt h/k 87.12-13: 4 *hu-ub-ku-ú ra-bi4-ú-tum* 4 *hu-ub-ku-ú ṣa-hu-ru-tum*, '4 large and 4 small *h.*-s'
huruppum - 'a dish'; bronze; AHw, p.360b (1) 'eine Metallschüssel')
  kt c/k 1517.8: 6 *hu-ru-pá-tum*
hutūlum - 'a vessel?'; bronze; AHw, p.1562b
  ICK 2, 344.15: 4 *hu-tù-lá-tum*
  kt h/k 87.17: 5 *hu-tù-lá-tum*

ištittum - 'nails?'; bronze

kt a/k 527.10-14: ... 5 *ma-na* URUDU SIG_5 *ù* 77 *iš-tí-na-tim ša Wa-áš-ha-ni-a a-na E.* DUMU *A. a-na ší-a-ma-tim a-dí-in*, 'I gave 5 mins of fine copper and 77 nails of Wašhania to E. son of A. to make purchases with'

kt n/k 36.13-17: 1 GÍN KB *ší-im iš-tí-na-tí-kà ša* ZABAR *i-na Ha-tù-uš a-dí-šu* , 'I gave him 1 shekel of silver in Hattuš as proceeds of your bronze nails' (level Ib; by courtesy of V. Donbaz)

itqurum - 'ladle'; bronze; CAD I, p.300

CCT 4, 19b.7-12: ... *qá-áb-li-tám ša a-bi-ni zi-ra-am ù it-qú-ra-am pu-ṭu-ur-ma a-na Wa-ah-šu-ša-na a-ṣé-ri-a šé-bi-lam*, 'redeem the *q.*-container of our father, the cauldron, and the ladle, and send it to me in Wahšušana'

ICK 2, 344.4: 2 *it-qú-ra*-an* (collation by V. Donbaz)

kt n/k 228.24-27: *it-qú-ra-am ša sí-pá-r[i-im] šé-bi$_4$-lam ù zi-ra-am ša* 30 *ma-na ù-ul* 40 *ma-na šé-bi$_4$-lam*, 'send me a bronze ladle and also a cauldron of 30 or 40 minas weight' (by courtesy of S. Bayram)

kt 91/k 154.19-21: *it-qú-ru-um ša* 2 *ma-na* 10 GÍN *sí-pá-ri-im*, 'a ladle of 2 minas 10 shekels of bronze' (by courtesy of K.R. Veenhof)

itqurum ša qātim - 'spoon'; bronze; Balkan 1965, p.161: 'mit Stiel'

kt h/k 87.22: 1 *it-qú-ru-um ša qá-tim*

kalappum - 'axe'; bronze?; CAD K, p.66a ('an ax')

CCT 4, 20a.9: ... 1 *kà-lá-pu*

kāsum - 'cup'; metal and stone, here bronze; CAD K, p.253b

kt c/k 1517.11-13: 4 *kà-sá-tum* 2 *li-bu* . 1 *kà-sú-um ša Ší-mu-ri-im*

kt m/k 1.26-27: ... *kà-sà-tum ša qá-áb-li-a-tim* (by courtesy of K. Hecker)

kt t/k 76+79 V.4-9: URUDU DIŠ ZABAR GAL ZABAR URUDU *ša ša-du-i*[!]*-šu* URUDU *zi-ra-tum* URUDU *e-li-t[um]*, 'copper, bronze, a bronze cup, native copper, copper cauldrons, a copper *elītum*' (school text, Hecker 1993, p.287)

KUG 25.10-12: ... 1 *ma-na* 17 GÍN / KB GAL ZABAR *ša* 1 1/3 *ma-na* KB *ša-wi-ru*, '1 mina 17 shekels of silver, a bronze cup, silver rings worth 1 1/3 mina'

kulium - '?'; bronze?; CAD K, p.526b (*kūlu*, a metal object)

kt n/k 172.15-18: ... 23 *ma-na ša-nu-um ù ku-li-um šu-qú-ul-ta-šu-nu* É *nu-a-im nu-šé-ṣí*, 'we redeemed from the Anatolian creditor's house 23 minas (of bronze): the weight of a kettle and a *k*.' (by courtesy of S. Bayram)

kunakkium - '?'; bronze; (related with *kanakku*, AHw, p.434b?)

kt h/k 87.15: 5 *ku-na-ki-ú*

libbum - 'a cup'; bronze; CAD L 167b (c) 'a cup shaped like a heart')

kt c/k 1517.12: 2 *li-bu*

makraktum - '?'; bronze

kt c/k 1517.7: 9 *ma-ak-ra-kà-tum*

marrum - 'spade'; copper; CAD M/I, p.287a ('spade, shovel')

BIN 4, 10.29-30: URUDU *ša a-na ma*-ri ta-dí-nu-/šu-ni mì-ma lá i-dí-nam*, 'he gave me nothing of the copper that you had given him for the spades' (collation by K.R. Veenhof)

maslaqtum - 'a metal pot used for cooking'; copper; CAD M/I, p.325b

ICK 2, 344.2-3: 2 *ša-pì-il*$_5$*-ta-an ma-as-la*$^!$*-aq-ta-an* (collation by V. Donbaz confirmed L. Matouš's copy of line 3 as *ma-az-za-a-ak-ta-an*)

mašqaltum - 'measuring cup'; bronze; CAD M/I, p.382a (a container)

ICK 2, 344.9-12: [1 *ma-áš*]*-qá-al-/tum ša* 3 SÌLA 1 *ma-áš-qá-al-tum ša* 2 SÌLA 1* *ma-áš-qal-/tum ša S.*, '1 *m.* of 3 litres, 1 *m.* of 2 litres, 1 *m.* of S.' (collation by V. Donbaz)

kt c/k 1517.10: 10 *ma-áš-qá-lá-tum*

kt h/k 87.7-8: *maš-qal-tum ša* 2 SÌLA *maš-qal-tum ša* 1 SÌLA, 'a *m.* of 2 litres, a *m.* of 1 litre'

mazlagum - 'fork'; bronze?; CAD M/I, p.438b (a fork or hook)

CCT 4, 20a.5-6: *a-ša-lá-ší-šu šu-ga-ri-a-ú a-ša-lá-ší-šu ma-az-lu-/gu*$_5$, 'three *š.*-s, three forks'

OIP 27, 6.33-34: 10 *ma-na šu-qú-ul-tám ma-az-li-gi*$_5$ *ù šu-ga-ri-a-e*, '10 minas, the weight of forks and *š.*-s'

mulum - '?'; bronze

kt h/k 87.18: 2 *áš-hu-lu* 2 *mu-lá-/tum*

mušālum - 'mirror'; bronze; CAD M/II, p.256b

kt c/k 1517.15: 1 *mu-ša-lu-um*

kt 91/k 348.7-8: *mu-ša-lá-am ša* ZABAR *ù ma-az-li-gi*$_5$, 'a bronze mirror and forks' (by courtesy of K.R. Veenhof)

nakkapum - 'hammer'; bronze?; CAD N/I, p.186a

AKT 3, 80.37: *pá-ší na-ki-pí*, 'axes, hammers'

BIN 4, 198.1-3: ⌜2⌝ *pá-ší* 1 *na-ku-pu-um šu-qú-ul-ta-áš-nu* 2 *ma-na*, '2 axes (and) 1 hammer, weighing 2 minas'

napraṣum - 'chisel?'; bronze?; CAD N/I, p.313b; AHw, p.740b

VS 26, 12.13-17: *áb-na-am i-na ša-ra-mì-im ki-ma ša-ar-ma-at-ni-ma li-ik-šu-du-ší i-na na-áp-ri-ṣí-im lá i-na-dí-ú-ší*, 'when the stone (a block of *husārum* of 12 minas) is cut, they should receive it just as it has been cut; they must not work it with a *n.*'

niggallum - 'sickle'; copper; CAD N/II, p.213a; frequently used as a metal concentrator, see § 3.1.6

TC 3, 178.9-10: 30 *ma-na ni-ga-lu lá-bi-ru-tum*, '30 minas of old sickles'

VS 26, 19.10-12: 40 *ma-na ni-ga-lu* 2 1/2 *ma-na* AN.NA

pāšum - 'axe'; copper; AHw, p.846a

AKT 3, 80.37: *pá-ší na-ki-pí*, 'axes, hammers'

BIN 4, 198.1-7, 13-16: ⌜2⌝ *pá-ší* 1 *na-ku-pu-um šu-qú-ul-ta-áš-nu* 2 *ma-na a-na K. a-dí-in* 1 1/3$^!$ *ma-na* 5 GÍN *pá-ší a-na Z. a-dí-in* (...) 1 GÍN LÁ 7 1/2 ŠE KÙ.GI *ù* 2 *ma-na pá-ší a-na N. a-dí-in*, 'I gave 2 axes and 1 hammer - their weight is 2 minas - to K. I gave 1 mina 25 shekels in axes to Z. (...) 1 shekel minus 7 1/2 grains of gold and 2 minas in axes I gave to N.'

kt m/k 1.32: 1 *ma-na pá-šu-um* (by courtesy of K. Hecker)

RA 58, Sch. 4.17-22: 1 *pá-šu-um ša* 2 *ma-na* 1 *pá-šu-um ša* 2/3 *ma-na* 5 GÍN *Z. na-ší* 1 *šé-šé-ru-um ša* 5 *ma-na*, 'Z. is bringing 1 axe weighing 2 minas and 1 axe weighing 45 shekels; 1 saw weighing 5 minas'

patrum - 'dagger'; bronze; AHw, p.848b

TC 3, 86.11-13: *ú ur-kà-tám-ma pá-at-ra-am ta-dí-na-ma ú-ta-e-/ra-ku-šu*, 'and afterwards you gave me a dagger, which I have returned to you'

TC 3, 101.23-24: *pá-at-ri ša ú-ša-hi-zu šé-pì-iš*, 'have the daggers made concerning which I gave instructions'

ba-at-ra-am in CCT 1, 27a (= CCT 5, 48c).10-11: *ù* 1/2 *ma-na* URUDU *ba-at-ra-am* is with J. Lewy (1952, p.274) to be interpreted as *watrum*.

qablītum - '?'; copper; CAD Q, p.5b (a container)

BIN 4, 90.9-11: *a-ša-bi-šu qá-áb-li-a-tum ša* KB *ší-ta qá-áb-li-a-tum ša* URUDU, '7 silver *q.*-s and 2 copper *q.*-s'

ribum, 'a jar'; bronze; AHw, p.981a (*rību* IV, 'ein Gefäß')

kt c/k 1517.1: 1 *ri-bu-um*

sappum - 'a bowl'; bronze; CAD S, p.166b (a metal object or container)

kt a/k 479b.25: URUDU *ú sà-pè-e*

ICK 2, 344.14: 15 *sà-pu ša mu-sà-ri*, '15 bowls with metal band'

kt c/k 1517.5-6: 17 *sà-pu-ú ša mu-sà-ri*

kt h/k 87.14: 6 *sà-pu-ú ša mu-sà-ri*

kt h/k 87.9-10: 9 *ha-bu-ra-a-tum* ŠÀ.BA *sà-pu-um ša na-aṣ-bi$_4$-tim*, '9 *h.*-vessels among which a bowl with handle'

kt h/k 87.4: 1 *ša-pì-il$_5$-tum ša sà-pé-e*, '1 stand for a bowl'

kt c/k 1517.4-5: 10 *sà-pu-ú ša-ah-tù-tum*, '10 bowls without metal band'

kt h/k 87.19-20: 3 *sà-pu-ú ša-ah-tù-tum*

The translation of *sappū ša musārī* and *sappū šahtūtum* is based on the interptretation of *m.* as a functional or ornamental strip of metal attached to the bowl. These strips could be removed, leaving a 'stripped' (*šahtum*) bowl. K. Balkan (1965, p.161) interpreted *ša m.* as (a bowl) with and *šahtum* as one without inscription or engraving.

su'um - 'a disc-shaped object?'; bronze; CAD S, p.337a ('hammerstone')

kt m/k 1.25: 2 *ma-na sú-a-an ša* ZABAR, '2 minas (weight), two bronze *s.*-s' (by courtesy of K. Hecker)

The occurrence of a golden *su'um* as the proceeds of a *kutānum*-textile in TC 2, 54 LE.1 points to a piece of metal named after a specific form.

supannum - 'a container'; bronze; CAD S, p.391b

kt h/k 87.6: 3 *sú-pá-nu . ša Kà-ni-iš*[ki]

kt m/k 1.27-28: 2 *sú-pá-nu ša* 2 *ma-na*.TA . *ša* ZABAR, '2 bronze *s.*-s of 2 minas each' (by courtesy of K. Hecker)

CCT 2, 36a.10: 2 *sú-[pá]-na-an ša* ZABAR

CCT 3, 20.6-7: 1 *sú-pá-nam ša* ZABAR *ú it-qú-ra-am ša* ZABAR

CCT 4, 20a.8: 2 *sú-pá-nu . ma-mì-ri*

ṣurṣuppum - 'a container'; bronze; CAD Ṣ, p.261a ('a container provided with teat-shaped protuberances')

kt h/k 87.5: 2 *ṣú-ur-ṣú-pá-tum*

kt c/k 1517.3: 2 *ṣú-ur-ṣú-pá-an*

ICK 2, 344.1: 3 *ṣú-ur-ṣú-pá-tum*

šāhum - 'a vessel'; bronze; CAD Š/I, p.105b ('a drinking or cooking vessel, usually of metal')

ICK 2, 344.16: 3 *ša-ha-tim*

kt c/k 1517.8-9: 6 *hu-ru-pá-tum* 6 *ša-ha-tum*

kt h/k 87.11: 18 *ša-ha-tum*

šannum - 'a kettle'; copper, bronze; CAD Š/II, p.289a

AnOr 6, 15.22-23: (silver) ŠÀ.BA 2/3 *ma-na* 7 1/3 GÍN *ší-im ša-nim áš-qúl*, 'out of it I paid 47 1/3 shekels as the price of a kettle'

kt n/k 172.15-18: 23 *ma-na ša-nu-um ù ku-li-um šu-qú-ul-ta-šu-nu* É *nu-a-im nu-šé-ṣí*, 'we got back from the Anatolian creditor's house 23 minas (of bronze): the weight of a kettle and a *k*.' (by courtesy of S. Bayram)

kt 91/k 347.7-8: *ša-na-tim sí-pá-*⌜*ri*⌝, 'bronze kettles' (by courtesy of K.R. Veenhof)

KTS 33a.18-22: *ša-na-am . lu a-na Wa-ah-šu-ša-na-i-e lu a-na Kà-ni-ší-e a-na ig-ri . dí-na-ší-ma lu-ub-lu-ni-ší*, 'give the kettle to the men from Wahšušana or Kanish, and let them carry it here (by receiving) hire'

TMH 17a.1-3: 1/3 *ma-na* 5 1/4 GÍN *ší-im ša-nim ša* 20 LÁ 1 *ma-na*, '25 1/4 shekels (of silver), the price of a kettle of 19 minas (weight)'

šapiltum - 'stand'; bronze; CAD Š/I, p.451b ('lower or inner part')

kt h/k 87.1-2, 4: 10 *ša-áp-lá-tum ša tí-ra-*⌜*ni*⌝ 1 *ša-pí-il$_5$-tum ša-šu-ru-um* (...) 1 *ša-pì-il$_5$-tum ša sà-pé-e*, '10 grooved stands, 1 stand, a sieve (...) 1 stand for *s.*bowls'

ICK 2, 344.2-3: 2 *ša-pì-il$_5$-ta-an ma-as-la*[!]*-aq-ta-an*, '2 grinding slabs, 2 cooking pots' (CAD M/I, p.325a)

šaršarratum - 'chains'; copper; CAD Š/II, p.320b

BIN 6, 227.8-10: 5 1/3 *ma-na* URUDU *ša-ar-ša-ri-tí-/in a-sà-ap-tí-nim*, '5 1/3 minas of copper in the shape of two chains for a *saptinnum*-cloth'

ša(š)šānum - '?'; bronze?; CAD Š/II, p.174a

OIP 27, 62.37-38: 20 *ma-na hu-ša-*[*ú*] 1 *ša-ša-num* 7 *pá-šu* ⌜7⌝ *na-ku-*[*pu*], '20 minas of scrap (copper) and 1 *š*., 7 axes, 7 hammers'

šeššērum - 'saw'; bronze; CAD Š/II, p.174a

RA 58, Sch. 4.21-22: 1 *šé-šé-ru-um ša* 5 *ma-na*, '1 saw weighing 5 minas'

šugariāum - 'a small knive'; bronze?; CAD Š/III, p.196b

CCT 4, 20a.5-6: *a-ša-lá-ší-šu šu-ga-ri-a-ú a-ša-lá-ší-šu ma-az-lu-/gu$_5$*, '3 *š*.-s, 3 forks'

OIP 27, 6.33-34: 10 *ma-na šu-qú-ul-tám ma-az-li-gi$_5$ ù šu-ga-ri-a-e*, '10 minas, the weight of forks and *š*.-s'

TC 1, 78.1-3: 1 GÍN *a-na šu-ga-ri-a-/e ù* 2/3 *hu-ša-e a-na A. a-dí-in*, 'I gave 1 shekel (of silver) for *š*.-s and 2/3 (minas of) scrap (copper) to A.'

ušhium - 'a type of lamp'; bronze

kt c/k 1517.1: 1 *uš-hi-um*

kt h/k 87.3: 2 *uš-hi-ú ša bi-ṣí-ni*, '2 lampbowls for wicks'

zirum - 'cauldron'; copper, bronze

CCT 4, 20a.10-11: 1 *zi-ru ša* 20 *ma-na*, '1 cauldron of 20 minas'

CCT 3, 18b.4-6, 10: *zi-ra-am iš-tù* 20 *ma-na ú ša-áp-/li-iš ša Ma-ma-a* (...) *šé-bi₄-lá-nim*, 'send to me a cauldron of Mama-type,[565] from 20 minas or less'

kt h/k 87.26-28: 3 *zi-ra-tum ša* 30 *ma-na*.TA *mì-ší-it zi-ri-im ša hu-ur-ší-a*, '3 cauldrons of 30 minas each, the stock of cauldrons in my kitchen'

kt n/k 228.24-27: *it-qú-ra-am ša sí-pá-r*[*i-im*] *šé-bi₄-lam ù zi-ra-am ša* 30 *ma-na ù-ul* 40 *ma-na šé-bi₄-lam*, 'send me the bronze ladle and send me also a cauldron of 30 or 40 minas (weight)' (by courtesy of S. Bayram)

kt n/k 812.15-16: *zi-ra-am ša* 1 GÚ LÁ 2 1/3 *ma-na*, 'a cauldron of 57 2/3 minas' (by courtesy of E. Bilgiç)

kt t/k 76+79 V.4-7: URUDU DIŠ ZABAR GAL ZABAR URUDU *ša ša-du-i*[!]-*šu* URUDU *zi-ra-tum* URUDU *e-li-t*[*um*], 'copper, bronze, a bronze cup, native copper, copper cauldrons, a copper *elītum*' (school text, Hecker 1993, p.287)

CCT 4, 19b.6-12: *um-ma a-na-ku-ma qá-áb-li-tám ša a-bi-ni zi-ra-am ù it-qú-ra-am pu-ṭu-ur-ma a-na Wa-ah-šu-ša-na a-ṣé-ri-a šé-bi-lam*, 'I said: "Redeem our father's *q*., the cauldron, and the ladle, and send them to me in Wahšušana"'

kt 89/k 369.27-28: 1 *zi-ra-*[*am*] [*š*]*a* ZABAR (...), '1 bronze cauldron' (level Ib; Donbaz 1993, p.143)

kt 91/k 194.5: *zi-ra-am ša* 15 *ma-na*, 'a cauldron of 15 minas' (by courtesy of K.R. Veenhof)

zuršum - 'a cup'; bronze; CAD Z, p.167a; AHw, p.1115a *ṣuršu* 2)

ICK 2, 344.17: 2 *zu-ur-ša-*[*an*]

kt c/k 1517.14: 3 *zu-ur-ša-tum*

kt h/k 87.16: 2 *zu-ur-ša-tum*

565 Cf. Michel 1991a II, p.241. Add the references to Mama in lines 6 and 13 to RGTC 4, p.82.

APPENDIX 4
COPPER AND BRONZE OBJECTS: ARCHAEOLOGICAL EVIDENCE

I Weapons
II Implements
III Vessels
IV Jewellery, ornaments
V Ingots

Esin = analysis number according to Esin 1969.

I Weapons

Arrowheads

- kt h/k 113; level IV; Özgüç 1959, Pl. 69, 7-8; Esin 17646; 1.85%Sn; 0.89%As; ++Fe.
- kt h/k 107; level III; unpublished; Esin 17650; ~0.03%Sn; 0.51%As; ++Fe.
- kt h/k 111; level III; unpublished; Esin 17648; traces Sn; 0.48%As; ++Fe.
- level II; Özgüç 1959, p.109, Pl. 49, 4.
- kt h?/k 192; level II; unpublished; Esin 6815; 0%Sn; traces As; >0.5%Fe.
- level Ib; Özgüç 1953, p.195 Abb. 555. Length with shaft varies from 10-18 cm.
- level Ib; Özgüç 1959, p.109, Pl. 49, 5 and 6.
- kt p/k 135, kt k/k 38, found in some level Ib houses; Özgüç 1986, p.75, Pl. 129, 1-2 and 5-8.

Axes

- kt b/k 104, lugged axe from level Ib, court of House 3; L. 13 cm, W. 4.5 cm, Th. 3 cm; Özgüç 1953, p.195, Abb. 572; Esin 17639; ~0.01%Sn; 0.97%As; ++Fe.
- kt c/k 199; shaft-hole axe from level II tomb; Özgüç 1986, p.45, Pl. 89, 3a-b; Esin 6806; ~10%Sn; 2.7%As; >0.5%Fe.
- shaft-hole axe from a cist grave from level II; L. 18.3 cm, W. 2 cm; Özgüç 1959, p.110 and Özgüç 1955, 70, Fig. 41a-b and 43.
- kt f/k 393; shaft-hole axe found in a grave from level Ib; Özgüç 1959, p.110 and Özgüç 1955, p.70, Fig. 40a-b.
- kt 82/k 271; fragment of bronze shaft-hole axe; level II workshop; Özgüç 1986, p.50.
- kt 84/k 20, shaft-hole axe outside a looted cist grave from level Ib; 'Syrian import'; L. 11.8 cm, W. 3.2 cm, Th. 2.3 cm; Özgüç 1986, p.74, Pl. 128, 5.
- kt ş/k 65, crescentic axe from level Ib tomb, imported; 15x6.5 cm; Özgüç 1986, p.74, Fig. 59 and Pl. 128, 6.

- kt 83/k 201; fenestrated axe from level Ib cist-grave; L. 5,2 cm, W. 3 cm, Th. 1 cm; Özgüç 1986, p.46, Pl. 90, 4.
- fenestrated axe found in house of Adad-şulūlī's neighbour (level II); L. 8.9 cm, W. 7.6 cm, Th. 2 cm; Özgüç 1986, p.46, Pl. 89, 5.
- fenestrated axe found in a cist grave from level II; Özgüç 1959, p.109, Fig. 64 and Pl. 49, 1.

Daggers

- from a level II grave; Özgüç 1954, p.389, Abb. 27.
- from a level II grave; L. 18 cm; Özgüç 1954, p.389, Abb. 28.
- kt a!/k 105; from cist grave 3, level Ib; Özgüç 1950, p.201, Abb. 368; Esin 17728; ~0.03%Sn; 1.25%As; ++Fe.
- from a level II grave; Özgüç 1954, p.389, Abb. 26; Esin 6825; traces Sn; 0.71%As; >0.5%Fe.
- kt e/k 3; level II; Özgüç 1954, p.371, Fig. 27; Esin 6824; ~10%Sn; 1.15%As; 0.16%Fe.
- kt a/k 381; level Ia; Özgüç 1950, p.201, Abb. 378, L. 22 cm; Esin 6796 (level Ia); ~10%Sn; traces As; traces Fe.
- from cist grave 3, level Ib; Özgüç 1950, p.201, Abb. 367; Esin 6800; 0%Sn; traces As; 0.05%Fe.
- kt b/k 242; level I; L. 16.5 cm, W. 4 cm; Özgüç 1953, p.194, Abb. 530.
- kt b/k 140; level I; L. 19 cm, W. 5 cm; Özgüç 1953, p.194, Abb. 531.
- kt b/k 325; level I; L. 10.3 cm, W. 3 cm; Özgüç 1953, p.194, Abb. 535.
- kt b/k 292; level I; L. 14.5 cm, W. 3.2 cm; Özgüç 1953, p.194, Abb. 536.
- kt b/k 244; from pot grave 11, level I; L. 27 cm, W. 4.3 cm; Özgüç 1953, p.194, Abb. 589.
- kt e/k 101; level Ib; Özgüç 1954, p.389, Fig. 24; Esin 6830; 0.6%Sn; 2.15%As; >0.5%Fe.
- kt e/k 157; level Ib; Özgüç 1954, p.389, Fig. 25; Esin 6831; 0.12%Sn; 3.0%As; >0.5%Fe.
- kt h/k 215; level Ib; unpublished; Esin 17654; ~0.03%Sn; 0.55%As; ++Fe.
- kt n/k 115; unpublished; Esin 17647; ~10%Sn; 0.24%As; ++Fe.
- kt h?/k 108; Özgüç 1953, Fig. 56, 242??; Esin 17649; ~0.03%Sn; 0.48%As; ++Fe.
- kt i/t 291; unpublished; Esin 6790; 0%Sn; 0.58%As; >0.5%Fe.
- kt h/k 19; unpublished; Esin 17735; ~8.2%Sn; 1.1%As; ++Fe.
- Anitta's dagger; Özgüç 1956, pp.33-36 Fig. 2-3; Esin 6821; ~10%Sn; 0%As; 0.04%Fe.

Knives

- kt b/k 63; (level II?) Özgüç 1953, p.195, Abb. 553, L. 9 cm, W. 2.7 cm; Esin 6812; ~9.6%Sn; traces As; traces Fe.
- kt b/k 243; (level II?) Özgüç 1953, p.195, Abb. 554, L. 8.6 cm, W. 2.5 cm; Esin 6819; 0.055%Sn; 0.8%As; >0.5%Fe.
- with curved point; level Ib; Özgüç 1959, p.109, Fig. 62, Pl. 48, 2. L. 20.5 cm; Esin 17659; ~8.3%Sn; 1.7%As; ++Fe.

Lanceheads

- from cist grave 3; Özgüç 1950, p.200, Abb. 366.
- from cist grave 3; L. 12 cm.; Özgüç 1950, p.200, Abb. 369.
- from cist grave 2; L. 18 cm.; Özgüç 1950, p.200, Abb. 373.
- 2 lanceheads; Özgüç 1954, p.389, Abb. 29.
- kt a/k 674; level Ia; Özgüç 1950, p.200 Abb. 390; Esin 6797; 0%Sn; 0.34%As; >0.5%Fe.
- level Ia; Özgüç 1950, p.200, Abb. 374; Esin 6820; ~7.8%Sn; 0.73%As; 0.3%Fe.

Macehead

- kt e/k 6; level II; unpublished; Esin 6829; ~10%Sn; 1.4%As; >0.5%Fe.

Spearheads

- from level IV; L. 13.5 cm; Özgüç 1959, p.110, Pl. 50, 4.
- from level III; L. 19 cm; Özgüç 1959, p.110, Fig. 65 and Pl. 49, 9 and 10.
- from level II; L. 27 cm; Özgüç 1959, p.110, Fig. 67 and Pl. 49, 8.
- kt b/k 47; from level Ib, L. 6.8 cm, W. 2.4 cm; Özgüç 1953, p.195, Abb. 529.
- from level Ib; L. 15.5 cm; Özgüç 1959, p.110, Fig. 68 and Pl. 49, 7.
- from level Ib; Özgüç 1959, p.110, Fig. 69 and Pl. 48, 4; Fig. 66 and Pl. 48, 1; L. 29 cm; Fig. 70 and Pl. 49, 2; Fig. 71 and Pl. 49, 3; Fig. 72, 73, and 74.
- (Esin 6821: see Dagger of Anitta).

Swords

From the Old Palace on the citadel, contemporary with *kārum* level II:

- kt s/t 58; Preserved L. 41.7 cm, W. 5 cm; Özgüç 1986, p.75, Pl. 129, 4.
- kt s/t 59; Preserved L. 32.8 cm, W. 5.2 cm; Özgüç 1986, p.75, Pl. 129, 3.
- kt b/k 46; only half preserved; level Ib; L. 14 cm, W. 3 cm; Özgüç 1953, p.194, Abb. 532; Esin 6802; 0.33%Sn; 2.75%As; >0.5%Fe.

Tridents

- from level Ib, found together with knive with curled point (Özgüç 1959, Fig. 62). Has a tubular handle which fitted a long wooden shaft. L. 70.2 cm, W. 13 cm; Özgüç 1959, p.109, Fig. 63, Pl. 49, 11.
- kt h/k 14, from level Ib tomb; L. 29 cm, W. 11.5 cm; Özgüç 1986, p.75, Pl. 128, 8.
- kt ş/k 64, without central prong, from level Ib tomb; L. 30.6 cm, W. 10.1 cm; Özgüç 1986, p.75, Pl. 128, 7.

Two-pronged fork

- kt 83/k 292; from a house; L. 40 cm, W. 10 cm; Özgüç 1986, p.75, Pl. 128, 9.

II Implements

Awls

- kt a/k ...; ?; level II; Esin 6804; ~8.9%Sn; 0.8%As; >0.5%Fe.

- kt c/k 165; level II; unpublished; Esin 6811; 0.092%Sn; 0.92%As; ~ 0.5%Fe.
- kt e/k 52; level II; unpublished; Esin 6826; 0.079%Sn; 3.6%As; 0.022%Fe.
- kt e/k 102; level II; unpublished; Esin 6827; traces Sn; 1.3%As; >0.5%Fe.
- kt e/k 234; level I; unpublished; Esin 6833; traces Sn; 4.05%As; >0.5%Fe.

Chisels
- kt c/t 281; level III *tepe*; unpublished; Esin 6818; traces Sn; 0.71%As; 0.058%Fe.
- kt c/t 284; level III *tepe*; unpublished; Esin 6816; 0%Sn; traces As; ~0.5%Fe.
- kt b/k 35; level Ib; Özgüç 1953, p.195, Abb. 557, L. 6 cm.
- kt i/k 20; unpublished; Esin 6794; 0%Sn; 1.2%As; 0.096%Fe.
- kt i/t 305; unpublished; Esin 6788; 0.069%Sn; 1.6%As; traces Fe.

Nails
- kt a/k 183 or 381; level I; unpublished; Esin 6798; ~10%Sn; 0.88%As; >0.5%Fe.
- kt a/k 171; level II; Özgüç 1950, p.201, Abb. 389; Esin 6799; 0.45%Sn; 0.9%As; 0.17%Fe.
- kt b/k 304; unpublished; Esin 6805; ~5.1%Sn; 1.14%As; >0.5%Fe.

Needles
- kt b/k 324; Özgüç 1953, p.196.
- from level II; Özgüç 1955, p.69, Fig. 92.

Scraper
- kt a/k 71; from level Ib; unpublished; Esin 17655; 0%Sn; 0.25%As; ++Fe.

Sickles
- kt 82/k 272; bronze sickle; worn by frequent use; level II workshop; Özgüç 1986, p.50, Pl. 95, 8.
- kt b/k 105; in the court of house 3: a 'Hittite' bronze sickle; level Ib; L. 11.5 cm, W. 2.5 cm; Özgüç 1953, p.195, Abb. 537; Esin 6817; traces Sn; 1.35%As; 0.054%Fe.
- Özgüç 1954, p.371 Fig. 30; Esin 17656; 0%Sn; 0.59%As; ++Fe.
- 'sickle-shaped hand-weapon'; L. 16 cm, W. 2.4 cm; Özgüç 1955, p.80, Fig. 38a-b; Esin 17737; 0%Sn; 0.39%As; ++Fe.
- kt h/k 30; unpublished; Esin 17657; 0.05%Sn; 0.66%As; ++Fe.
- kt n/k 190; unpublished; Esin 17643; 0%Sn; 0.27%As; ++Fe.

III Vessels

Bowls, jars
- Metal Vessels no. 45; bronze bowl; H. 13.5 cm, D. at rim 24 cm; from a cist grave from level II; 19th century B.C.; Museum inventory no. 116-5-64 (This bowl might be the same as the bronze bowl published in Özgüç 1955, p.68, Fig. 22, which was found in 1950 in a level II cist grave).

- Metal Vessels no. 46; bronze bowl with base perforated with sieve holes. On body, below shoulder, is hole for threading string for suspension; H. 4.5 cm, D. at rim 9.5 cm; from a limestone cist; 18th century B.C.; Museum inventory no. 117-79-64.
- kt e/k 2; bronze jar with flutes on shoulder and near bottom; from level II grave; H. c. 9 cm; Özgüç 1954, p.388, Abb. 35.
- bronze bowl with basket handle attached to hook; level II(?) tomb; H. c. 5 cm, D. at rim 8 cm; Özgüç 1954, p.388, Abb. 36.
- 2 bowls with thick rim from cist grave 3: Özgüç 1950, p.200, Abb. 438 = 533, H. 8.5 cm, D. 25 cm; Abb. 439 = 535, D. 10.5 cm, H. 3.1 cm.
- kt b/k 250; bowl from cist grave 2; H. 6.5 cm, D. at rim 21 cm; Özgüç 1953, p.194, Abb. 456 and 573.
- kt b/k 245; bowl from pot grave 11; H. 6 cm, D. at rim 20 cm; Özgüç 1953, p.194, Abb. 457 and 574.
- kt h/k 15; unpublished fragment of rim; level Ib; Esin 17660; >10%Sn; 0.94%As; +Fe.
- kt t/k 34, copper bowl of omphalos type, level Ib; D. 13.7 cm; Özgüç 1986, p.73, Pl. 126, 1a-b.
- bowl, level Ib; Özgüç 1986, p.73, Fig. 55 and Pl. 126, 2.
- kt 78/k 252, a globular vase with cylindrical neck standing on a ring-base, level Ib; H. 7.7 cm, W. 7 cm; Özgüç 1986, p.73, Pl. 126, 3.
- vessel on a ring base, round-bodied with neck, level Ib; H. 9 cm, W. 10.6 cm; Özgüç 1986, p.73, Pl. 126, 4.
- kt f/k 384, two vases found inside each other, level Ib; Özgüç 1955, p.68, Fig. 20a-b, Özgüç 1986, p.73, Pl. 126, 5a-b: One (n?/k 129 ??) with an upper part consisting of a strainer with built-in sieve. Height 19 cm, sieve at depth of 7 cm from top; Özgüç 1955, p.68, Fig. 65; Esin 17645; 4.3%Sn; 0.32%As; +Fe; the other with ring-base and carinated neck and body; Özgüç 1955, Fig. 62.
- kt 84/k 160, jar with thick rim from level II tomb with basket handle, round body, on a ring base. Ends of the handle are hooks. H. 8.5 cm, D. 12.7 cm; Özgüç 1986, p.73, Fig. 58 and Pl. 127, 1.
- bowl with smooth base; Özgüç 1955, p.68, Fig. 59.
- bowl with ring base; Özgüç 1955, p.68, Fig. 18 and 58.
- bowl with ring base; Özgüç 1955, p.68, Fig. 19 and 60.

From Assur:

- jar Ass. 21809n; Grab 21; H. c. 9.5 cm.; Haller 1954, p.10, Pl. 10g.

Bucket with basket-handle

- bucket with basket-handle from Ib tomb, handle fastened to the body with rivets and hooks; Özgüç 1959, p.109, Fig. 60 and Pl. 48, 5. Height of hooks c. 6 cm.

Cauldrons

- kt f/k 1, with two riveted handles, found in level Ib tomb; Measures 33x33 cm; Özgüç 1955, p.68, Fig. 21 and 57a-c, and Özgüç 1986, p.73, Pl. 127, 2.

- kt 74/t 40, one of two cauldrons found on the *tepe* in a house of *kārum* level II type. Coarsely made of metal sheets joined by rivets. With two riveted band handles, showing signs of wear, repaired; H. 45 cm, W. max. 46.4 cm; Özgüç 1986, p.74, Pl. 127, 3a-b.
- kt 74/t 41, cauldron on ring-base with tall cylindrical neck and thick rim, with two handles riveted to the body; H. 32 cm, W. 36.4 cm; Özgüç 1986, p.74, Pl. 127, 4.

Cups
- cup with handle from cist grave 3; H. 8 cm; Özgüç 1950, p.200, Abb. 440 = 534.
- cup with handle; level Ib grave; Özgüç 1955, p.68, Fig. 17 and 61, 63.

Handles
- kt f/k 5; copper basket handle of wooden vessel; found in a level II tomb; L. 10.6 cm; Özgüç 1955, p.68, Fig. 68 and Özgüç 1986, p.73, Pl. 127, 5.
- kt f/k 35; copper basket handle of wooden vessel; found in a level II tomb; L. 8.2 cm; Özgüç 1955, p.68, Fig. 67 and Özgüç 1986, p.73, Pl. 127, 6.
- bucket with basket handle from a Ib tomb, handle fastened to the body with rivets and hooks; Özgüç 1959, p.109, Fig. 60 and Pl. 48, 5. Height of hooks c. 6 cm.
- kt n/k 131; unpublished handle, level Ib; Esin 17642; ~6.2%Sn; 1.2%As; ++Fe.
- kt h/k 112; handle, from B IV; Özgüç 1955, p.68, Fig. 67; Esin 17727; traces Sn; 1.4%As; ++Fe.
- kt h!/k 170; handle, level II; Özgüç 1955, p.68, Fig. 68; Esin 6813; ~10%Sn; 0.86%As; traces Fe.

Hooks
- kt b/k 97; level Ia, house 1; Özgüç 1953, p.196, Abb. 533. H. c. 5.5-6 cm. Esin 6832; ~6.2%Sn; 1.7%As; 0.036%Fe.
- Hooks in which a handle was placed were fastened to the body of a bucket from a level Ib tomb, the handle was fastened to the body with rivets and hooks; Özgüç 1959, p.109, Fig. 60 and Pl. 48, 5. Height of the hooks c. 6 cm.

Pans (cf. Calmeyer 1977)
- kt n/k 130; frying pan with a long handle found in a tomb from level II; Diameter at rim c. 7.2 cm, H. c. 3 cm, L. 15 cm (Özgüç 1986, p.73), total L. c. 28 cm; Özgüç 1959, p.109, Fig. 61 and Pl. 48, 6. Esin 17637; ~0.02%Sn; 0.84%As; +Fe.

From Assur:
- Ass. 20504ac; *Grab* 20; L. c. 52 cm, D. c. 25 cm; Haller 1954, p.10, Pl. 10b.
- Ass. 21410f; *Gruft* 21; L. c. 38 cm, D. c. 21.5 cm; Haller 1954, p.104, Pl. 21a.

IV Jewellery, ornaments

Bracelets
- kt c/k 197; level II; unpublished; Esin 6810; 0.53%Sn; 2.0%As; >0.5%Fe.

- piece of bracelet (four-edged half 'ring') from cist grave 3; Özgüç 1950, p.201, Abb. 370.
- kt b/k 7; a half ring-shaped object found in a level Ib house; Özgüç 1953, p.197, Abb. 551.
- kt h/k 22; level Ib; unpublished; Esin 17661; 0.07%Sn; 1.6%As; ++Fe.
- kt h/k 23; level Ib; unpublished; Esin 17662; 0.09%Sn; 2.0%As; ++Fe.
- kt h/k 66; level Ib; unpublished; Esin 17734; traces Sn; 0.31%As; ++Fe.
- kt h/k 67; level Ib; unpublished; Esin 17733; traces Sn; 0.28%As; ++Fe.
- kt h/k 68; level Ib; unpublished; Esin 17664; >10%Sn; 1.15%As; +Fe.
- kt h/k 69; level Ib; unpublished; Esin 17729; ~8.0%Sn; 0.81%As; ++Fe.
- kt h/k 91; level Ib; unpublished; Esin 17730; 4.0%Sn; 0.16%As; ++Fe.
- kt h/k 92; level Ib; unpublished; Esin 17731; 4.0%Sn; traces As; ++Fe.
- kt i/k 25; level Ib; unpublished; Esin 6789; 1.45%Sn; 2.9%As; 0.053%Fe.
- kt k/k 166; bronze bracelet wrapped in silver wire; level Ib; D. 6 cm; Özgüç 1986, p.30, Pl. 69, 4.
- kt n/k 132; level Ib; unpublished; Esin 17644; 4.3%Sn; 1.0%As; ++Fe.

Buckles

- kt f/k 8; level II, imported; Özgüç 1955, p.72, Fig. 37a-b; Esin 6823; ~10%Sn; traces As; 0.064%Fe.
- kt 84/k 141; cast in a mould; imported; H. 9.5 cm, W. 6.4 cm; Özgüç 1986, p.76, Pl. 124, 18.
- Ankara 12919; unpublished; Esin 6822; ~10%Sn; traces As; 0.12%Fe.

Cymbals

- kt d/k 112 and 113, a pair of cymbals pierced in the middle was found in the house of Peruwa; D. 12.8 and 13 cm; Özgüç 1953c, p.301, Fig. 14 and Özgüç 1986, p.74, Pl. 128, 3-4.
- kt 73/k 62 and 63, pair of cymbals from level Ib tomb; D. 6.4 and 6.2 cm; Özgüç 1986, p.74, Pl. 128, 1-2.

Figurines

- kt a/k 48; lion, unpublished surface find, Hittite?; Esin 6795; traces? Sn; ~4.0%As; >1%Fe.
- kt b/k 51; Head of bull, consisting of solid bronze; L. 2.2 cm, Th. 1.7 cm; Özgüç 1953, p.196, Abb. 571.
- kt b/k 251; In grave 14 statuette of a dog(?); Özgüç 1953, p.196, Abb. 575; L. 8.3 cm., H. 6.2 cm. Consists of a clay core with thin layer of bronze.

Pectorals

- kt b/k 247 and 248; a pair in pithos burial 5; with 4 holes in rim at regular interval; D. 8.3 cm, Th. 1.5 cm; Özgüç 1953, p.196, Abb. 540 (b/k 247) and 541 (b/k 248).
- kt b/k 249; a pair in grave 6; with 4 holes in rim at regular intervals; D. 8.3 cm, Th. 1.5 cm; Özgüç 1953, p.196, Abb. 539.

- kt j/k 162 and 163, two thin discs of bronze; D. 7 cm; Özgüç 1986, p.27, Pl. 66, 1-2.

Pins (*tudittum*)

Many pins for fastening clothes were found in graves. They are listed in Klein 1992, pp.339-340 ('Kültepe/Kaneš (Türkei')). Only those analysed by U. Esin are presented here.

- Fluted pins from level II tombs, imported, Özgüç 1959, p.110, Fig. 75, 76, and Pl. 50, 2 and 3.
- kt c/k 164 = Klein 20; level II; Özgüç 1955, Fig. 75; Klein Taf. 115/2; Esin 6814; >10%Sn; 0.67%As; 0.17%Fe.
- kt a/k 156?; level II; ?; Esin 6828; 4.8%Sn; 2.3%As; >0.5%Fe.
- kt h/k 26; level Ib; unpublished; Esin 17651; ~6.8%Sn; 1.3%As; ++Fe.
- kt h/k 27; level Ib; unpublished; Esin 17663; ~5.8%Sn; 0.96%As; +Fe.
- kt h/k 28; level Ib; unpublished; Esin 17732; 0.07%Sn; 1.25%As; ++Fe.
- kt h/k 90 = Klein 10; level Ib; L. 7.6 cm.; Özgüç 1955, Fig. 74; Esin 17653; ~7.6%Sn; 0%As; ++Fe; Özgüç 1953, Pl. 56, no. 549; Klein, Pl. 113/6.
- level Ib; Özgüç 1959, Fig. 50, 3??; Esin 17736; 4.3%Sn; 1.65%As; +Fe.
- h/k 29; level I; Özgüç 1959, ??; Esin 17652; ~5.0%Sn; 1.95%As; ++Fe.
- kt i/t 303; unpublished; Esin 6793; 0%Sn; 3.0%As; traces Fe.
- kt i/t 304; unpublished; Esin 6791; 0.56%Sn; traces As; traces Fe.
- kt i/t 308; unpublished; Esin 6792; 0.74%Sn; 2.95%As; 0.074%Fe.
- kt n/t 202; unpublished; Esin 17640; 0%Sn; 0.91%As; ++Fe.
- kt n/t 205; unpublished; Esin 17641; 0%Sn; 1.2%As; ++Fe.

Rings

Special rings (chain?):
- kt c/k 182; level II; unpublished; Esin 6807; >10%Sn; 1.35%As; 0.088%Fe.
- kt c/k 183; level II; unpublished; Esin 6808; >10%Sn; 1.15%As; 0.33%Fe.

Earring:
- Özgüç 1950, p.201, Abb. 380, 382.

Finger rings:
- Özgüç 1950, p.201, Abb. 386.
- kt b/k 49; Özgüç 1953, p.197, Abb. 561.
- kt b/k 266; Özgüç 1953, p.197, Abb. 560.

Esin 6809 (comprises Özgüç 1950, Abb. 380-387); 0.06%Sn; 0.77%As; >0.5%Fe.

Spools

three bronze spools found in level Ib tombs, probably imports
- kt c/k 184; H. 8.8 cm, W. 7.2 cm; Özgüç 1986, p.75, Pl. 124, 16.
- kt n/k 131; H. 9.5 cm, W. 5 cm; Özgüç 1986, p.76, Pl. 124, 17.
- kt v/k 179; H. 10.8 cm, W. 8.6 cm; Özgüç 1986, p.76, Pl. 96, 4a-b.

Stamp seal

- kt h/k 162; level Ib; unpublished; Esin 17658; ~0.04%Sn; 0.16%As; ++Fe.

V Ingots

Bars, rods

- nail-like rods: Özgüç 1953, p.196, Abb. 534 (kt b/k 48), 547 (kt b/k 310), 556, 548 (kt b/k 312), 549 (kt b/k 294), 552 (kt b/k 89) in majority from level I, rest from level II.
- kt b/k 296; rod-shaped; level I; Özgüç 1953, p.196; Esin 6801; 0.087%Sn; 1.15%As; >0.5Fe.

Disc

- kt b/k 246; a disc-shaped item found in pot grave 11 was interpreted as a weight by the excavator; D. 9.2 cm., Th. 1 cm; Özgüç 1953, p.197, Abb. 538.

APPENDIX 5
THE OLD ASSYRIAN WEIGHTS

The number between brackets after the excavation number is the number in the list in Özgüç 1986, pp.78-81; if an asterisk is added, bars are visible on the weight (not always on the photo). The weight in grams is given in the second column; the third gives the unit the weight-stone was used for with the aberration from the mean value of one shekel calculated on the basis of the 19 one-shekel weights: 1 še = 0.04587426 gr, 1 shekel = 8.257368 gr, and 1 mina = 495.44208 gr. The 'h' as material is haematite.

excavation-no.	grams	weight	length	thick	mater.	level
kt f/k 144 (10)	0.65	15 še, -5.5%	1.3	–	h	I B
kt 85/k 194 (73)	1.1	30 še, -20.1%	1.6	0.5	h	I A
kt 84/k 97 (63)*	1.1	30 še, -20.1%	2.3	0.4	h	II tomb
kt y/k 95 (38)	1.15	30 še, -16.4%	1.5	0.5	h	I B
kt 82/k 138 (56)*	1.3	30 še, -5.5%	2.5	0.4	h	surface, Pl. 131,41
kt 85/k 81 (72)	1.6	30 še, +16.3%	2.9	0.5	h	II
kt y/k 33 (35)*	1.8	45 še, -12.8%	2.9	0.4	h	surface, Pl. 131,20
kt 84/k 21 (59)*	1.8	45 še, -12.8%	2.2	0.5	h	I A, Pl. 131,38
kt y/k 93	1.9	45 še, -8%	1.3	0.9	h, duck	II, Pl. 132,1
kt 76/k 183 (49)	2.2	45 še, +6.6%	1.4	0.6	h	I B, Pl. 131,30
kt i/k 287 (19)	2.68	60 še, -2.6%	1.8	0.7	h	I B, Pl. 131,14
kt 84/k 96 (62)	2.8	60 še, +1.7%	2.7	0.5	h	II, tomb, Pl. 131,43
kt 84/k 150 (65)	2.8	60 še, +1.7%	2.5	0.6	h	II
kt y/k 94 (37)	2.85	60 še, +3.6%	2.1	0.7	h	I A
kt m/k 226 (21)	3.05	60 še, +10.8%	2.2	0.8	h	II
kt ş/k 55 (32)*	3.22	60 še, +17%	3.6	0.6	h	I B, Pl. 131,32
kt 82/t 22 (55)	3.3	60 še, +20%	3	0.6	h	II
kt 74/k (39)	3.7	90 še, -10.4%	2.2	0.8	h	I B
kt f/k 205 (11)*	3.8	90 še, -8%	2.8	0.8	h	II
kt o/k 155 (24)	3.95	90 še, -4.3%	1.8	0.8	h	II
kt 76/k 182 (48)	5.12	120 še, -7%	2.9	1.9	h	I B, Pl. 131,27
kt 76/k 179 (47)	5.6	120 še, +1.7%	2.4	1	h	I B, Pl. 131,29
kt 85/k 14 (71)	6.3	135 še, +1.7%	3.8	0.8	h	II
kt ı/k 58[566] (16)*	6.3	135 še, +1.7%	2.5	1	h	I B, Pl. 131,12
kt ./. 123 (67)	6.35	135 še, +2.5%	2.4	1	h	

566 Apparently wrong number, compare p.79 with p.81.

kt p/t 10 (25)	6.75	135 še, +9%	2.4	1	h	I B
kt 82/k 266 (57)	7.8	1, -5.5%	3.1	0.9	h	II, Pl. 131,36
kt ş/k 54 (31)*	8	1, -3.1%	3.3	0.9	h	II, Pl. 131,24
kt 82/k 1 (54)	8	1, -3.1%	3	0.9	h	I B, Pl. 131,34
kt e/k 280 (8)	8.05	1, -2.5%	2.8	1	h	II
kt g/k 114 (14)*	8.05	1, -2.5%	3.5	1	h	I B, Pl. 131,10
kt f/k 400 (12)	8.1	1, -1.9%	3.4	0.9	h	tomb I B, Pl. 131,8
kt n/k 256 (23)	8.1	1, -1.9%	3.1	0.9	h	I B
kt 81/k 41 (52)	8.1	1, -1.9%	3.4	0.8	h	tomb I B
kt 84/k 117 (64)	8.1	1, -1.9%	3.6	0.8	h	I B, Pl. 131,44
kt a/k 690 (1)	8.17	1, -1.1%	4	0.9	h	I B, Pl. 131,7
kt 83/k 109 (58)	8.2	1, -0.7%	4.1	0.8	h	I B, tomb
kt 84/k 77 (60)*	8.2	1, -0.7%	3.5	0.9	h	I A, Pl. 131,42
kt 74/k 8	8.22	1, -0.5%	2.2	1.4	h, duck	II, Pl. 132,1 ?
kt 84/k 80 (61)*	8.4	1, +1.7%	3.8	0.9	h	I B, tomb
kt 74/k 50 (41)*	8.45	1, +2.3%	4.9	0.8	h	I B, Pl. 131,25
kt g/k 44 (13)	8.5	1, +2.9%	3.3	1	h	I B
kt t/k 77 (33)	8.6	1, +4.2%	3.6	0.8	h	II
kt 76/k 178 (45)	8.85	1, +7.2%	3	0.9	h	I B
kt 85/k 79 (70)	9	1, +9%	3.1	1.3	h	I B
kt r/k 125 (29)	10.2	1 1/4?, -1.2%	3.5	1.5	rock crystal	IB, Pl. 131,18
kt ı/k 60 (17)	12.5	1.5, +0.9%	4.1	1	h	I A
kt 76/k 177 (46)*	13.2	1 2/3?, -4.1%	4.4	1.1	h	I B, Pl. 131,37
kt r/k 124 (28)	14.5	1 2/3?, +5.4%	4.2	1.6	rock crystal	Pl. 131,17
kt 85/k 193 (69)	16.1	2, -2.5%	5.5	1.1	h	I A
kt 85/k 80 (68)	16.6	2, +0.5%	4.4	1.2	h	II; IB
kt n/k 205 (22)	16.6	2, +0.5%	4.1	1.3	h	I A, Pl. 131,11
kt 76/k 175, 2 notches	19.4	2, +17.5%			lead	Pl. 130,5
kt e/k 50 (7)*	24.35	3, -1.7%	5.7	1.4	h	I B, Pl. 131,4
kt ş/k 53 (30)*	24.5	3, -1.1%	3.3	0.9	h	II, Pl. 131,23
kt v/k 74 (34)	24.5	3, -1.1%	4.9	1.5	stone	I B
kt 74/t 21 (40)*	24.62	3, -0.6%	4.4	1.5	h	I B, Pl. 131,33
kt 81/k 56 (53)	41	5, -0.7%	4.5	2	stone	II
kt y/k 37 (36)	41	5, -0.7%	4.9	2	stone	purchased
kt b/k 472 (6)	42	5, +1.7%	4.4	1.9	stone	I A
kt 76/k 178 (44)*	45	5, +9%	2.8	0.7	h	I B
kt j/k 152	50.75	6, +2.4%			lead	Pl. 130,1
kt j/k 153	59.95	7, +3.7%			lead	Pl. 130,2
kt f/k 131 (9)	62.7	8?, -5.1%	5.8	2	h	II
kt b/k 317 (3)	68	8, +2.9%	4.6	2.4	h	I B
kt u/k 95	80	10, -3.1%			lead	Pl. 130,7
kt b/k 323 (5)	81	10, -1.9%	6.6	2.2	stone	tomb
kt p/k 172 (27)	82.5	10, -0.1%	6	2.3	h	I B

kt b/k 322 (4)	89	10, +7.8%	5.3	2.5	h	tomb
kt ./. 122 (66)*	95	10, +15.1%	4.6	3	h	Pl. 131,46
kt 77/k 131 (50)	98.4	10?, +19.1% 12?, -0.7%	4.6	3	h	I B
kt 75/k 20 (42)*	109	15, -12%	8.5	3	h	II, Pl. 131,35
kt ı/k 58	124.4	15, +0.4%	3.4	4.2	stone	Pl. 132,3, worn
kt p/k 165 (26)	165	20, -0.1%	8.8	2.5	h	I B
kt 79/k 29 (51)	173	20, +4.8%	9.3	3.5	stone	I A
kt i/k 59	172.9	20, +4.7%	5.9	3.3	stone	Pl. 132,4, worn
kt i/k 185 (18)	178.5	20, +8.1%	6.3	3.6	h	II
kt ı/k 47 (15)*	180	20, +9%	6.9	3.2	h	II, Pl. 131,2
kt l/k 184 (20)	248	30, +0.1%	10	2.9	h	I B
kt 82/k 180	405	50, -1.9%			lead	Pl. 130,6
kt 75/k 21 (43)*	430	50, +4.2%	3.4	0.7	h	II, Pl. 131,31
kt f/k 103 (75)	475[567]	1 mina, -4.1%	6.4	8.1	stone	II, Pl. 132, 2
kt e/k 142	575	1 m?, +16.1% 1 1/4 m?, -7.2%			lead	Pl. 130,3
kt e/k 137	1012	2 m, +2.1%			lead	Pl. 130,4
kt n/k 244 (76)	4250	10 m?, -14.2%	18	13	stone	II
kt u/k 108 (74)	5250	10 m, +6%	21	12	stone, animal	I B, Pl. 119,6
kt b/k 271 (2)	–	—	3.7	1.3	h	I B, Pl. 131,1

Several haematite weights have been found in the *kārum* area of Boğazköy:

3732	3	60 še, +9%	duck, Boehmer 1979, p.55, Pl. 33
Bo 109/o	5.76	120 še, +4.6%	Bittel 1957, pp.31-32, Pl. 28,1
Bo 107/o	513.0	1 m, +3.5%	Bittel 1957, pp.31-32, Pl. 28,1

(see for the lead disc weighing 115.7 gr. p.88)

567 According to p.81 it weighs 440 gr.

MAP A

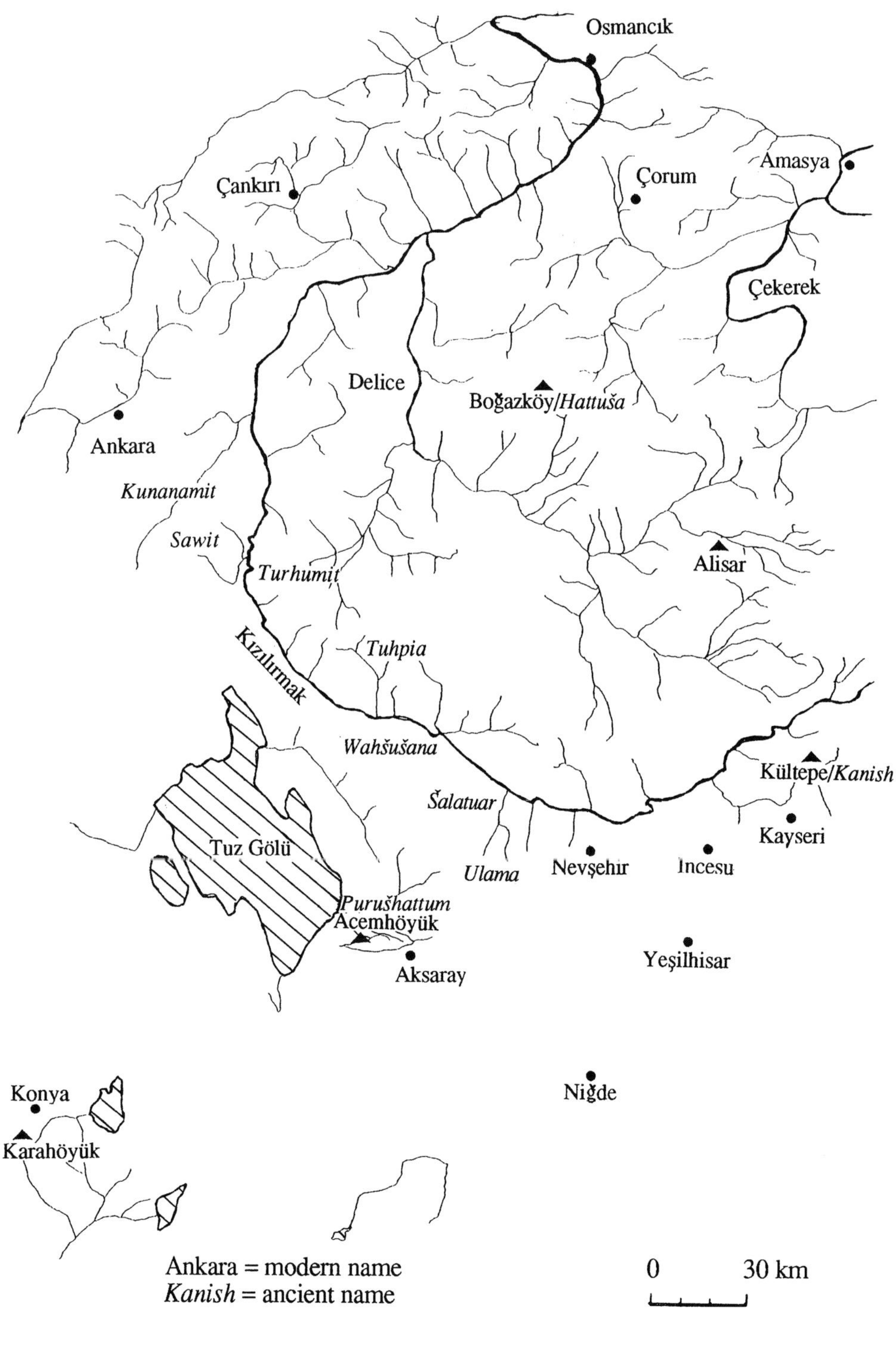

Map A: Geography of Old Assyrian activites in Anatolia

MAP B

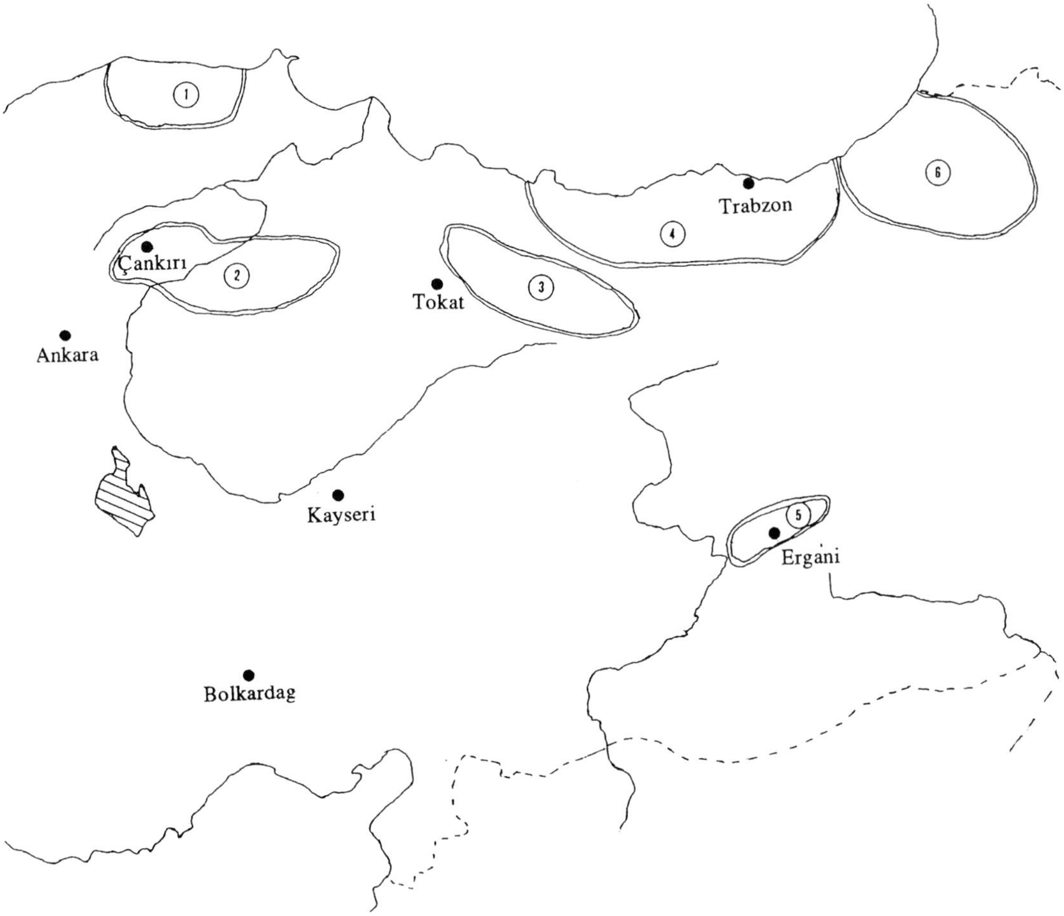

Areas of intensive exploitation in Antiquity as suggested by De Jesus (see pp.9-10):

1. Küre
2. Yapraklı
3. Pontic
4. Giresun-Trabzon
5. Ergani
6. Murgul-Kuvarshan

Map B: The main copper deposits in Turkey (after MTA 1972)

BIBLIOGRAPHY

The abbreviations used in this book are those of the CAD (*The Chicago Assyrian Dictionary*) and AHw (W. von Soden, *Akkadisches Handwörterbuch*)

Additional abbreviations are:
AbB 12 - W.H. van Soldt, *Altbabylonische Briefe Heft XII. Letters in the British Museum*, Leiden 1990.
AJA - *American Journal of Archaeology.*
AKT 1 - E. Bilgiç, H. Sever, C. Günbattı, and S. Bayram, *Ankara Kültepe Tabletleri (Ankaraner Kültepe-Tafeln) I*, Ankara 1990.
AKT 2 - E. Bilgiç and S. Bayram, *Ankara Kültepe Tabletleri II*, Ankara 1995.
AKT 3 - E. Bilgiç and C. Günbattı, *Ankaraner Kültepe-Texte III. Texte der Grabungskampagne 1970. Deutsche Übersetzung von Karl Hecker*, Stuttgart 1995.
AMM - *Anadolu Medeniyetleri Müzesi. Yıllığı*, Ankara.
DTCFD - *Ankara Üniversitesi, Dil ve Tarih-Cografya Fakültesi Dergisi*, Ankara.
JbRGZM - *Jahrbuch des Römisch-Germanischen Zentralmuseums Mainz*, Mainz.
MTA - Maden Tektit ve Arama Enstitüsü, Ankara.
N.A.B.U. - *Nouvelles Assyriologiques Brèves et Utilitaires*, Paris.
PACT - *Pact: revue du groupe d'études pour les techniques physiques, chimiques et mathématiques appliquées à l'archéologie*, Strasbourg.
RDAC - *Report of the Department of Antiquities Cyprus*, Nicosia.
RGTC 4 - Kh. Nashef, *Die Orts-und Gewässernamen der altassyrischen Zeit*, Wiesbaden 1991.
RGTC 6 - G.F. del Monte and J. Tischler, *Die Orts- under Gewässernamen der hethitischen Texte*, Wiesbaden 1978.
RGTC 6/2 - G.F. del Monte, *Die Orts- und Gewässernamen der hethitischen Texte. Supplement*, Wiesbaden 1992.
RlA - *Reallexikon der Assyriologie und Vorderasiatischen Archäologie*, Berlin.

Agricola - Georg Agricola, *De Re Metallica Libri XII, (Basel 1556), Zwölf Bücher vom Berg- und Hüttenwesen (...)*, Übersetzt und bearbeitet von Carl Schiffner unter Mitwirkung von Ernst Darmstaedter, Paul Knauth, Wilhelm Pieper, Friedrich Schumacher, Victor Tafel, Emil Treptow, Erich Wandhoff. Fünfte Auflage, Düsseldorf 1978.

Alp 1993 - S. Alp, Eine *kārum*-zeitliche Gußform und die Siegel von Karahöyük, *Istanbuler Mitteilungen* 43, 185-193.

Bachmann 1976 - H.G. Bachmann, Pontus - Studien zur Geschichte einer Metallprovinz, *Der Anschnitt* 28, 14-24.

Bachmann 1982 - H.G. Bachmann, *The Identification of Slags from Archaeological Sites*, London.

Balkan 1957 - K. Balkan, *Letter of King Anum-hirbi of Mama to King Warshama of Kanish*, Ankara.

Balkan 1965 - K. Balkan, review of L. Matouš (ed.), *Inscriptions cunéiformes du Kültépé. Vol. II*, *OLZ* 60, 146-162.

Balkan 1967 - K. Balkan, Contributions to the Understanding of the Idiom of the Old Assyrian Merchants of Kanish, *OrNS* 36, 393-415.

Balkan 1974 - K. Balkan, Cancellation of Debts in Cappadocian Tablets from Kültepe, in: K. Bittel, Ph.H.J. Houwink ten Cate, E. Reiner (eds.), *Anatolian Studies Presented to Hans Gustav Güterbock on the Occasion of his 65th Birthday*, Istanbul, 29-41.

Balkan 1979 - K. Balkan, '*makriš*' and '*ašiš*', Component-parts of Wagon and Ploughs Respectively, in a Cappadocian Tablet from Kültepe, in: *Florilegium Anatolicum. Mélanges offerts à Emmanuel Laroche*, Paris, 49-58.

Bamberger et al. 1986 - M. Bamberger, P. Wincierz, H.-G. Bachmann, B. Rothenberg, Ancient Smelting of Oxide Copper Ore. Archaeological Evidence at Timna and Experimental Approach, *Metall* 40, 1166-1174.

Bayram 1990 - S. Bayram, Kültepe tabletlerinde geçen yeni bir vâde ifâdesi ve çıkan neticeler, in: *X. Türk Tarih Kongresi; Ankara (22-26 Eylül 1986); Kongreye sunulan bildiriler, II. cilt*, Ankara, 453-462.

Berthoud et al. 1980 - T. Berthoud, S. Bonnefous, M. Dechoux, J. Françaix, Data Analysis. Towards a Model of Chemical Modification of Copper from Ores to Metal, in: P.T. Craddock (ed.), *Scientific Studies in Early Mining and Extractive Metallurgy. 19. International Symposium on Archaeometry and Archaeological Prospection*, London, 87-102.

Bilgiç 1945/51 - E. Bilgiç, Die Ortsnamen der 'kappadokischen' Urkunden im Rahmen der alten Sprachen Anatoliens, *AfO* 15, 1-37.

Bilgiç 1954 - E. Bilgiç, *Die einheimischen Appellativa der kappadokischen Texte und ihre Bedeutung für die anatolischen Sprachen*, Ankara.

Bilgiç 1964 - E. Bilgiç, Three Tablets from the City Mound of Kültepe, *Anatolia* 8, 145-163.

Bilgiç 1992 - E. Bilgiç, 'Ebla' in Cappadocian Inscriptions, in: H. Otten, H. Ertem, E. Akurgal, A. Süel (eds.), *Hittite and other Anatolian and Near Eastern Studies in Honour of Sedat Alp*, Ankara, 61-66.

Bittel 1957 - K. Bittel, R. Naumann, Th. Beran, R. Hachmann, G. Kurth, *Bogazköy III. Funde aus den Grabungen 1952-1955*, Berlin.

Boehmer 1979 - R.M. Boehmer, *Die Kleinfunde aus der Unterstadt von Bogazköy. Grabungskampagnen 1970-1978*, Berlin.

Brill and Wampler 1967 - R.H. Brill, J.M. Wampler, Isotope Studies of Ancient Lead, *AJA* 71, 63-77.

Calmeyer 1977 - P. Calmeyer, Das Grab eines Altassyrischen Kaufmanns, *Iraq* 39, 87-97.

Craddock and Meeks 1987 - P.T. Craddock, N.D. Meeks, Iron in Ancient Copper, *Archaeometry* 29, 137-204.

Crawford 1993 - H. Crawford, London-Bahrain Archaeological Expedition: Excavations at Saar 1991, *Arabian Archaeology and Epigraphy* 4, 1993, 1-19

Çambel 1974 - H. Çambel, The Southeast Anatolian Prehistoric Project and its Significance for Culture History, *Belleten* 38/151, 361-377.

Çeçen 1990 - S. Çeçen, Yeni Kültepe metinleri'ne göre yerli-Asurlu münâsebetleri, in: *Uluslararası 1. Hititoliji kongresi bildirileri (19-21 Temmuz 1990)*, 140-153.

Çeçen 1991 - S. Çeçen, Koloni devri Anadolu'sunda, tarihi ve sosyal olayların Asurlu tüccarlar tarafından târihleme olarak kullanılışı, *DTCFD* 35, 51-59.

Çeçen and Hecker 1995 - S. Çeçen and K. Hecker, *ina mātīka eblum*. Zu einem neuen Text zum Wegerecht in der Kültepe-Zeit, in: M. Dietrich and O. Loretz eds., *Vom Alten Orient zum Alten Testament. Festschrift für Wolfram Freiherrn von Soden zum 85. Geburtstag am 19. Juni 1993*, Neukirchen-Vluyn, 31-41.

Çukur and Kunç 1990 - A. Çukur and Ş. Kunç, Acem Höyük bakır buluntu analizleri, *V. Arkeometri sonuçları toplantısı* (Antalya, 18-23 Mayıs 1989), 33-39.

De Jesus 1978 - P. de Jesus, Metal Resources in Ancient Anatolia, *AnSt* 28, 97-102.

De Jesus 1980 - P.S. de Jesus, *The Development of Prehistoric Mining and Metallurgy in Anatolia*, Oxford.

Dercksen 1992 - J.G. Dercksen, review of C. Michel, *Innāya dans les tablettes paléo-assyriennes*, *BiOr* 49, 790-800.

Donbaz 1985a - V. Donbaz, New Evidence on the Reading of the Old Assyrian Month-name *kanwarta*; with an Edition of the Memorandum kt c/k 839, *JEOL* 28, 3-10.

Donbaz 1985b - V. Donbaz, More Old Assyrian Texts from Assur, *Akkadica* 42, 1-23.

Donbaz 1988a - V. Donbaz, The Business of Ašēd, an Anatolian Merchant, *AfO* 35, 48-63.

Donbaz 1988b - V. Donbaz, review of B. Kienast, *Das altassyrische Kaufvertragsrecht*, *AfO* 35, 190-191.

Donbaz 1993 - V. Donbaz, Some Remarkable Contracts of 1-B Period Kültepe Tablets II, in: M.J. Mellink, E. Porada, and T. Özgüç (eds.), *Aspects of Art and Iconography: Anatolia and its Neighbors. Studies in Honor of Nimet Özgüç*, Ankara, 131-154.

Donbaz and Veenhof 1985 - V. Donbaz and K.R. Veenhof, New Evidence for Some Old Assyrian Terms, *Anatolica* 12, 131-155.

Donbaz and Horasanlı 1976 - V. Donbaz and E. Horasanlı, Bursa arkeloji müzesinde bulunan Kültepe tabletleri, *Belleten* 40/158, 175-183

Eaton and McKerrell 1976 - E.R. Eaton, H. McKerrell, Near Eastern Alloying and Some Textual Evidence for the Early Use of Arsenical Copper, *World Archaeology* 8, 169-191.

Edzard 1989 - D.O. Edzard, Altassyrisch *nuwā'um*, in: K. Emre, B. Hrouda, M. Mellink, and N. Özgüç (eds.), *Anatolia and the Ancient Near East. Studies in Honor of Tahsin Özgüç*, Ankara, 107-109.

Eidem 1991 - J. Eidem, An Old Assyrian Treaty from Tell Leilan, in: D. Charpin and F. Joannès (eds.), *Marchands, diplomates et empereurs. Etudes sur la civilisation mésopotamienne offertes à Paul Garelli*, Paris, 185-207.

Elitzsch et al. 1983 - Ch. Elitzsch, E. Pernicka, G.A. Wagner, Thermoluminescence Dating of Archaeometallurgical Slags, *Pact* 9, 271-286.

Emre 1971 - K. Emre, *Anatolian Lead Figurines and their Stone Moulds*, Ankara.

Esin 1969 - U. Esin, *Kuantitatif spektral analiz yardımıyla Anadolu'da başlangıcından Asur kolonileri çasgına kadar bakır ve tunç madenciligi*, Istanbul.

Forlanini 1985 - M. Forlanini, Remarques géographiques sur les textes cappadociens, *Hethitica* 6, 45-67.

Forlanini 1987 - M. Forlanini, Toponymie antique d'origine hattie?, *Hethitica* 8, 105-122.

Forlanini 1990 - M. Forlanini, Uda, un cas probable d'homonymie, *Hethitica* 10, 109-127.

Frangipane et al. 1993 - M. Frangipane, H. Hauptmann, M. Liverani, P. Matthiae, M. Mellink (eds.), *Between the Rivers and over the Mountains; archaeologica anatolica et mesopotamica Alba Palmieri dedicata*, Rome.

French 1993 - D. French, Acemhöyük, Colonia Archelais and Road-knots, in: M.J. Mellink, E. Porada, and T. Özgüç (eds.), *Aspects of Art and Iconography: Anatolia and its Neighbors. Studies in Honor of Nimet Özgüç*, Ankara, 201-207.

Gale et al. 1985 - N.H. Gale, Z.A. Stos-Gale, G.R. Gilmore, Alloy Types and Copper Sources of Anatolian Copper Alloy Artefacts, *AnSt* 35, 143-173.

Garelli 1963 - P. Garelli, *Les Assyriens en Cappadoce*, Paris.

Garelli 1989 - P. Garelli, Le marché de Burušhattum, in: K. Emre, B. Hrouda, M. Mellink, and N. Özgüç (eds.), *Anatolia and the Ancient Near East. Studies in Honor of Tahsin Özgüç*, Ankara, 149-152.

Garstang and Gurney 1959 - J. Garstang, O.R. Gurney, *The Geography of the Hittite Empire*, London.

Gelb 1969 - I.J. Gelb, A Cappadocian Tablet Supposedly from the Neighborhood of Carchemish, *Athenaeum* 47, 119-124.

Giles and Kuijpers 1974 - D.L. Giles, E.P. Kuijpers, Stratiform Copper Deposit, Northern Anatolia, Turkey: Evidence for Early Bronze I (2800 B.C.) Mining Activity, *Science* 186, 29 November 1974, 823-825.

Gstrein and Lippert 1987 - P. Gstrein, A. Lippert, Untersuchung bronzezeitlicher Pingen am Hochmoos bei Bischofshofen, Salzburg, *Archaeologia Austriaca* 71, 89-100.

Günbattı 1987 - C. Günbattı, Yeniden işlenen bir Kültepe tableti (Kt. g/t 42+z/t 11), *Belleten* 51, 1-8 and Pl. 1-5.

Günbattı 1989 - C. Günbattı, Ev satışı ile ilgili beş Kültepe tableti ve bunlardan çıkan bazı sonuçlar, *Belleten* 53, 51-59 (with two plates).

Günbattı 1994 - C. Günbattı, Asur ticaret kolonileri devrinda Anadolu kralları ile Asurlu tüccarlar arasındaki ilişkilere ışık tutan iki yeni belge, congress hand-out with hand-copies of kt 85/k 27 (no. 1) and kt n/k 388 (no. 2).

Güterbock 1961 - H.G. Güterbock, The North-Central Area of Hittite Anatolia, *JNES* 20, 85-97.

Haller 1954 - A. Haller, *Die Gräber und Grüfte von Assur*, Berlin.

Hauptmann 1985 - A. Hauptmann, *5000 Jahre Kupfer in Oman. Band 1: Die Entwicklung der Kupfermetallurgie vom 3. Jahrtausend bis zur Neuzeit*, Bochum.

Hauptmann 1987 - A. Hauptmann, Kupfer und Bronzen der südostarabischen Halbinsel, *Der Anschnitt* 39, 1987, 209-218.

Hauptmann 1989 - A. Hauptmann, The earliest periods of copper metallurgy in Feinan, Jordan, in: Hauptmann et al. 1989, 119-135.

Hauptmann et al. 1985 - A. Hauptmann, G. Weisgerber, E.A. Knauf, Archäometallurgische und bergbauarchäologische Untersuchungen im Gebiet von Fenan, Wadi Arabah (Jordanien), *Der Anschnitt* 37, 163-195.

Hauptmann et al. 1989 - A. Hauptmann, E. Pernicka, G.A. Wagner (eds.), *Archäometallurgie der Alten Welt: Beiträge zum Internationalen Symposium 'Old World Archaeometallurgy', Heidelberg 1987.* Bochum.

Hauptmann et al. 1993 - A. Hauptmann. J. Lutz, E. Pernicka, Ü. Yalçın, Zur Technologie der frühesten Kupferverhüttung im östlichen Mitelmeerraum, in: M. Frangipane et al. 1993, 541-572.

Hauptmann and Roden 1988 - A. Hauptmann and C. Roden, Archäometallurgische Untersuchungen zur Kupferverhüttung der Frühen Bronzezeit in Fenan, Wadi Arabah, Jordanien, in: Symposion Archäometallurgie von Kupfer und Eisen in Westeuropa, *JbRGZM* 35, 1988 [1991], 510-516.

Hecker 1993a - K. Hecker, Schultexte vom Kültepe, in: M.J. Mellink, E. Porada, T. Özgüç (eds.), *Aspects of Art and Iconography: Anatolia and its Neighbors. Studies in Honor of Nimet Özgüç*, Ankara, 281-291.

Hecker 1993b - K. Hecker, Zur Herkunft der hethitischen Keilschrift, *Uluslararası 1. Hititoloji Kongresi Bildirileri (19-21 Temmuz 1990)*, 53-63.

Heimpel 1987 - W. Heimpel, Das Untere Meer, *ZA* 77 (1987), 22-91.

Heimpel 1994 - W. Heimpel, Towards an Understanding of the *siKKum* Institution, *RA* 88, 5-31.

Heinrich 1973 - E. Heinrich, 'Kaufmannshaus' und *bīt kārim*, in: E. Neu and Chr. Rüster (eds.), *Festschrift Heinrich Otten*, Wiesbaden, 89-98.

Heskel and Lamberg-Karlovsky 1980 - D. Heskel and C.C. Lamberg-Karlovsky, An Alternative Sequence for the Development of Metallurgy: Tepe Yahya, Iran, in: Th.A. Wertime and J.D. Muhly (eds.), *The Coming of the Age of Iron*, New Haven, 229-265.

Hoffner 1974- H.A. Hoffner, Jr., *Alimenta Hethaeorum; Food Production in Hittite Asia Minor*. New Haven.

Houwink ten Cate 1967 - Ph.H.J. Houwink ten Cate, Mursilis' North-Western Campaigns - A Commentary, *Anatolica* 1, 44-61.

Hrozny 1927 - F. Hrozny, Rapport préliminaire sur les fouilles tchécoslovaques du Kultépé, *Syria* 8, 1-12.

Hurtel and Tallon 1990 - L. Hurtel, F. Tallon, Le métal en provenance du Tell F 6. Description des objets et analyses, *Failaka: 1986-1988*, Lyon, 149-154.

Ichisar 1981 - M. Ichisar, *Les archives cappadociennes du marchand Imdilum*, Paris.

Joannès 1989 - F. Joannès, La culture matérielle à Mari (IV): les méthodes de pesée, *RA* 83, 113-152.

Joannès 1993 - F. Joannès, article 'Metalle und Metallurgie. A. I. In Mesopotamien', *RlA* 8, 96-112.

Kammenhuber 1977 - A. Kammenhuber, Die Arier im Vorderen Orient und die historischen Wohnsitze der Hurriter, *OrNS* 46, 129-144.

Kaptan 1986, E. Kaptan, Ancient Mining in the Tokat Province, *Anatolica* 13, 19-36.

Kaptan 1990, E. Kaptan, Findings related to the History of Mining in Turkey, *Bulletin of the Mineral Research and Exploration Institute*, (= MTA Bulletin) 111, 75-84.

Kienast 1984 - B. Kienast, *Das altassyrische Kaufvertragsrecht*, Stuttgart.

Klein 1992 - H. Klein, *Untersuchung zur Typologie bronzezeitlicher Nadeln in Mesopotamien und Syrien*, Saarbrücken.

Klengel 1989 - H. Klengel, Syrischer Handel und die Texte aus Kültepe/Kaniš, in: K. Emre, B. Hrouda, M. Mellink, and N. Özgüç (eds.), *Anatolia and the Ancient Near East. Studies in Honor of Tahsin Özgüç*, Ankara, 263-268.

Laroche 1966 - E. Laroche, *Les noms des Hittites*, Paris.

Larsen 1967 - M.T. Larsen, *Old Assyrian Caravan Procedures*, Istanbul.

Larsen 1976 - M.T. Larsen, *The Old Assyrian City-state and its Colonies*, Copenhagen.

Larsen 1977 - M.T. Larsen, Partnerships in the Old Assyrian trade, *Iraq* 39, 119-145.

Larsen 1982 - M.T. Larsen, Your Money or your Life! A Portrait of an Assyrian Businessman, in: J.N. Postgate (ed.), *Societies and Languages of the Ancient Near East. Studies in Honour of I.M. Diakonoff*, Warminster, 214-245.

Larsen 1987 - M.T. Larsen, Commercial Networks in the Ancient Near East, in: M. Rowlands, M. Larsen, K. Kristiansen (eds.), *Centre and Periphery in the Ancient World*, 47-56.

Larsen and Møller 1991 - M.T. Larsen and E. Møller, Five Old Assyrian texts, in: D. Charpin and F. Joannès (eds.), *Marchands, diplomates et empereurs. Etudes sur la civilisation mésopotamienne offertes à Paul Garelli*, Paris, 227-252.

Leemans 1960 - W.F. Leemans, *Foreign Trade in the Old Babylonian Period as revealed by Texts from Southern Mesopotamia*, Leiden.

Lewin and Hauptmann 1984 - B. Lewin, A. Hauptmann, *Kodō-zuroku, 'Illustrierte Abhandlung über die Verhüttung des Kupfers', 1801. Zur Geschichte der Kupfergewinnung in Japan*, Bochum.

Lewy 1947 - J. Lewy, Naram-Sin's Campaign to Anatolia in the Light of the Geographical Data of the Kültepe texts, *Halil Edhem hâtıra kitabı* 1, Ankara, 11-18.

Lewy 1950 - J. Lewy, Studies in Old Assyrian Grammar and Lexicography, *OrNS* 19, 1-36.

Lewy 1950b - J. Lewy, Hatta, Hattu, Hatti, Hattuša and 'Old Assyrian' Hattum, *ArOr* 18/3, 366-441.

Lewy 1952 - J. Lewy, Studies in the Historic Geography of the Ancient Near East, *OrNS* 21, 393-425.

Lewy 1955 - J. Lewy, Old Assyrian *husârum* and Sanchunyâtôn's Story about Chusor, IEJ 5, 154-162.

Lewy 1956 - J. Lewy, On some Institutions of the Old Assyrian Empire, *HUCA* 27, 1-79.

Lewy 1958 - J. Lewy, Some Aspects of Commercial Life in Assyria and Asia Minor in the Nineteenth Pre-Christian Century, *JAOS* 78, 89-101.

Lewy 1964 - H. Lewy, The Assload, the Sack, and other Measures of Capacity, *RSO* 39, 181-197.

Lechtman 1980 - H. Lechtman, The Central Andes: Metallurgy without Iron, in: T.A. Wertime and J.D. Muhly (eds.), *The Coming of the Age of Iron*, New Haven, 267-334.

Littauer and Crouwel 1979 - M.A. Littauer and J.H. Crouwel, *Wheeled Vehicles and Ridden Animals in the Ancient Near East*, Leiden.

Lorenzen 1965 - W. Lorenzen, *Helgoland und das früheste Kupfer des Nordens; Ein Beitrag zur Aufhellung der Anfänge der Metallurgie in Europa*, Otterndorf/Niederelbe.

Lutz et al. 1994 - J. Lutz, E. Pernicka, G.A. Wagner, Chalkolitische Kupferverhüttung in Murgul, Ostanatolien, in: R.-B. Wartke (ed.), *Handwerk und Technologie im Alten Orient*, Mainz, 59-66.

Maddin et al. 1980 - R. Maddin, T. Stech Wheeler, J.D. Muhly, Distinguishing Artefacts Made of Native Copper, *Journal of Archaeological Science* 7, 211-225.

Maddin 1989 - R. Maddin, The Copper and Tin Ingots from the Kaş Shipwreck, in: A. Hauptmann et al. 1989, 99-105.

Matouš 1957/58 - L. Matouš, Zwei 'kappadokische' Tontafeln im Nationalmuseum zu Kraków, *Journal of Juristic Papyrology* 11-12, 111-118.

Matouš 1969 - L. Matouš, Der Streit um den Nachlass des Puzur-Aššur, *ArOr* 37, 156-180.

McKerrell 1977 - H. McKerrell, The Use of Tin-bronze in Britain and the comparative Relationship with the Near East, in: A.D. Franklin (ed): *The Search for Ancient Tin*, Washington, 7-24.

Merkel 1990 - J.F. Merkel, Experimental Reconstruction of Bronze Age Copper Smelting based on archaeological Evidence from Timna, in: Rothenberg (ed.), *The Ancient Metallurgy of Copper* [= Researches in the Arabah 1959-1984 volume 2], London, 78-122.

Merrillees 1982 - R.S. Merrillees, Archaeological Symposium 'Early Metallurgy in Cyprus 4000-500 B.C.' Historical summary, in: J.D. Muhly, R. Maddin, V. Karageorghis (eds.), *Early Metallurgy in Cyprus, 4000-500 B.C.*, Nicosia, 373-376.

Metal Vessels - *Museum of Anatolian Civilizations: Metal Vessels*. Prepared by A. Toker and revised and edited by J. Öztürk, Ankara 1992.

Michel 1986 - C. Michel, Réédition des trente tablettes 'cappadociennes' de G. Contenau, *RA* 80, 105-140.

Michel 1991a - C. Michel, *Innāya dans les tablettes paléo-assyriennes*, Paris.

Michel 1991b - C. Michel, Durhumid, son commerce et ses marchands, in: D. Charpin and F. Joannès (eds.), *Marchands, diplomates et empereurs. Etudes sur la civilisation mésopotamienne offertes à Paul Garelli*, Paris, 253-273.

Michel 1992 - C. Michel, Transporteurs, responsables et propriétaires de convois dans les tablettes paléo-assyriennes. Réflexions sur les expresssions *šēp* NP et *ellat* NP, in: D. Charpin, F. Joannès (eds.), *La circulation des biens, des personnes et des idées dans le Proche-Orient ancien*, 137-156.

Michel and Foster 1989 - C. Michel and B.R. Foster, Trois textes paléo-assyriens de New York et les affaires confuses d'Iddin-Ištar, *JCS* 41, 34-56.

Moesta et al. 1989 - H. Moesta, R. Rüffler, G. Schnau-Roth, Zur Verfahrenstechnik der bronzezeitlichen Kupferhütten am Mitterberg. Mößbauer- und mikroskopische Studien, in: Hauptmann et al. 1989, 141-153.

Moosleitner 1988 - F. Moosleitner, Spangenbarrendepots aus Obereching, *Germania* 66, 29-67.

MTA 1972 - Maden Tektik ve Arama Enstitüsü, *Lead, Copper and Zinc Deposits of Turkey*, MTA Publication No. 133, Ankara.

Müller-Karpe 1989 - M. Müller-Karpe, Neue Forschungen zur frühen Metallverarbeitung in Mesopotamien, *JbRGZM* 36 [1992], 179-192.

Müller-Karpe 1993 - M. Müller-Karpe, *Metallgefäße im Iraq I* (*Von den Anfängen bis zur Akkad-Zeit*), Stuttgart.

Muhly 1973-1976 - J.D. Muhly, *Copper and Tin. The Distribution of Mineral Resources and the Nature of the Metals Trade in the Bronze Age*, New Haven. Supplement published in 1976.

Muhly 1980 - J.D. Muhly, The Bronze Age setting, in: T.A. Wertime and J.D. Muhly (eds.), *The Coming of the Age of Iron*, New Haven, 25-67.

Muhly 1983 - J.D. Muhly, article 'Kupfer. B. Archäologisch', *RlA* 6, 348-364.

Muhly 1984 - J.D. Muhly, Timna and King Solomon, *BiOr* 41, 275-292.

Muhly 1985 - J.D. Muhly, Lead Isotope Analysis and the problem of lead in copper, *RDAC* 1985, 78-82.

Muhly 1993a - J.D. Muhly, Early Bronze Age tin and the Taurus, *AJA* 97, 239-253.

Muhly 1993b - J.D. Muhly, article 'Metalle. B. Archäologisch', *RlA* 8, 119-136.

Muhly et al. 1990 - J.D. Muhly, F. Begemann, Ö. Öztunalı, E. Pernicka, S. Schmitt-Strecker, G.A. Wagner, The Bronze Metallurgy of Anatolia and the Question of Local Tin Sources, in: E. Pernicka (ed.), *Archaeometry '90*, Zürich, 209-220.

Nashef 1987 - Kh. Nashef, *Rekonstruktion der Reiserouten zur Zeit der altassyrischen Handelsniederlassungen*, Wiesbaden.

Neve 1965 - P. Neve, Die Grabungen auf Büyükkale im Jahre 1962, *MDOG* 95, 6-68.

Northover 1989 - J.P. Northover, Properties and Use of Arsenic-copper Alloys, in: Hauptmann et al. 1989, 111-118.

Özgüç 1950 - T. Özgüç, *Kültepe Kazısı Raporu* 1948, *Ausgrabungen in Kültepe. Bericht über die im Auftrag der Türkischen Historischen Gesellschaft, 1948 durchgeführten Ausgrabungen*, Ankara.

Özgüç 1953 - T. Özgüç und N. Özgüç, *Kültepe Kazısı Raporu 1949. Ausgrabungen in Kültepe. Bericht über die im Auftrage der Türkischen Historischen Gesellschaft, 1949 durchgeführten Ausgrabungen*, Ankara.

Özgüç 1953b - T. Özgüç, Vorläufiger Bericht über die Grabungen von 1950 in Kültepe ausgeführt im Auftrag des Türk Tarih Kurumu, *Belleten* 17, 109-118.

Özgüç 1953c - N. Özgüç, Preliminary Report on the 1951 Excavations at Kültepe, *Belleten* 17, 289-305.

Özgüç 1954 - T. Özgüç, Die Grabungen von 1953 in Kültepe, *Belleten* 18, 373-390.

Özgüç 1955a - T. Özgüç, Excavations at Kültepe 1954 - Finds of Level Ib, *Belleten* 19, 64-72.

Özgüç 1955b - T. Özgüç, Report on a Work-shop belonging to the Late Phase of the Colony Period (Ib), *Belleten* 19, 77-88.

Özgüç 1956 - T. Özgüç, The Dagger of Anitta, *Belleten* 20, 33-36.

Özgüç 1959 - T. Özgüç, *Kültepe-Kanish. New Researches at the Center of the Assyrian Trade Colonies*, Ankara.

Özgüç 1965 - N. Özgüç, *The Anatolian Group of Cylinder Seal Impressions from Kültepe*, Ankara.

Özgüç 1986 - T. Özgüç, *Kültepe-Kaniş II. New researches at the trading center of the Ancient Near East*, Ankara.

Öztan 1979 - A. Öztan, Two Stone Plates from Sarıkaya Palace at Acemhöyük, *Belleten* 170, 381-388.

Oppenheim 1938 - A.L. Oppenheim, review of J. Lewy (ed.), TC 3/I, *OrNS* 7, 132-134.

Otten 1959 - H. Otten, Ritual bei Erneuerung von Kultsymbolen hethitischer Schutzgottheiten, *Festschrift Johannes Friedrich* , Heidelberg, 351-359.

Otto 1973 - H. Otto, Über die Auswertung der Metallanalysen in Band 1 und 2 der 'Studien zu den Anfängen der Metallurgie' (SAM), *Acta Praehistorica et Archaeologica* 4, 1-9.

Palmieri et al. 1993 - A.M. Palmieri, K. Sertok, E. Chernykh, From Arslantepe Metalwork to Arsenical Copper Technology in Eastern Anatolia, in: Frangipane et al. 1993, 573-599.

Pernicka 1984 - E. Pernicka, Instrumentelle Multi-Elementanalyse archäologischer Kupfer- und Bronzeartefakte: ein Methodenvergleich, *JbRGZM* 31, 517-531.

Pernicka 1987 - E. Pernicka, Erzlagerstätten in der Ägäis und ihre Ausbeutung im Altertum: geochemische Untersuchungen zur Herkunftsbestimmung archäologischer Metallobjekte, *JbRGZM* 34 [1989], 607-714.

Pernicka et al. 1984 - E. Pernicka, T.C. Seeliger, G.A. Wagner, u.a., Archäometallurgische Untersuchungen in Nordwestanatolien, *JbRGZM* 31, 533-599.

Powell 1979 - M.A. Powell Jr., Ancient Mesopotamian Weight Metrology, in: M.A. Powell Jr. and R.H. Sack (eds.), *Studies in Honor of Tom B. Jones* (= AOAT 203), Neukirchen-Vluyn, 71-109.

Projektgruppe Plinius 1984 - *Plinius der Ältere über Kupfer und Kupferlegierungen; eine Übersetzung der chemischen, technischen und medizinischen Abschnitte mit Fachkommentaren aus heutiger Sicht*, Düsseldorf.

Otten 1983 - H. Otten, 'Brücken' im hethitischen Schrifttum, in: R.M. Boehmer and H. Hauptmann eds., *Beiträge zur Altertumskunde Kleinasiens, Festschrift für Kurt Bittel, Band I*, Mainz, 433-434.

Rapp 1977 - G. Rapp Jr., Trace Elements as a Guide to the Geographical Source of Tin Ore: Smelting Experiments, in: A.D. Franklin (ed): *The Search for Ancient Tin*, Washington, 59-63.

Riederer 1994 - J. Riederer, Die frühen Kupferlegierungen im Vorderen Orient, in: R.-B. Wartke (ed.), *Handwerk und Technologie im Alten Orient*, Mainz, 85-93.

Röllig 1983 - W. Röllig, article 'Kupfer. A. Philologisch', *RlA* 6, 345-348.

Roman 1990 - I. Roman, Copper ingots, in: Rothenberg (ed.) 1990 - B. Rothenberg, *The Ancient Metallurgy of Copper* [= Researches in the Arabah 1959-1984 volume 2], London, 176-181.

Rosen 1977 - B.L. Rosen, *Studies in Old Assyrian Loan Contracts*, Ann Arbor.

Rothenberg 1990 - B. Rothenberg, Copper Smelting Furnaces, Tuyeres, Slags, Ingot-moulds and Ingots in the Arabah: the Archaeological Data, in: Rothenberg (ed.),*The Ancient Metallurgy of Copper* [= Researches in the Arabah 1959-1984 volume 2], London, 1-77.

Rothenberg and Glass 1992 - B. Rothenberg, J. Glass, The Beginnings and the Development of Early Metallurgy and the Settlement and Chronology of the Western Arabah, from the Chalcolithic Period to Early Bronze Age IV, *Levant* 24, 141-157.

Schmitt-Strecker et al. 1992 - S. Schmitt-Strecker, F. Begemann, E. Pernicka, Chemische Zusammensetzung und Bleiisotopenverhältnisse der Metallfunde vom Hassek Höyük, in: M.R. Behm-Blancke (ed.), *Hassek Höyük; Naturwissenschaftliche Untersuchungen und lithische Industrie*, Tübingen, 108-123.

Schulz 1983 - H.D. Schulz, Zuordnung von Kupfer-Metall zum Ausgangserz. Möglichkeiten und Grenzen der Methode, *Prähistorische Zeitschrift* 58, 1-14.

Seeliger et al. 1985 - T. Seeliger, E. Pernicka, G.A. Wagner, F. Begemann, S. Schmitt-Strecker, C. Eibner, Ö. Öztunalı, I. Baranyi, Archäometallurgische Untersuchungen in Nord- und Ostanatolien, *JbRGZM* 32, 597-659.

Sever 1990a - H. Sever, Kültepe metinlerine göre '*birtum*' kelimesinin yeni anlamı, in: *X. Türk Tarih Kongresi; Ankara (22-26 Eylül 1986); Kongreye sunulan bildiriler, II. cilt*, Ankara, 467-479.

Sever 1990b - H.Sever, Yeni Kültepe tabletlerinde geçen '*Kīma awāt naruā'im*' tabiri ve değerlendirilmesi, *DTCFD* 34, 251-265.

Stech 1985 - T. Stech, Copper and Society in Late Bronze Age Cyprus, in: Knapp and Stech (eds): *Prehistoric Production and Exchange; the Aegean and Eastern Mediterranean*, Los Angeles, 100-105.

Tafel 1951 - V. Tafel, *Lehrbuch der Metallhüttenkunde. 2. Auflage hrsg. von K. Wagenmann, Band 1. Gold, Silber, Platinmetalle, Kupfer, Quecksilber, Wismut*, Leipzig.

Teissier 1993 - B. Teissier, The Ruler with the Peaked Cap and other Syrian Iconography on Glyptic from Kültepe in the Early Second Millennium B.C., in: M.J. Mellink, E. Porada, and T. Özgüç (eds.), *Aspects of Art and Iconography: Anatolia and its Neighbors. Studies in Honor of Nimet Özgüç*, Ankara, 601-612.

Teissier 1994 - B. Teissier, *Sealing and Seals on Texts from Kültepe kārum Level 2*, Istanbul.

Tunca 1989 - Ö. Tunca, Cylinder Seal Insrciptions of Šamši-Adad I and his Officials from Acemhöyük, in: K. Emre, B. Hrouda, M. Mellink, and N. Özgüç (eds.), *Anatolia and the Ancient Near East. Studies in Honor of Tahsin Özgüç*, Ankara, 481-483.

Tylecote 1976 - R.F. Tylecote, *A History of Metallurgy*, London.

Tylecote 1980 - R.F. Tylecote, Furnaces, Crucibles, and Slags, in: T.A. Wertime and J.D. Muhly (eds.), *The Coming of the Age of Iron*, New Haven, 183-228.

Tylecote et al. 1977 - R.F. Tylecote, H.A. Ghaznavi, P.J. Boydell, Partitioning of Trace Elements between the Ores, Fluxes, Slags and Metal during the Smelting of Copper, *Journal of Archaeological Science* 4, 305-333.

Tylecote 1982 - R.F. Tylecote, The Late Bronze Age: Copper and Bronze Metallurgy at Enkomi and Kition, in: J.D. Muhly, R. Maddin, V. Karageorghis (eds.), *Early Metallurgy in Cyprus, 4000-500 B.C.*, Nicosia, 81-100.

Veenhof 1967 - K.R. Veenhof, review of K. Hecker, *Die Keilschrifttexte der Universitätsbibliothek Giessen*, *BiOr* 24, 183-188.

Veenhof 1972 - K.R. Veenhof, *Aspects of Old Assyrian Trade and its Terminology*, Leiden.

Veenhof 1985 - K.R. Veenhof, Observations on Old Assyrian Memorandums; with particular Reference to Kt c/k 839, *JEOL* 28, 10-23.

Veenhof 1986 - K.R. Veenhof, Cuneiform Archives. An Introduction, in: K.R. Veenhof (ed.), *Cuneiform Archives and Libraries. Papers read at the 30e Rencontre Assyriologique Internationale. Leiden, 4-8 July 1983*, Istanbul, 1-36.

Veenhof 1987 - K.R. Veenhof, 'Dying Tablets' and 'Hungry Silver'; Elements of Figurative Language in Akkadian Commercial Terminology, in: M. Mindlin, M.J. Geller, and J.E. Wansbrough (eds.), *Figurative Language in the Ancient Near East*, London, 41-75.

Veenhof 1988 - K.R. Veenhof, Prices and Trade. The Old Assyrian Evidence, *AoF* 15, 243-263.

Veenhof 1989 - K.R. Veenhof, Status and Offices of an Anatolian Gentleman. Two Unpublished Letters of Huharimataku from *kārum* Kanish, in: K. Emre, B. Hrouda, M. Mellink, and N. Özgüç (eds.), *Anatolia and the Ancient Near East. Studies in Honor of Tahsin Özgüç*, Ankara, 515-525.

Veenhof 1991 - K.R. Veenhof, Private Summons and Arbitration amoung the Old Assyrian Traders, in: M. Mori (ed.), *Near Eastern Studies dedicated to H.I.H. Prince Takahoti Mikasa on the Occasion of his seventy-fifth Birthday*, Wiesbaden, 437-459.

Veenhof 1992 - K.R. Veenhof, *makkārum*, 'Trader'?, *N.A.B.U.* 1992, 4 (note 5).

Veenhof 1993 - K.R. Veenhof, On the Identification and Implications of some Bullae from Acemhöyük and Kültepe, in: M.J. Mellink, E. Porada, and T. Özgüç (eds.), *Aspects of Art and Iconography: Anatolia and its Neighbors. Studies in Honor of Nimet Özgüç*, 645-657.

Veenhof 1994 - K.R. Veenhof, article 'Miete. C. Altassyrisch', *RlA* 8, 181-184.

Veenhof 1995 - K.R. Veenhof, Old Assyrian *iṣurtum*, Akkadian *eṣērum* and Hittite GIŠ.HUR, in: Th.P.J. van den Hout and J. de Roos (eds.), *Studio Historiae Ardens. Ancient Near Eastern Studies Presented to Philo H.J. Houwink ten Cate on the Occasion of his 65th Birthday*, Istanbul, 311-322.

Waetzoldt and Bachmann 1984 - H. Waetzoldt and H.G. Bachmann, Zinn- und Arsenbronzen in den Texten aus Ebla und aus dem Mesopotamien des 3. Jahrtausends, *OA* 23, 1-18.

Wagner et al. 1983-1984 - G.A. Wagner, E. Pernicka, T.C. Seeliger, Ö. Öztunali, I. Baranyi, F. Begemann, S. Schmitt-Strecker, Geologische Untersuchungen zur frühen Metallurgie in Nordwest-Anatolien, *Bulletin of the Mineral Research and Exploration Institute of Turkey*, No. 101/102, 45-81.

Wagner et al. 1986 - G. A. Wagner, E. Pernicka, T.C. Seeliger, u.a., Geochemische und isotopische Charakteristika früher Rohstoffquellen für Kupfer, Blei, Silber und Gold in der Türkei, *JbRGZM* 33, 723-752.

Wagner et al. 1989a - G.A. Wagner, Ö. Öztunali, C. Eibner, Early Copper in Anatolia. Archaeometallurgical Field Evidence, in: Hauptmann et al. 1989, 299-305.

Wagner et al. 1989b - G.A. Wagner, F. Begemann, C. Eibner, J. Lutz, Ö. Öztunalı, E. Pernicka, S. Schmitt-Strecker, Archäometallurgische Untersuchungen an Rohstoffquellen des frühen Kupfers Ostanatoliens, *JbRGZM* 36 [1992], 637-686.

Wayman et al. 1985 - M.L. Wayman, R.R. Smith, C.G. Hickey, M.J.M. Duke, The Analysis of Copper Artefacts of the Copper Inuit, *Journal of Archaeological Science* 12, 367-375.

Weisgerber 1980 - G. Weisgerber, '... und Kupfer in Oman' – Das Oman-Projekt des Deutschen Bergbau-Museums, *Der Anschnitt* 32, 62-110.

Weisgerber 1981 - G. Weisgerber, Mehr als Kupfer in Oman. Ergebnisse der Expedition 1981, *Der Anschnitt* 33, 174-263.

Wilhelm 1989 - G. Wilhelm, *The Hurrians*, Warminster.

Wulff 1966 - H.E. Wulff, *The Traditional Crafts of Persia; Their Development, Technology, and Influence on Eastern and Western Civilizations*, Cambridge.

Yener and Vandiver 1993 - K.A. Yener and P.B. Vandiver, Tin Processing at Göltepe, an Early Bronze Age Site in Anatolia, *AJA* 97, 207-238.

Zaccagnini 1986 - C. Zaccagnini, Aspects of Copper Trade in the Eastern Mediterranean during the Late Bronze Age, in: *Traffici micenei nel Mediterraneo*, 413-424.

INDEXES

Texts

Nearly all kt-texts are unpublished Ankara. Those which have been published are listed at the end of kt 92/k.

TC 2

TC 3

TMH 1

TTC

UF 7

VS 26

Sumerograms and Sumerian words

Personal names

Place-names

Subject index